SpringerWienNewYork

Angel Fierros Palacios

The Hamilton-Type Principle in Fluid Dynamics

Fundamentals and Applications to Magnetohydrodynamics, Thermodynamics, and Astrophysics

SpringerWienNewYork

Dr. Angel Fierros Palacios
Instituto de Investigaciones Eléctricas, Cuernavaca, Morelos, México

SpringerWienNewYork is part of Springer Science Business Media
springeronline.com

Typesetting: Camera ready by the author

Printed on acid-free and chlorine-free bleached paper
SPIN: 11393856

Library of Congress Control Number: 2006920792

ISBN 978-3-211-24964-2 SpringerWienNewYork

This book is dedicated:

To my beloved wife and companion Rosa María who has fulfilled my life with hapiness. For her patience and understanding.
To the memory of my parents Angel and Catalina, and my beloved son Angel Arsenio.

What is important is not to be the one who knows the most, but the one who knows what to do with what he knows.

Contents

X. Magnetohydrodynamic Waves

XI. The Sunspots

XII. The Hamilton Equations of Motion

XIII. Thermodynamics

About the Author

Angel Fierros Palacios was born in March 1933 in Pochutla, Oaxaca, México. He studied Physics at the Universidad Nacional Autonoma de México (UNAM), where he also obtained a Master degree and Ph. D. in Science. For many years he was Professor at the UNAM and a researcher at the Instituto Mexicano del Petróleo and at the Instituto Nacional de Investigaciones Nucleares. He also worked at different areas of the public administration such as the Instituto Mexicano del Seguro Social, the Secretaría de Energía, as well as the Comisión Federal de Electricidad (CFE). He is the author of the project which originated the Instituto de Investigaciones Eléctricas (IIE) and also, he was General Director of the Museo Tecnológico de la CFE. Presently he is working at the IIE as Director de la División de Energías Alternas.

Foreword

The objective of this book is to contribute to specialized literature with the most significant results obtained by the author in Continuous Mechanics and Astrophysics. The nature of the book is largely determined by the fact that it describes Fluid Dynamics, Magnetohydrodynamics, and Classical Thermodynamics as branches of Lagrange's Analytical Mechanics; and in that sense, the approach presented in it is markedly different from the treatment given to them in traditional text books.

In order to reach that goal, a Hamilton-Type Variational Principle, as the proper mathermatical technique for the theoretical description of the dynamic state of any fluid is formulated. The scheme is completed proposing a new group of variations regarding the evolution parameter which is time; and with the demonstration of a theorem concerning the invariance of the action integral under continuous and infinitesimal temporary transformations. With all that has been mentioned before and taking into account the methods of the calculus of variations and the adequate boundary conditions, a general methodology for the mathematical treatment of fluid flows characteristic of Fluid Dynamics, Magnetohydrodynamics, and also fluids at rest proper of Classical Thermodynamics is presented.

On the other hand, and as it is well known, to completely characterize any fluid, five functions are needed: the three components of the velocity field, the total energy of the system,and the mass density. For the mathematical scheme to be consistent, the theoretical frame of Continuous Mechanics has to dispose of a set of five partial differential equations for these five field variables. However, a scalar equation for mass density is missing, so strictly speaking, there is a serious inconsistency in the theoretical frame of Fluid Dynamics. That inconsistency is solved obtaining the scalar equation for the mass density within the scheme of Newton´s Vectorial Mechanics, and the field equation for the mass density within the theoretical frame of Lagrange´s Analytical Mechanics. Moreover, chapters XI and XIV are devoted to the treatment of some subjects in the scope of Astrophysics. One of them concerning the problem of the origin, permanency, and disappearance in the solar disk, and properties of the Sunspots; and the other one, referring to the role played by the general magnetic field in the structure and stability of gaseous stars. In that last chapter, the problem of the origin, magnitude, and structure of the Geomegnetic Field is also studied.

References have been listed at the end of each chapter, after the sections called *Selected Topics*. The list is not intended to be in any way complete; many of the good books on the subject have been omitted. Nevertheless, the list contains the refrences used in writing this book, and must therefore serve also as an acknoledgement of my debt to these sources.

My appreciation and thanks go to Lady Elizabeth Casarín Corpus, who deciphered my handwriting and converted it into neat work. In the preparation of the English edition, I´m grateful also to my English Professor Luis Alberto Viades Valencia for his very valuable aid in the translation of this book.

Angel Fierros Palacios
Ciudad de México.

Acknowledgements

I want to express my gratitude for the support and affection received from many people throughout the conceivement and writing of the present work.

To my wife Rosa María, and my children Rosa María, Fernando, Angel Arsenio, Ara-antz-azu, Luis Javier, and my granddauther Mariana, who deserve a very special mention. Not least do I have to say of my brothers and sisters, nephews, and teachers. I found in all of them warmth, understanding and ecouragment for my ideas and projects as well as the motivation which helps me continue along the hard path of scientific research. To all of them, thanks.

Angel Fierros Palacios
Ciudad de México.

Notation

$\boldsymbol{v}$	Velocity field
J	Jacobian of the transformation
$\square_{ijk}$, $\square^{ijk}$	Totally antisymmetrical density of Levi-Civita
$\dfrac{d}{dt}$	Hidrodynamics derivative
$\mathcal{D}$	Reynolds´ differential operator
W	Action integral
L	Classical lagrangian. Luminosity
$\mathscr{L}, \circ$	Lagrangian density
$\boldsymbol{u}$	Displacement vector
u_{ik} , u^{ik}	Components of the small deformation tensor
g_{ik} , g^{ik}	Components of the fundamental metric tensor
δ_{ik}	Components of Kronecker´s delta
$\Gamma^{i}_{ki};\Gamma^{i}_{ik}$	Contracted Christoffel´s simbol
λ	Specific lagrangian
p	Hydrostatic pressure
ρ	Mass density
$\square$	Specific internal energy
$\boldsymbol{F}_H$	Hydrodynamic force
$\mathcal{H}$	Hamiltonean density

$\phi(\boldsymbol{x})$	Conservative potential
$\boldsymbol{f}$	External force per unit mass
t	Kinetic energy density. Evolution parameter
u	Potential energy density
T	Temperature
s	Specific entropy
$w = \mathcal{E} + p/\rho$	Specific enthalpy
$\phi(\boldsymbol{x},t);\varphi(\boldsymbol{x},t)$	Velocity potential. Gravitational potential
c_o	Sound velocity
$\square^2$	Laplacian operator
σ_{ij}	Components of the stress tensor
σ'_{ij}	Components of the viscosity stress tensor
η	Dynamic viscosity
$\vartheta = \eta/\rho$	Kinematics viscosity
$\boldsymbol{q}$	Vectorial heat flux density
κ	Thermal conductivity
α	Thermal coefficient
R'	Gas constant for the air
$\mathcal{R}, \mathcal{R}^o$	Work done by the internal stresses.
	Universal gas constant
$\chi = \kappa/\rho c_p$	Thermometric conductivity
c_p, c_V	Specific heats at a pressure and volume constant
$\boldsymbol{g}$	Aceleration of gravity
Δ	Adiabatic thermal gradient
$\boldsymbol{H}$	Magnetic field

$\boldsymbol{B}$	Magnetic induction
μ	Magnetic permability. Mean molecular weigth
σ	Electrical conductivity. Fractional change in the volume
c	Light velocity in the empty space
$\widetilde{m}$	Maxwell´s magnetic stress tensor
$\tilde{\sigma}^{o}$	Generalized stress tensor
$\boldsymbol{j}$	Conduction current density
$\boldsymbol{E}$	Electric field
$\boldsymbol{S}$	Pointing´s vector
j^2/σ	Joule´s heat per unit volume
$\varepsilon = \frac{1}{V}\left(\frac{\partial V}{\partial T}\right)$	Volume coefficient of expansion
$k = -\frac{1}{V}\left(\frac{\partial V}{\partial p}\right)$	Isothermal compresibility
$\boldsymbol{h}$	Perturbation in the magnetic field
$a_{\circ}$	Phase velocity of a MHD longitudinal wave
$\boldsymbol{v_a}$	Alfven´s vectorial velocity
$\boldsymbol{f} = \boldsymbol{j} \times \frac{\boldsymbol{H}}{c}$	Magnetic force
$\frac{H^2}{8\pi}$	Hydrostatic magnetic pressure
$\frac{H^2}{4\pi}$	Magnetic tension
$\boldsymbol{F_v}$	Viscous force
$\boldsymbol{F_{MHD}}$	Magnetohydrodynamics force

q	Thermal gradient. Effective cross section for collisions
v'_c	Velocity of the convective movements
τ	Mean time life of the Sunspots
E	Total energy
S, σ	Total entropy
R	Mechanical work. Stellar radius
Q	Heat quantity
$SdT - pdV = 0$	Gibbs-Duhem´s relationship
$E = TS - pV$	Total internal energy
Θ	Absolute temperature
$F = E - TS$	Helmholtz free energy
$W = E + pV$	Entalpy
$\Phi = E - TS + pV$	Gibbs´ free energy
$\hat{F} = F + \frac{B^2}{8\pi}$	Generalized Helmholtz free energy
$\hat{E} = E + \frac{B^2}{8\pi}$	Generalized total internal energy
$\hat{W} = W + \frac{B^2}{8\pi}$	Generalized enthalpy
$\hat{\Phi} = \Phi - \frac{B^2}{8\pi}$	Generalized Gibbs´ free energy
G	Granvitational constant
τ_o	Expected life
τ^*	Relative age
k, k_c	Coefficient of opacity

h	Specific hamiltonian. Hydrogen atom
$\overline{X}$	Average hydrogen content
E_{cc}^{*}	Conversion factor which convert grams burns into ergs
γ	Thermodynamic Grüneisen parameter

Chapter I

General Principles

§1. Introduction

Classical Fluid Dynamics is a branch of Mechanics of Continuous Media dedicated to the analysis of the motion of gases and liquids. Since the phenomena considered are macroscopic, a fluid is regarded as a continuous medium. The study of those systems is based on the hypothesis that their structure can be considered from a practical point of view as continuous and homogeneous. In Fluid Mechanics, it is assumed that the macroscopic properties are a continuous function of position and time. The fundamental property which differentiates a fluid from other continuous media is that it can not be in equilibrium in a state of stress such that the reciprocal action between two of its adjacent parts shows itself sharply over their common boundary. There are two kinds of forces that act over a fluid: long range forces that affect all the volume penetrating in the interior of the system, acting over all the fluid elements; forces such as gravity, centrifugal force or electromagnetic forces which act over a conducting fluid. They are called body forces and have the property of being proportional to the size of the volume element. The second type of forces already mentioned are the stress or contact forces since they are of short range and they have their origin in molecular interactions. The surface total force which acts over a volume element of fluid is determined by its superficial area. The above mentioned property is basic to solve problems in Hydrostatic and Hydrodynamics but it is not enough to make a complete description of a fluid motion. In order to characterize the dynamical behaviour of fluids, it is necessary to have a set of five partial differential equations which must satisfy the relevant variables of the system. In general, a fluid can be completely characterized with the aid of the three components of the velocity field, the mass density, and the total energy of the system. Also, there are two theoretical methods to deduce the equations which describe the dynamic state of a fluid. One of them consists of facing the problem from the molecular point of view considering the system as formed by material particles whose motions are ruled by Classical Mechanics laws. Thus, with the aid of Statistical Mechanics, the macroscopic behaviour of a fluid can be predicted from fundamental principles of Mechanics and the Theory of Probability.

The other approaching method adopts a continuous model for the fluid. In consequence, in Fluid Mechanics the average properties of gases and liquids considered as continuous substances are studied. In the theoretical treatment given to those systems it is assumed that the number of particles that form them is so big that in any volume element it is neither possible nor desirable to distinguish individual particles; that is to say, they are macroscopic systems. As in Classical Theory of Fields and regardless of the molecular structure of the system, the dynamical state of the fluid is described by means of certain functions of coordinates and time. The field variables used, refers to fixed points in space and given instants, and not to fixed particles of the fluid. This implies that the description this refers to is only valid for local equilibrium.

Throughout this book, a continuous model to describe the dynamic state of a fluid will be adopted and its molecular structure will only be referred to when incidentally it becomes necessary to analyse some microscopic behaviour of the system being studied.

§2. Kinematics

Fluid flow is a physically intuitive notion which can be represented mathematically by means of a continuous transformation of the three-dimensional Euclidean Space in itself. Time is the evolution parameter which describes the transformation and it is such that $-\infty < t < \infty$, where $t = 0$ is an arbitrary initial instant. To describe the transformation analytically let us introduce an inertial frame of reference fixed in space. *Be* $\{\boldsymbol{x}\}$ a set of n linearly independent vectors in a vectorial space R of dimension n which defines the inertial frame of reference above mentioned. From it and according to given initial conditions, the motion of a discrete system formed by N individual particles with mass m_i $(i = 1,2,\ldots,N)$, can be described, associating to them N paths such that in every point of them the respective particles move with velocities defined as the derivative of position with respect to time.

If the initial conditions are changed in such a way that the N particles of the previous mechanical system describe at each nstant all the possible paths such that the space is totally covered by them, and imposing the addi-tional condition that they do not cross each other; then, to each point of the space a velocity vector and only one can be associated. It is said that the velocity field is stationary and of the form $\boldsymbol{v}=\boldsymbol{v}(\boldsymbol{x})$ if the defined velo-city vectors are not explicit functions of time. Let us consider now the case of a continuous medium such that it can be thought that to a volume element of fluid passing by any given point of space, a velocity equal in magnitude and direction to the vector velocity must be associated at said point. If the velocity field is not only a function of position but also of time, each field vector depends on time according to some definite relationship. In general the velocity field will have the following form

$$\boldsymbol{v}(\boldsymbol{x},t)=\frac{d\boldsymbol{x}(t)}{dt}. \tag{2.1}$$

The velocity field defined this way is an operation which instantly associates each fluid element at each point of space a velocity equal in magnitude and direction to the vector velocity linked to that point in that instant.

The coordinates of every point of space are functions of time, that is to say, $\boldsymbol{x}=\boldsymbol{x}(t)$.

§3. Deformation

Let us consider a region of three-dimensional Euclidean Space occupied by a moving continuous medium in such a way that at changing its position occupies a new region of space. Be R_o the region initially occupied by the fluid and R the region that subsequently is occupied by the system. Be P_o a point in R_o located by a position vector $\boldsymbol{x}_o=\boldsymbol{i}x_o^1+\boldsymbol{j}x_o^2+\boldsymbol{k}x_o^3$; and let us suppose that by the effect of the motion of the continuous medium this point has been translated to the position P in R, located by another position vector $\boldsymbol{x}=\boldsymbol{i}x^1+\boldsymbol{j}x^2+\boldsymbol{k}x^3$, where $(\boldsymbol{i},\boldsymbol{j},\boldsymbol{k})$, is a triad of unity vectors along the coordinate axes of the inertial reference system used to make the description of the process. The transformation of all points $\{\boldsymbol{x}_o\}$ of R_o in the points $\{\boldsymbol{x}\}$ of R is what is known as a deformation of the medium; that is to say

$$\boldsymbol{x}=\boldsymbol{x}(\boldsymbol{x}_o). \tag{3.1}$$

Be (3.1) a continuous and uniform transformation that can be inverted in such a way that it is possible to state that the following inverse transformation exists

$$\boldsymbol{x}_o = \boldsymbol{x}_o(\boldsymbol{x}) \ . \tag{3.2}$$

As a continuous model for the fluid has been assumed and a continuous transformation has been considered, the relations (3.1) and (3.2) define, mamathematically speaking, the deformation process of continuous medium.

In that case it is said that R_o is the undeformed initial region; whereas and due to the fluid flow effect, R is the deformed final region occupied by the continuous medium at the end of the process.

§4. Transformation of a volume element

To determine the deformation that a fluid experiments when it moves, it is necessary to isolate a volume element and determine how it changes during the flow process by means of the analytical application of the transformation (3.1).

From the physical view point, one wishes to find the correspondence existing between a volume element of fluid in its undeformed initial state to some later situation when it changes its position and has suffered a deformation process. Be a volume element of fluid located around the $\{\boldsymbol{x}_o\}$ point in R_o. For simplicity it is assumed that it is a rectangular parallepiped with sides dx_o^i , dx_o^j, dx_o^k In this case, the volume element in the initial undeformed state can be written as

$$dV_o = dx_o^i dx_o^j dx_o^k \ . \tag{4.1}$$

Under transformation (3.1) we have that

$$\begin{aligned} dx^1 &= \frac{\partial x^1}{\partial x_o^i} dx_o^i \\ dx^2 &= \frac{\partial x^2}{\partial x_o^j} dx_o^j \\ dx^3 &= \frac{\partial x^3}{\partial x_o^k} dx_o^k \ ; \end{aligned} \tag{4.2}$$

where dx^1, dx^2, dx^3 are the sides of the new volume element which due to the change of position of the fluid, has been deformed and clearly it is no longer a rectangular parallelepiped as it is easy to see form relations (4.2). Then the volume element in the already deformed region is expressed as follows

$$dV = \frac{\partial x^1}{\partial x_o^i} \frac{\partial x^2}{\partial x_o^j} \frac{\partial x^3}{\partial x_o^k} dx_o^i dx_o^j dx_o^k \ . \tag{4.3}$$

This expression can be written in terms of (4.1) to finally obtain that

$$dV = JdV_o \ ; \tag{4.4}$$

where

$$J = \frac{\partial x^1}{\partial x_o^i} \frac{\partial x^2}{\partial x_o^j} \frac{\partial x^3}{\partial x_o^k} \tag{4.5}$$

is the Jacobian of the transformation.

§5. Rate of change of a volume element

To obtain this rate of change it is necessary to use the tensor notation to express the Jacobian in the following manner

$$\Box_{abc} J = \Box_{ijk} \frac{\partial x^i}{\partial x_o^a} \frac{\partial x^j}{\partial x_o^b} \frac{\partial x^k}{\partial x_o^c} \ ; \tag{5.1}$$

where $\Box_{abc}$ and $\Box_{ijk}$ are respectively, the components of the totally antisymmetrical densities of Levi-Civita. In general, their properties are the following

$\Box_{ijk} = +1$; if i, j, k is a cyclic permutation of 1,2, and 3;
 i.e. if $(i, j, k) = (1,2,3), (2,3,1)$ or $(3,1,2)$.
$\Box_{ijk} = -1$; if i, j, k is an anticyclic permutation of 1,2, and 3; $\qquad$ (5.2)
 i.e. , if $(i, j, k) = (1,3,2), (2,1,3)$ or $(3,2,1)$.
$\Box_{ijk} = 0$; otherwise.

Let δ be a differential operation with respect to α ; where $\{\alpha\}$ is a set of continuous and time independent geometric parameters, such that

$$\delta J = \underset{\delta\alpha \to o}{\circ im} \frac{J(\alpha+\delta\alpha)-J(\alpha)}{\delta\alpha} \quad . \tag{5.3}$$

On applying this operation to the expression (5.1) the following is obtained

$$\square_{abc}\,\delta J = 3\square_{ijk}\frac{\partial x^i}{\partial x_o^a}\frac{\partial x^j}{\partial x_o^b}\frac{\partial}{\partial x_o^c}(\delta x^k) \ .$$

If now all is multiplied by $\square^{abc}$ it can be seen that

$$\square^{abc}\square_{abc}\,\delta J = 3\square^{abc}\left[\square_{ijk}\frac{\partial x^i}{\partial x_o^a}\frac{\partial x^j}{\partial x_o^b}\right]\square\frac{\partial}{\partial x_o^c}(\delta x^k) .$$

On the other hand and given that

$$\frac{\partial}{\partial x_o^c}\left(\delta x^k\right)=\frac{\partial x^p}{\partial x_o^c}\frac{\partial}{\partial x^p}\left(\delta x^k\right) \ , \tag{5.4}$$

it is immediately obtained

$$\frac{1}{3}\square_{abc}\,\delta J = \left[\square_{ijk}\frac{\partial x^i}{\partial x_o^a}\frac{\partial x^j}{\partial x_o^b}\frac{\partial x^p}{\partial x_o^c}\right]\frac{\partial}{\partial x^p}\left(\delta x^k\right).$$

Finally and according to (5.1) the rate of change looked for is

$$\delta J = J\,div\left(\delta \boldsymbol{x}\right). \tag{5.5}$$

Particularly, when one wishes to obtain the change dJ/dt when the attention is centered on the motion of a volume element of fluid, the last demonstration is equally valid since it is only enough to suppose that α is t and δ, d/dt, in order to obtain the following result

$$\frac{dJ}{dt} = J\,div\,\boldsymbol{v}. \tag{5.6}$$

This is the well known Euler´s relationship. If we observe that *div* $\boldsymbol{v} = 0$, it is said that the flux is isochoric; that is to say, that a constant volume occurs and it is possible to assure that the fluid is incompressible.

§6. The hydrodynamics derivative

From elemental Classical Mechanics it is known that the acceleration of a particle is equal to the time rate of change of its velocity $\boldsymbol{a} = d\boldsymbol{v}/dt$, where $\boldsymbol{v} = d\boldsymbol{x}/dt$ is the vector velocity. However, within the theoretical frame of Fluid Dynamics, the total derivative $d\boldsymbol{v}/dt$ does not denote the rate of change of the fluid velocity in a fixed point of space; but rather, the rate of change of the velocity of a fluid elemental volume when it moves in space. This derivative must be expressed in terms of quantities referring to points fixed in space. To do so, notice that the change $d\boldsymbol{v}$ in the velocity of a given fluid elemental volume during the time dt is made of two parts, namely, the change during dt in the velocity at a point fixed in space, and the difference between the velocities (at the same instant) at two points $d\boldsymbol{r}$ apart, where $d\boldsymbol{r}$ is the distance in which the fluid volume element has moved during the time dt. The first part is$(\partial \boldsymbol{v}/\partial t)dt$, where partial derivative is taken at a fixed point in space whose coordinates are (x, y, z). The second part is

$$dx\frac{\partial \boldsymbol{v}}{\partial x} + dy\frac{\partial \boldsymbol{v}}{\partial y} + dz\frac{\partial \boldsymbol{v}}{\partial z} = (d\boldsymbol{r} \cdot \boldsymbol{grad})\boldsymbol{v} \; .$$

Thus,

$$d\boldsymbol{v} = \left(\frac{\partial \boldsymbol{v}}{\partial t}\right) dt + (d\boldsymbol{r} \cdot \boldsymbol{grad})\boldsymbol{v} \; ;$$

or, dividing both sides by dt, finally one obtains

$$\frac{d\boldsymbol{v}}{dt} = \frac{\partial \boldsymbol{v}}{\partial t} + (\boldsymbol{v} \cdot \boldsymbol{grad})\boldsymbol{v} \; .$$

The operator

$$\frac{d}{dt} = \frac{\partial}{\partial t} + \boldsymbol{v} \cdot \boldsymbol{grad} \tag{6.1}$$

is known as the convective or hydrodynamics derivative.

§7. Reynolds´ transport theorem

An important kinematical theorem can be derived from the so-called Euler´s formula; that is the equation (5.6). The above mentioned theorem is due to Reynolds and it concerns the rate of change not of an infinitesimal volume element but of any volume integral. Be R any region of the three-dimensional Euclidean Space and let $F(\boldsymbol{x}, t)$ be any scalar or vectorial function of position and time. Its volume integral

$$\int_R F(\boldsymbol{x},t)dV \ ,$$

with dV the volume element in region R is a well defined function of time. If one wants to know the time rate of change of that integral, it is necessary to use the results obtained in paragraphs 4 and 5, since the integral is taken over a region $R(t)$ which varies with time and it is not possible to make the derivation with respect to the time of the quantity inside it. However, the result (5.6) makes it possible to carry out this operation. In fact, if the formula for the change of the volume, deduced in paragraph 4 is used, the following is obtained

$$\frac{d}{dt}\int_R FdV = \frac{d}{dt}\int_{R_o} FJdV_o \ ;$$

where R_o is a fixed region independent of time so that now the integral is not a function of time and it is possible to interchange differentiation and integration operators. In this case we have that

$$\begin{aligned} \frac{d}{dt}\int_R F(\boldsymbol{x},t)dV &= \int_{R_o}\left(J\frac{dF}{dt} + F\frac{dJ}{dt}\right)dV_o \\ &= \int_{R_o}\left(\frac{dF}{dt} + F\,div\,\boldsymbol{v}\right)JdV; \end{aligned}$$

where Euler´s formula has been used. Clearly, it is possible to return to region R that varies with time by means of the use of relation (4.4); in which case the following is obtained

$$\frac{d}{dt}\int_R F(\boldsymbol{x},t)dV = \int_R \mathcal{D}F(\boldsymbol{x},t)dV.$$

This result is known as Reynolds´ transport theorem; whereas

$$\mathcal{D} \equiv \frac{d}{dt} + div\,\boldsymbol{v} \tag{7.1}$$

is Reynolds´ differential operator

§8. Hamilton´s principle

The balance equations of Fluid Dynamics can be obtained from an extremal action principle and with the aid of a lagrangian density.

However, the route to follow presents some difficulties. In fact, the classical lagrangian of a system of particles is a function that depends on time and on two sets of linearly independent variables which are the generalized coordinates and the generalized velocities as functions of the evolution parameter that is the time. For the case of a fluid considered as a continuous and homogeneous dynamical system, the velocity field is a function of coordinates and time, so that coordinates and velocities are not independent between them. This fact complicates the use of lagrangian formalism and Hamilton´s principle to deal with the problem. Within the theoretical frame of Classical Mechanics, it is known that to obtain the differential equations of motion of a mechanical system a lagrangian function is used which depends on time and on the kinematical entities $q(t)$ and $\dot{q}(t)$ which are respectively the generalized coordinates and velocities. Here the dot means the total derivative with respect to time and j is an index that goes from 1 to N. Both the generalized coordinates $q_j\,(t)$ as well as the generalized velocities $\dot{q}_j(t)$ are taken as functions of some set of continuous and time independent geometrical parameters. Then the action integral W is defined as follows

$$W = \int_{t_1}^{t_2} L dt \quad ; \tag{8.1}$$

where the integral must be calculated between two fixed values of time. Hamilton´s principle proposes that the action be an extremal with respect to the above mentioned set of parameters. This condition leads directly to the motion differential equations known as Euler-Lagrange´s equations.

If on the other hand one wishes to obtain the so called field differential equations of some continuous mechanical system, the use of a lagrangian density $\mathcal{L}$ depending on certain field functions, on its first gradients, on the coordinates, and time is proposed so that the classical lagrangian be

$$L = \int_V \mathcal{L} dV \quad ; \tag{8.2}$$

where dV is the volume element in the region considered. The action integral is defined again as before, but in this case the field functions and their first gradients are parametrized with respect to a set of continuous parameters independent of position and time. Then, and according to Hamilton´s principle, the action integral must be an extremal under a variation with respect to said parameters. According to this, the field differential equations of Euler-Lagrange are obtained.

In the case of Fluid Mechanics and provided that velocity field depends on coordinates and time, a hybrid Lagrange formalism to obtain the field differential equations for any fluid has to be developed. Those circumstances compel to use a lagrangian density as one of the Classical Field Theory and a Hamilton-Type Extremal Action Principle. However, for this case $\mathcal{L}$ is a function of certain field functions, and of the coordinates, and time, only. Then and according to the Hamilton-Type Variational Principle the action integral defined as

$$W = \int_{t_1}^{t_2} L dt = \int_{t_1}^{t_2} \int_V \mathcal{L} dV dt \tag{8.3}$$

must be an invariant under a continuous and infinitesimal variation with respect to a set of continuous and time independent geometrical parameters. Field variables and coordinates depend on the above mentioned parameters, obtaining this way the momentum balance equation for any fluid.

Selected Topics

The velocity field

From a mathematical view point, the velocity field is an analytical operation known by the name of *mapping*. In general, a *mapping* is a geometrical transformation which assigns every point of the space to another position or point in this same space. To translate the above said to language of Fluid Dynamics, let us consider the following definitions:

- **A deformed body is a differentiable manifold with boundaries.** It is represented with the symbol $\mathcal{B}$. By a manifold a collection of objects is understood.
- **The elements of $\mathcal{B}$ are known by the name of material points.** Thus, the body $\mathcal{B}$ can be represented as follows

$$\mathcal{B} \equiv \{X, Y, Z, \ldots\};$$

where X, Y, Z… are the material points of $\mathcal{B}$. Be Ψ a set of *isomorphic mappings*[†] of $\mathcal{B}$ in regions of the three-dimensional Euclidean Space E_3, such that if ψ is an element of the set Ψ; that is to say, if $\psi \in \Psi$, then

$$\psi : \mathcal{B} \rightarrow B \subset E_3,$$

where B is some region of E_3. Graphically we have the following

---

† *Isomorphism: Correspondence one to one.*
It is said that two groups are isomorphic if the elements of one of them can be associated with the elements of the other such that the product of two arbitrary elements of the first, corresponds to the product of the respective elements of the second.

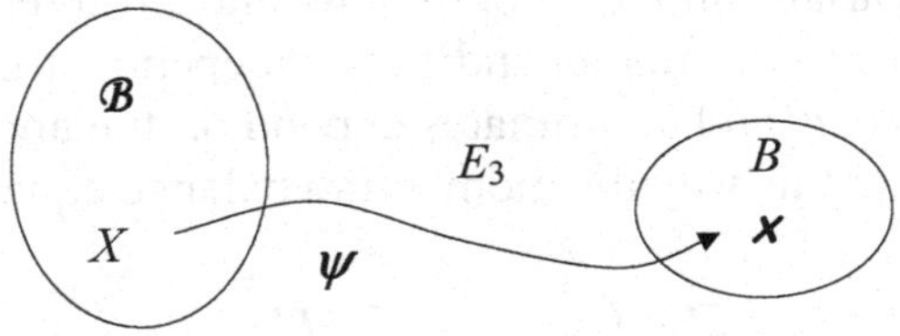

In the language of the theory of functions, the former means that as a consequence of *mapping* ψ, $\boldsymbol{x}$ is the image of X. That is to say

$$\boldsymbol{x} = \psi(X) \text{ for all } \boldsymbol{x} \in B \text{ and } X \subset \mathcal{B}.$$

The *mapping* ψ is an operation that associates every material point of $\mathcal{B}$ with a point $\boldsymbol{x}$ in region B on E_3. In other words, $\boldsymbol{x}$ is the coordinate of X in the three-dimensional Euclidean Space. As the inverse *mapping* exists; that is $\psi^{-1} \in \Psi$ it is clear that

$$\psi^{-1} : B \to \mathcal{B}.$$

In consequence

$$X = \psi^{-1}(\boldsymbol{x})$$

because it is an isomorphism.

On the other hand, there is a strictly monotonic parameter called the time t which has the property of establishing an order relation of the mappings of $\mathcal{B}$ in E_3, at least into a closed interval $[a,b]$ such that

$$\psi^{-1} = \psi(t) \quad \text{and} \quad \psi : [a,b] \times \mathcal{B} \to B.$$

In the language of the theory of functions the former means that

$$\psi(t) = \psi(X,t).$$

Let us consider the next figure

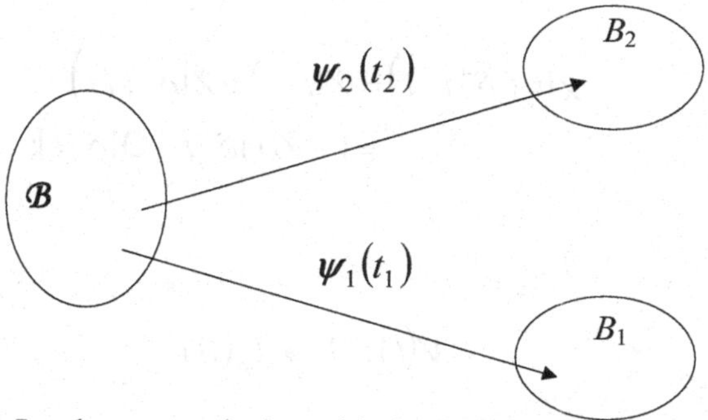

Be $t_1 < t_2$. In that case it is said that $B_1(t_1)$ occurs before $B_2(t_2)$. By definition be

$$\chi(t_2/t_1) \equiv \psi(t_2) \square \psi^{-1}(t_1).$$

It is said that $\chi(t_1/t_2)$ is a displacement of the region image B from $B_1(t_1)$ to B_2 (t_2) which has the following properties

$$1)\ \ \chi(t_3/t_2) \square \chi(t_2/t_1) \equiv \chi(t_3/t_1)$$
$$2)\ \ \chi(t_2/t_1) \equiv [\chi(t_1/t_2)]^{-1}$$
$$3)\ \ \chi(t_1/t_1) = 1\,.$$

In consequence, if the last properties are carried out it can be stated that $\chi(t/t')$ is a group since it satisfies the following requirements

a) It is commutative.
b) The inverse exists.
c) The unit exists.

It is said that $\{\chi(t/t')\}$ is the displacements group of the images of B. In fact, it is a one-parametric group and since it is commutative and according to Slur´s lemma, it is said that it is abelian; in other words, all of its elements commute one to one. Moreover, as the t-parameter is continuous, the group is continuous; that is to say, it is a Lie group. All that has been said before implies the existence of a group generator. In fact, be

$$t' = t + \delta^+ t\,,$$

in such a way that

$$\chi(t+\delta^+ t/t) = \chi(t/t) + \chi(\delta^+ t/t)$$
$$= 1 + \mathbf{V}(t)\delta^+ t + \mathbf{O}(\delta^+ t);$$

where

$$\mathbf{V}(t): B \to T_p(B)$$

is the group generator, with $T_p(B)$ the tangent space of B. Moreover, $O(\delta^+ t)$ represents the $\delta^+ t$ higher order terms. In consequence, it is clear that

$$\mathbf{x}'(t+\delta^+ t) = \mathbf{x}(t) + \delta^+ \mathbf{x}(t)$$

with

$$\delta^+ \mathbf{x}(t) = \frac{\partial \mathbf{x}}{\partial t}\delta^+ t = \mathbf{v}(\mathbf{x},t)\delta^+ t .$$

Clearly, the mapping $\mathbf{v}(\mathbf{x}, t)$ is the generator of the displacement group in the tangent space. On the other hand, it is evident that

$$\mathbf{v}(\mathbf{x},t) = \lim_{\delta^+ t \to 0} \frac{\mathbf{x}'(t+\delta^+ t) - \mathbf{x}(t)}{\delta^+ t} .$$

That is to say

$$\mathbf{v}(\mathbf{x},t) \equiv \frac{d\mathbf{x}(t)}{dt};$$

which is no other thing than the relation (2.1). In the language of the theory of functions, $T_p(B)$ is the tangent space of B whose elements are the velocities of the points $\mathbf{x}(t)$ from B.

Jacobian variation

In order to obtain the equation (5.5) from the following relation

$$\frac{1}{3}\square_{abc}\delta J = \left[\square_{ijk}\frac{\partial x^i}{\partial x_o^a}\frac{\partial x^j}{\partial x_o^b}\frac{\partial x^p}{\partial x_o^c}\right]\frac{\partial}{\partial x^p}\left(\delta x^k\right);$$

the procedure is as follows. The square parenthesis is multiplied by δ_p^k which are the components of Kronecker's delta, in such a way that we have the following

$$\left[\square_{ijk}\frac{\partial x^i}{\partial x_o^a}\frac{\partial x^j}{\partial x_o^b}\frac{\partial x^k}{\partial x_o^c}\right]\delta_p^k\frac{\partial}{\partial x^p}\left(\delta x^k\right) = \square_{abc}J\delta_p^k\frac{\partial}{\partial x^p}\left(\delta x^k\right);$$

where equation (5.1) was used. In this case,

$$\frac{1}{3}\square_{abc}\delta J = \square_{abc}J\frac{\partial}{\partial x^p}\left(\delta x^p\right).$$

In consequence and since

$$\frac{\partial}{\partial x^p}\left(\delta x^p\right) = \frac{1}{3}div(\delta \mathbf{x}),$$

it is finally obtained that

$$\delta J = J\,div(\delta \mathbf{x}),$$

which is the result looked for.

Euler's relation

Be

$$J = \frac{\partial\left(x^1, x^2, x^3\right)}{\partial\left(x_o^a, x_o^b, x_o^c\right)} = det\left(\frac{\partial x^i}{\partial x_o^\alpha}\right);$$

such that $0 < J < \infty$. This means that a volume element can not be compressed to zero or expanded to infinity; that is to say, it must remain finite. Let $A^{\alpha}{}_{i}$ be the cofactor of $\partial x^{i}/\partial x^{\alpha}{}_{o}$ in the expansion of Jacobian's determinant, in such a way that

$$J\delta^{i}_{j} = \frac{\partial x^{i}}{\partial x^{\alpha}_{o}} A^{\alpha}_{j} .$$

Then

$$\frac{dJ}{dt} = \frac{d}{dt}\left(\frac{\partial x^{i}}{\partial x^{\alpha}_{o}}\right) A^{\alpha}_{i} = \frac{\partial}{\partial x^{\alpha}_{o}}\left(\frac{dx^{i}}{dt}\right) A^{\alpha}_{i}$$
$$= \frac{\partial v^{i}}{\partial x^{\alpha}_{o}} A^{\alpha}_{i} = \frac{\partial v^{i}}{\partial x^{j}} \frac{\partial x^{j}}{\partial x^{\alpha}_{o}} A^{\alpha}_{i}$$
$$= \frac{\partial v^{i}}{\partial x^{j}} \left(J\delta^{j}_{i}\right).$$

In that case it is fulfilled that

$$\frac{dJ}{dt} = J \, div \, \mathbf{v} ,$$

which is no other thing than the equation (5.6).

Hamilton's principle in fluid dynamics

In the monumental article ***Mathematical Principles of Classical Fluid Mechanics*** of James Serrin published in the Handbuch der Physic, VIII/1 of Springer-Verlag editorial house (Berlin, 1959), it is stated that it is not possible to obtain motion equations for any viscous fluid from Hamilton's Principle, fundamentally due to the dissipative character of these systems. Apparently, the continuous mechanical systems are not conservative so and according to the last statement, Hamilton's Principle least action can not be applied to them. In fact, Serrin is right. Let us see why. In the first place, Hamilton's Principle is based on the energy conservation law. The action integral

$$W = \int_{t_1}^{t_2} \int_R \mathcal{L} \, dV dt ,$$

contains a lagrangian density that according to the scheme of Classical Field Theory must be a function of the relevant field variables and their first gradients. In the case of Fluid Mechanics, two theoretical problems to construct the lagrangian density must be faced, that is to say:

1. The relevant field variable is the velocity field $\boldsymbol{v}(\boldsymbol{x}, t)$; so that their first gradients are intimately linked to the energy dissipation process known by the generic name of viscosity. In consequence, the classical Hamilton's Principle can not be used because the action integral will contain explicitly the dissipative process and this is obviously not compatible with the energy conservation principle.
2. The velocity field is a function of coordinates and time such that velocities and coordinates are not independent between them. That complicates the problem due to the fact that we do not have any more a scheme like we do in Analytical Mechanics of particles where the generalized coordinates and the generalized velocities as functions of the evolution parameter which is time, forming a set of independent variables between them. In this case, it is always possible to express the variation of the generalized velocities in terms of the variation of generalized coordinates and extend the validity of Hamilton's Principle even when the $\dot{q}_1$ are considered as a second set of independent variables.

In the case of Fluid Dynamics and despite the fact that velocities and coordinates are not independent between them, the validity of Hamilton-Type Principle can be extended even when the velocity field can be considered as another set of independent variables since the variation of that field can always be expressed in terms of the variation of coordinates. This is so because fortunately, the velocity field is a mapping that has the form given in (2.1) of the text.

Finally and related to the first theoretical problem, in Hamilton-Type Principle an integral action containing a lagrangian density that is independent of the gradients of field function is used. In the proposed theory, the dissipative processes are effects that are derived from the relevant field functions in such a way that in Hamilton-Type Variational Principle are neither important nor appear explicitly in it, such that the energy conservation principle can be fulfilled. In other words and according to that theoretical scheme, the fluids are conservative continuous mechanical systems.

References

1. Aris R., "Vector, Tensors, and the Basic Equations of Fluid Mechanics" Prentice Hall, Inc. (1966).
2. C. Lanczos. "The Variational Principles of Mechanics" University of Toronto Press, 4th edition (1970).
3. Fierros Palacios, Angel. "Una formulación lagrangiana de la hidrodinámica clásica" Tesis Doctoral (1973).
4. Fierros Palacios, Angel y Viniegra H., Fermín. "Notas para un curso de Mecánica de Fluidos". Fac. de Ciencias, UNAM (1977).
5. Landau, L.D. and Lifshitz E. M. "The Classical Theory of Fields". Addison-Wesley Publishing Co. (1962).
6. Serrin J. "Mathematical Principles of Classical Fluid Mechanics". Handbuch der Physic, VIII/I, Springer-Verlag, Berlin (1959).
7. Viniegra H., Fermín, Salcido, Alejandro y Fierros Palacios, Angel. "Las ecuaciones de balance de un fluido perfecto a partir de un principio variacional tipo Hamilton", Rev. del IMP. Vol. XVI, Núm.1 (enero, 1984).

Chapter II
Mass Density

§9. The continuity equation

To completely characterize any fluid, five field functions are needed, that is to say: the three components of the velocity field, the total energy of the system, and the mass density. The latter is due to the fact that mass density not only is a fundamental characteristic of any substance, but in the theoretical frame of Fluid Dynamics plays a very important role. As it will be seen further on, when the case of a perfect fluid is studied, mass density is the argument of specific internal energy of the system.

As a consequence of the latter and so that the mathematical scheme will be consistent, the theoretical frame of Fluid Mechanics has to dispose of a set of five partial differential equations for the five characteristic field functions of any continuous medium.

Until not long ago in specialized literature, there were only four basic equations for the five variables above mentioned. Even at present, in the text books concerning the topic, there is only Canchy´s motion differential equation for the case of a viscous fluid or L. Euler´s for a perfect fluid and the corresponding energy conservation equations for both types of fluids.

A scalar equation for mass density is missing, so strictly speaking, there is a serious inconsistency in the theoretical frame of Classical Fluid Mechanics. As a consequence of this fact, that part of physics has been divided into two branches: one that is the Classical Hydrodynamics which has evolved very little and the other one the Hydraulics, which has made very important advances. This has been achieved thanks to engineers skill to construct models but at the expense of not having a complete theory. In Hydraulics each problem is solved empirically and as an isolated case. In the scope of Classical Hydrodynamics, the lack of a fundamental relationship that ends the inconsistency before mentioned has been traditionally covered by the continuity equation which is certainly not a fundamental relationship to describe the dynamic state of any fluid. This equation appears in the theory as a consequence of imposing a restriction on the motion of fluids; imposition which consists of demanding that their mass be constant along the flux process. Evidently, there are many cases of flow of fluids where that does not occur. On the other hand, before the inconsistency of the theory and with the purpose of simplifying the equations; mass density is usually considered as a constant in some terms and as a variable in others. Under these conditions it has been very difficult to obtain accurate solutions for the problems in the scope of Classical Hydrodynamics.

In order to obtain the continuity equation let us consider a region R in three-dimentional Euclidean Space which contains any fluid whose mass is given by the following relation

$$m = \int_R \rho(\boldsymbol{x},t)\, dV \,, \tag{9.1}$$

where $\rho(\boldsymbol{x}, t)$ is a function of position and time that denotes the quantity of matter contained in region R. This function is know by the name of mass density; whereas dV is the volume element. As it was said before, in Classical Hydrodynamics it is always considered that the mass of the fluid is preserved in such a way that $dm/dt=0$; and in (9.1) we have that

$$\frac{d}{dt}\int_R \rho(\boldsymbol{x},t)\, dV = 0 \,.$$

If Reynolds′ transport theorem is used, it is possible to interchange the differential and integral operators to obtain the following result

$$\int_R \left(\frac{d\rho}{dt} + \rho \, div \, \boldsymbol{v} \right) dV = 0 \,.$$

As the volume element was arbitrarily selected, it is clear that it is different from zero. Then, so that the last relation is fulfilled it is necessary that the integrand becomes null; that is to say

$$\frac{d\rho}{dt} + \rho \, div \, \boldsymbol{v} = 0 \,, \tag{9.2}$$

and this is the well known continuity equation whose role within the theoretical frame of Classical Fluid Dynamics must be maintained in its right dimention, that is as an additional condition which is imposed on the motion of the fluids.

Aiming to remove the above mentioned undetermination and from purely geometric considerations, next the scalar equation for mass density will be obtained.

§10. The small deformation strain tensor as a fundamental metric tensor

Experience indicates that when a region occupied by a continuous medium is deformed and the quantity of matter present there kept unchanged, the result of this deformation is translated into an increase or decrease in the mass density. It is evident that if the process is a compression, the mass density increases and if it is an expansion, its value decreases. Then, it is reasonable to suppose that if the quantity of matter remains fixed, there is a closed relation between the changes in form of the region that contains the fluid and its mass density. In order to determine that relation let us consider a region R in three-dimensional Euclidean Space that contains any fluid. If over the boundary surface of R an external force is applied, the geometry of the region changes in form and size; that is to say, it is deformed. If the process is made in such a way that the quantity of matter remains unchanged, the mass density of the fluid present here, changes its value.

In order to mathematically describe the deformation suffered by the region R, the procedure is as follows: before deformation, the position of any point in this region is described by its radius vector $\boldsymbol{r}$, with components x_i; with i=1,2,3; draught from the origin of some coordinate system fixed in space. After the deformation of the considered region, each point is displaced in such a way that now its position vector is $\boldsymbol{r}'$ with components x_i'. Thus, the displacement of any point due to deformation is then given by the vector $\boldsymbol{u} = \boldsymbol{r}' - \boldsymbol{r}$, known by the name of displacement vector whose components are

$$u_i = x_i' - x_i \; . \tag{10.1}$$

According to what was said before, the coordinates x_i' of the displaced point are functions of the coordinates x_i of the point before displacement, in such a way that the displacement vector is therefore only a function of the coordinates x_i ; that is to say $\boldsymbol{u} = \boldsymbol{u}(\boldsymbol{x})$. This simple fact is enough so that from a mathematical point of view the deformation process of the region R is entirely determined. Now, let us consider two points very close together. Be dx_i the i-component of the radius vector joining them before the deformation and $dx_i' = dx_i + du_i$, the radius vector joining the same two points in the deformed region. The distance between the points before and after deformation respectively are $ds = (dx_i^2)^{1/2}$ and $ds' = (dx'^2_i)^{1/2}$. In this case

$$ds^2 = dx_i^2 \tag{10.2}$$

and

$$\begin{aligned} ds'^2 = (dx_i + du_i)^2 = ds^2 + 2\frac{\partial u_i}{\partial x_k} dx_i dx_k \\ + \frac{\partial u_i}{\partial x_k}\frac{\partial u_i}{\partial x_m} dx_k dx_m \end{aligned} \tag{10.3}$$

where the following expansion was used

$$du_i(x_k) = \frac{\partial u_i}{\partial x_k} dx_k \, . \tag{10.4}$$

Since the summation is taken over both suffixes i and k; the second term on the right of (10.3) can be written as

$$\frac{\partial u_i}{\partial x_k} dx_i dx_k = \frac{\partial u_k}{\partial x_i} dx_i dx_k \; .$$

In the third term on the right of (10.3), the suffixes i and m can be interchanged, in such a way that finally it is obtained that

$$ds'^2 = ds^2 + 2U_{ik} dx_i dx_k \; . \tag{10.5}$$

In that result we have that

$$U_{ik} = \frac{1}{2}\left(\frac{\partial u_i}{\partial x_k} + \frac{\partial u_k}{\partial x_i} + \frac{\partial u_m}{\partial x_k}\frac{\partial u_m}{\partial x_i}\right), \tag{10.6}$$

are the components of the strain tensor. From its definition, it is clear that it is a symmetrical tensor; that is to say

$$U_{ik} = U_{ki} \, . \tag{10.7}$$

For the case of small deformations it is possible to neglect the last term in the general expression (10.3) and write that

$$ds'^2 = ds^2 + 2u_{ik}dx_i dx_k \, , \tag{10.8}$$

where

$$u_{ik} = \frac{1}{2}\left(\frac{\partial u_i}{\partial x_k} + \frac{\partial u_k}{\partial x_i}\right) \tag{10.9}$$

are the components of the small deformations strain tensor.

On the other hand, after deformation the distance between two neighbouring points in the region *R*, can be written as follows

$$ds'^2 = dx'_m\, dx'_m = \frac{\partial x'_m}{\partial x_i}\frac{\partial x'_m}{\partial x_k} dx_i\, dx_k = g_{ik}\, dx_i\, dx_k \, ; \tag{10.10}$$

where

$$g_{ik} \equiv \sum_{m=1}^{3} \frac{\partial x'_m}{\partial x_i}\frac{\partial x'_m}{\partial x_k} \tag{10.11}$$

are the components of the fundamental metric tensor. That tensor describes the metrical properties of the already deformed region *R*. If the equations (10.8) and (10.10) are compared we have that

$$u_{ik} = \frac{1}{2}\left(g_{ik} - \delta_{ik}\right) , \tag{10.12}$$

due to the fact that ds^2 can be written as follows

$$ds^2 = dx_i\, dx_i = \delta_{ik}\; dx_i\; dx_k \;;$$

where δ_{ik} are the components of Kronecker´s delta.

From this last expression it can easily be seen that δ_{ik} plays the role of the fundamental metric tensor for undeformed region R whose geometry corresponds to a flat Euclidean Space. When the region is deformed it can be said that the geometry of R is similar to a curved Euclidean Space whose metrical properties are described by the fundamental metric tensor g_{ik}.

Now, if instead considering that the points are separated and we make them coincide in the undeformed initial situation, it is clear that $ds^2 = dx_i^2 = 0$, and then

$$u_{ik} = \frac{1}{2} g_{ik} \quad ; \tag{10.13}$$

in such a way that the small deformation strain tensor becomes the fundamental metric tensor that describes the metrical properties of the deformed region R. This means that u_{ik} has all the properties of the tensor g_{ik}; that is to say, both tensors are symmetrical. In particular, if the contravariant metrical tensor g^{ik} is the reciprocal of the covariant tensor g_{ik}, it is fulfilled that

$$g_{ik}\; g^{kn} = \delta_i^n \;. \tag{10.14}$$

In this case and according to (10.13) and (10.14) it is easy to see that

$$u^{kn} = 2\, g^{kn} \;; \tag{10.15}$$

in such a way that

$$u_{ik}\, u^{kn} = \delta_i^n \;. \tag{10.16}$$

On the other hand, if g is the determinant of g_{ik} it is known that

$$\sqrt{g} = \frac{1}{J} \quad . \tag{10.17}$$

In this case and in similar form we have that

$$\sqrt{2U} = \frac{1}{J} \; ; \tag{10.18}$$

where U is the determinant of u_{ij}. In the relations (10.17) and (10.18), J is the Jacobian of the transformation. Moreover, if

$$\Gamma^{i}_{ik} = \frac{1}{2} g^{im} \frac{\partial g_{im}}{\partial x^{k}} \tag{10.19}$$

it is fulfilled that

$$\Gamma^{i}_{ik} = \frac{1}{2} u^{im} \frac{\partial u_{im}}{\partial x^{k}} \; ; \tag{10.20}$$

where $\Gamma^{i}_{ik} = \Gamma^{i}_{ki}$ is the contracted Christoffel´s symbol.

§11. The scalar equation for the mass density

The equations (10.19) and (10.20) can be written in terms of the corresponding determinants g and U. In fact, (10.19) can be expressed as follows

$$\Gamma^{i}_{ik} = \frac{1}{2g} \frac{\partial g}{\partial x^{k}} = \frac{\partial \ln \sqrt{g}}{\partial x^{k}} \; ; \tag{11.1}$$

whereas (10.20) remains as

$$\Gamma^{i}_{ik} = \frac{1}{2U} \frac{\partial U}{\partial x^{k}} = \frac{\partial \ln \sqrt{U}}{\partial x^{k}} \quad . \tag{11.2}$$

Now, if the equation (10.18) is used in the relation (11.2), the following result is obtained

$$\Gamma^{i}_{ik} = -\frac{1}{J}\, grad_k\ J. \tag{11.3}$$

From the last relation and since Γ^{i}_{ik} depends on coordinates only, it is clear that in this case J is not an explicit function of time. Thus, if (11.3) is multiplied by dx^k/dt we have that

$$\Gamma^{i}_{ik}\frac{dx^k}{dt} = -\frac{1}{J}v^k\frac{\partial J}{\partial x^k} = -\frac{1}{J}\frac{dJ}{dt} = -div\ \boldsymbol{v}. \tag{11.4}$$

However and according to (10.1)

$$\Gamma^{i}_{ik}\left(\frac{dx'^k}{dt} - \frac{du_k}{dt}\right) = -div\ \boldsymbol{v}; \tag{11.5}$$

where $d\boldsymbol{u}/dt = (\boldsymbol{v}\cdot\boldsymbol{grad})\boldsymbol{u}$ only. Thus, expanding the last equation and taking into account the result (11.4) the following is obtained

$$\frac{d}{dt}\left(\Gamma^{i}_{ik}u_k\right) = 0\ ; \tag{11.6}$$

where an integration by parts was made and the fact that Γ^{i}_{ik} is not a function of time was taken into account, so that

$$\frac{d\Gamma^{i}_{ik}}{dt} = 0.$$

The integration of the relation (11.6) gives the following result

$$\Gamma^{i}_{ik}\ u_k = f(\boldsymbol{x},t)\ ; \tag{11.7}$$

where $f(\boldsymbol{x},t)$ is some undimensional scalar function of position and time. Be

$$f(\boldsymbol{x},t) \equiv -\frac{\rho(\boldsymbol{x},t)}{\rho_o} \; ; \tag{11.8}$$

in such a way that in (11.7) we finally obtain

$$\rho(\boldsymbol{x},t) = \frac{\rho_o}{J}\,\boldsymbol{u}\cdot \boldsymbol{grad}\; J \; , \tag{11.9}$$

where the relation (11.3) was used. The last relationship is the scalar equation for the mass density of the fluid contained in the region R. As it was not specified at any moment what kind of fluid it was, that relationship is valid for any perfect or real fluid. In (11.9) ρ_o is the mass density corresponding to undeformed situation. On the other hand, $J(\boldsymbol{x})$ is a scalar field that can be represented by means of contour-lines on a plane. The field lines point out the direction of it in any point of the plane, whereas the magnitude of $\boldsymbol{grad}\, J$ is proportional to the distance between the contour-lines in such a way that the magnitude of this terms is a measurement of the rate of change of J.

The direction in which the term $\boldsymbol{u}\cdot \boldsymbol{grad}\, J$ points out is the same in which the increase or decrease in the mass density is produced. In fact, if the region R contracts, an increase in $\rho(\boldsymbol{X},t)$ is produced and clearly the term $\boldsymbol{u}\cdot\boldsymbol{grad}\, J$ points in the same direction in which the increase is produced. If R is expanded, $\rho(\boldsymbol{x},t)$ is diminished and $\boldsymbol{u}\cdot\boldsymbol{grad}\, J$ points in the same direction of this decrease.

When the last equation is derived with respect to time the following is obtained

$$\begin{aligned}\frac{d\rho}{dt} &= -\frac{\rho_o}{J^2}\,\boldsymbol{u}\cdot \boldsymbol{grad}\; J\left(\frac{dJ}{dt}\right) + \frac{\rho_o}{J}\left(\frac{d\boldsymbol{u}}{dt}\right)\cdot \boldsymbol{grad}\; J \\ &= -\frac{\rho}{J}\frac{dJ}{dt} + \frac{\rho_o\,\boldsymbol{v}_{rel}}{J}\cdot \boldsymbol{grad}\; J \; ;\end{aligned} \tag{11.10}$$

where the equation (11.9) was used again and

$$\boldsymbol{v}_{rel} \equiv \frac{d\boldsymbol{x}'}{dt} - \frac{d\boldsymbol{x}}{dt} \tag{11.11}$$

is the relative velocity of two points in *R*, when because of the effect of some external agent, the region is deformed. As it is easy to see if the relation (10.1) is derived with respect to time, the deformation velocity is the same for two near points in such a way that $\boldsymbol{v}_{rel} = 0$. In this case in (11.10) the continuity equation is obtained; that is to say

$$\frac{d\rho}{dt} + \rho \, div \, \boldsymbol{v} = 0,$$

where Euler´s relationship (5.6) was used.

If the fluid is incompressible the mass density is a constant equal to ρ_o, so that the relation (11.9) is transformed as follows

$$J = \boldsymbol{u} \cdot \boldsymbol{grad} \, J \,. \tag{11.12}$$

If the last relation is derived with respect to time and the results (5.6) and (11.11) are used we only obtain that

$$div \, \boldsymbol{v} = 0; \tag{11.13}$$

clearly, due to the fact that dJ/dt=0. This last relationship is the form of the continuity equation when the mass density of the fluid is considered as a constant.

§12. The dilute gas

Even when in the theoretical treatment of a gas considered as a fluid, the statistical approach is not used, since in all cases to be studied, only the continuous model discussed in paragraph 1 is used, then the case of a dilute gas formed by molecules of mass *m* will be dealt with; since the continuity equation rises in the Transport Theory as the zero moment of Boltzmann´s equation. Then be $X(\boldsymbol{r}, \boldsymbol{v}, t)$ some property of a molecule located in $\boldsymbol{r}$ that moves with velocity $\boldsymbol{v}$ to time t. Its mean value is defined as follows

$$\langle X(\boldsymbol{r},t) \rangle \equiv \frac{1}{n(\boldsymbol{r},t)} \int d^3\boldsymbol{v} \; g(\boldsymbol{r},\boldsymbol{v},t) \; X(\boldsymbol{r},\boldsymbol{v},t), \tag{12.1}$$

where integration is extended over all the possible velocities; $n(\boldsymbol{r},t)$ is the mean number of molecules for unit volume and $g(\boldsymbol{r}, \boldsymbol{v},t)$ is the distribution function. The variation of this mean value as a function of $\mathbf{r}$ and t is calculated in the following manner. Be $d\mathbf{r}^3$ an elemental fixed volume located in $d\mathbf{r}$ which contains $n(\mathbf{r},t)d\mathbf{r}^3$ molecules. The mean value nX of the property X for all molecules in $d\mathbf{r}^3$ calculated in the time dt, increases its value according to the relation

$$\frac{\partial}{\partial t}\langle nX\rangle d^3\mathbf{r}dt = A_{int} + A_{flux} + A_{col} \tag{12.2}$$

where

$$A_{int} = nd^3\mathbf{r}dt\langle DX\rangle \tag{12.3}$$

with

$$DX = \frac{\partial X}{\partial t} + \mathbf{v}\cdot\frac{\partial X}{\partial \mathbf{r}} + \frac{\mathbf{F}}{m}\cdot\frac{\partial X}{\partial \mathbf{v}}. \tag{12.4}$$

Moreover

$$A_{flux} = -\frac{\partial}{\partial x_\alpha}\langle nv_\alpha X\rangle d^3\mathbf{r}dt\ ; \tag{12.5}$$

and finally

$$A_{col} = \frac{1}{2}d^3\mathbf{r}dt\iiint\int d^3\mathbf{v}\,d^3\mathbf{v}_1 d^3\mathbf{v}'\,d^3\mathbf{v}_1' gg_1 V\sigma'\Delta X. \tag{12.6}$$

Be $X=m$ the mass of a molecule. In this case: $Dm=\Delta m=0$ in such a way that $A_{int} = A_{col} = 0$, so that in (12.2) we only have that

$$\frac{\partial}{\partial t}\langle nm\rangle + \frac{\partial}{\partial x_\alpha}\langle nv_\alpha m\rangle = 0. \tag{12.7}$$

However

$$\langle nm\rangle = \langle\rho(\mathbf{r},t)\rangle, \tag{12.8}$$

with $\langle\rho(\mathbf{r},t)\rangle$ the average mass density of the dilute gas. In this case in (12.7) the following is obtained

$$\frac{\partial}{\partial t}\langle \rho \rangle + \frac{\partial}{\partial x_\alpha}\langle \rho v_\alpha \rangle = 0 . \qquad (12.9)$$

For the case of a dilute gas formed by molecules of mass m and whose distribution function is $g(\boldsymbol{r}, \boldsymbol{v}, t)$, the above is the continuity equation. The rate of change of g due to molecular collisions satisfies all the usual hypothesis of the Transport Theory. The relation (12.9), in the specialized literature is known as the zero moment of Boltzmann´s equation.

Now, suppose that in region R of the three-dimensional Euclidean Space considered, the fluid we really have is a dilute gas formed by molecules of mass m. As it has been assumed, the deformations suffered by the region which contains the system are small, the molecules move with small velocities too, in such a way that only the binary collisions between them will be relevant. In this case in equation (11.8) we have that

$$f(\boldsymbol{x},t) = -\frac{\langle nm \rangle}{\langle n_o m \rangle} = -\frac{n}{n_o}; \qquad (12.10)$$

where $n(\boldsymbol{r}, t)$ is the mean number of molecules in R after deformation process whereas $n_o(\boldsymbol{r}, t)$ is the mean number of molecules in R before the deformation process. Thus, the fundamental relationship (11.9) is transformed into the following

$$n(\boldsymbol{x},t) = \frac{n_o}{J}\boldsymbol{u} \cdot \boldsymbol{grad}\ J . \qquad (12.11)$$

If the last relationship is now derived with respect to time, the continuity equation for the mean number of molecules of mass m contained in the region R is obtained; that is to say

$$\frac{\partial n}{\partial t} + div(n\boldsymbol{v}) = 0 .$$

In consequence, the scalar equation for the average mass density of a dilute gas as the one considered, becomes again the fundamental relationship (11.9); that is to say

$$\langle \rho(\boldsymbol{x},t) \rangle = \frac{\langle \rho_o \rangle}{J} \boldsymbol{u} \cdot \boldsymbol{grad}\ J .$$

Evidently, the fundamental relationship mentioned is the missing equation needed to complete the set of five basic equations to describe the dynamic state of any fluid. With this relationship the inconsistency holding back the evolution of Classical Fluid Dynamics has been removed.

Selected Topics

The continuity and motion equations†

The continuity equation is an expression of the mass conservation. It is a condition that is externally imposed on the system and in that sense is not a fundamental equation to make the description of dynamical state of any continuous medium. In order to demonstrate the last assertion let us consider Newton´s second law applied to the momentum of volume element of a fluid; that is to say

$$\frac{d}{dt} \int_R \rho \boldsymbol{v} dV = \boldsymbol{F}_{total} ,$$

where the total force is the resultant of all internal stresses. This force can be expressed as an integral over the surface only if

$$\left(\boldsymbol{F}_{total} \right)_i = \frac{\partial \sigma^o_{ij}}{\partial x^j} ,$$

in such a way that

$$\left(\boldsymbol{F}_{total} \right)_i = \int_R F_i dV = \int_R \frac{\partial \sigma^o_{ij}}{\partial x^j} dV = \oint_S \sigma^o_{ij} da_j .$$

† *Here, some results that will be obtained in chapters IV, VI and VIII are given in advance.*

Here S is the boundary surface of the region R, $\boldsymbol{da}$ is the surface element vector and $\sigma^{o}_{ij}=\sigma^{o}_{ji}$ is a symmetrical tensor that we call the generalized stress tensor. In this case it is clear that Newton's second law for any fluid can be expressed as follows

$$\frac{d}{dt}\int_R \rho\boldsymbol{v}\,dV = \oint_S \sigma^{o}_{ij}\,da_j\,.$$

According to Reynolds' transport theorem, on the left hand side of the last relation we have the following

$$\int_R \left[\rho\frac{d\boldsymbol{v}}{dt}+\boldsymbol{v}\left(\frac{d\rho}{dt}+\rho div\boldsymbol{v}\right)\right]dV = \int_R \frac{\partial\sigma^{o}_{ij}}{\partial x^j}\,dV\,.$$

So that the last relationship is fulfilled and since dV is arbitrary and then different from zero we need that

$$\rho\left[\frac{\partial\boldsymbol{v}}{\partial t}+(\boldsymbol{v}\cdot\boldsymbol{grad})\boldsymbol{v}\right]-\frac{\partial\sigma^{o}_{ij}}{\partial x^j}+\boldsymbol{v}\left[\frac{d\rho}{dt}+\rho div\boldsymbol{v}\right]=0\,.$$

In this case the motion equation

$$\rho\left[\frac{\partial\boldsymbol{v}}{\partial t}+(\boldsymbol{v}\cdot\boldsymbol{grad})\boldsymbol{v}\right]=div\tilde{\sigma}^{o}\,,$$

is only fulfilled if

$$\frac{d\rho}{dt}+\rho\,div\boldsymbol{v}=0\,.$$

This is the continuity equation, so the last relationship is the generalized Cauchy's equation for the MHD. If the external magnetic field is zero and due to the fact that

$$\sigma^{o}_{ij}=\sigma_{ij}+m_{ij}\,,$$

with $\widetilde{\sigma}(\boldsymbol{x},t)$ the mechanical stress tensor, whereas $\widetilde{m}(\boldsymbol{x},t)$ is Maxwell´s magnetic stress tensor whose components are

$$m_{ij} = \frac{1}{4\pi}\left[H_i H_j - \frac{1}{2}H^2\delta_{ij}\right];$$

with $\boldsymbol{H}(\boldsymbol{x},t)$ the external magnetic field, we have that

$$\rho\left[\frac{\partial \boldsymbol{v}}{\partial t} + (\boldsymbol{v}\cdot\boldsymbol{grad})\boldsymbol{v}\right] = div\,\widetilde{\sigma}\,.$$

This is no other than that Cauchy´s equation of motion for the newtonian viscous fluid. It is clear that for the case of volumetric compressions or volumetric expansions the stress tensor is the following

$$\sigma_{ij} = -p\delta_{ij}$$

with $p(\boldsymbol{x},t)$ the hydrostatic pressure, so it is fulfilled that

$$\rho\left[\frac{\partial \boldsymbol{v}}{\partial t} + (\boldsymbol{v}\cdot\boldsymbol{grad})\boldsymbol{v}\right] = -\boldsymbol{grad}\,p\,,$$

and this is Euler´s motion equation for the ideal fluid.

Bossinesq´s approaches

In the scope of Classical Hydrodynamics and with the objective of simplifying the equations, Bossinesq´s scheme, which is based on the formulation of some reasonable suppositions is used. To discuss thermal effects in gases, H. Bossinesq proposed a wise approach based on the following observations:

- **The coefficients of viscosity and diffusion vary slowly with temperature for most substances.**
- **When a thermal convection phenomenon is studied, the most important effects are expected to arise from the fact that warm air is**

lighter than cold air; that is to say, from the way in which the gas at different temperatures is affected by gravity.

These observations led Bussinesq to propose a theoretical scheme which contains the following approaches.

- **Ignore the variation of all quantities in the equations of motion with temperature except where they are concerned with gravitational field.**
- **Related to mass density, this must be considered in some terms concerned with the external force, as a variable with respect to temperature, but must be treated as a constant in all the other quantities.**

This scheme is known by the name of Bossinesq´s approaches. It is proposed as an effort to handle the fact that traditionally we only have four equations for the five variables needed to study the dynamical state of any fluid. That characteristic undetermination of the theory has been solved obtaining the scalar equation for the mass density in Newton´s theoretical frame, and of the field equation for mass density in the scope of Lagrange´s Analytical Mechanics.

The mass balance equation

For any deformable body $\mathcal{B}$ there is a non-negative measurement m called the mass distribution in the body $\mathcal{B}$, such that for each m of $\mathcal{B}$ a Lebesgue measure of B $\subset E_3$ is associated. By the Radon-Nikodym theorem, the mass of $\mathcal{B}$ may be expressed in terms of a mass density ρ_ψ. Thus, and for every region $R \subset$ B

$$m(R) = \int_{\psi(R)} \rho_\psi dV ,$$

with dV the volume element in the region. The mass density in region R is $\rho_\psi = \rho_\psi(\boldsymbol{x}, t)$. It is evident that $m(R)$ does not depend on the representation whereas $\rho_\psi(\boldsymbol{x}, t)$ certainly depends on it. Consequently, when an infinitesimal displacement from $B(t)$ to $B'(t + \delta^+ t)$ occurs it is fulfilled that

$$\delta^+ m(R) = 0 ,$$

since we have said that $m(R)$ does not depend on the representation. Then

$$\delta^+ \int_{\psi(R)} \rho_\psi(\boldsymbol{x},t)dV = 0 .$$

Be $dV = J(t)dV_o$, with dV_o a time independent volume element and $J(t)$ the Jacobian of the transformation, in such a way that the last relation is transformed into the following

$$\delta^+ \int_{\psi(R_o)} \rho_\psi(\boldsymbol{x},t)J(t)dV_o = 0 ,$$

where R_o is a time independent region. In this case, the symbols of variation and integration can be interchanged to obtain

$$\int_{\psi(R_o)} \left[J(t)\delta^+\rho_\psi(\boldsymbol{x},t) + \rho_\psi(\boldsymbol{x},t)\delta^+ J(t)\right]dV_o = 0 .$$

On the other hand, it is clear that

$$\delta^+ J(t) = \frac{dJ}{dt}\delta^+ t ,$$

and from the well known Euler´s relationship

$$\frac{dJ}{dt} = J\, div\, \boldsymbol{v}$$

it is evident that

$$\int_{\psi(R)} \left[\delta^+\rho_\psi + \rho_\psi div\, \boldsymbol{v}\delta^+ t\right]dV = 0 ,$$

where the formula for the change of a volume element was used again. Now,

$$\begin{aligned}\delta^+\rho_\psi &= \rho'_\psi\left(\boldsymbol{x}+\delta^+\boldsymbol{x};t+\delta^+ t\right) - \rho_\psi(\boldsymbol{x},t)\\ &= \cancel{\rho_\psi(\boldsymbol{x},t)} + \left(\frac{\partial \rho_\psi}{\partial x^i}\right)_{\delta^+\boldsymbol{x}=0}\delta^+ x^i + \left(\frac{\partial \rho_\psi}{\partial t}\right)_{\delta^+ t=0}\delta^+ t - \cancel{\rho_\psi(\boldsymbol{x},t)} .\end{aligned}$$

On the other hand, it is known that

$$\delta^{+}\boldsymbol{x} = \boldsymbol{v}(\boldsymbol{x},t)\delta^{+}t ,$$

in whose case the following result is finally obtained

$$\delta^{+}\rho_{\psi} = \delta^{+}t\left[\boldsymbol{v}\cdot\boldsymbol{grad}\,\rho_{\psi} + \frac{\partial\rho_{\psi}}{\partial t}\right] \equiv \delta^{+}t\left(\frac{d\rho_{\psi}}{dt}\right).$$

Therefore, it implies and it is implied that

$$\delta^{+}m = 0 \leftrightarrow \frac{d\rho}{dt} + \rho\, div\,\boldsymbol{v} = 0 .$$

This is the continuity equation which expresses the mass conservation law; that is to say, it is fulfilled that

$$\delta^{+}m = 0 \leftrightarrow \frac{dm}{dt}\delta^{+}t = 0 .$$

As $\delta^{+}t \neq 0$ we have that $m = constant$ and then, the continuity equation can be expressed as follows

$$\frac{\partial\rho}{\partial t} + div(\rho\boldsymbol{v}) = 0 ;$$

where the definition of the hydrodynamics derivative was used. The last relationship is the mass balance equation.

Demonstration of a few formulae

From equations (10.13) and (10.14) it is clear that

$$u_{ik}u^{kn} = \frac{1}{2}g_{ik}u^{kn} .$$

However it is fulfilled that

$$u_{ik}u^{kn} = \delta_i^n$$

so that

$$\delta_i^n = \frac{1}{2} g_{ik} u^{kn} .$$

On the other hand and due to the fact that

$$\delta_i^n = g_{ik} g^{kn}$$

it is evident that

$$u^{kn} = 2g^{kn};$$

and that is the relation to be demonstrated. On the other hand, if

$$U = det\, u_{ik} ,$$

it is fulfilled that

$$U\, det\, u^{kn} = det\left(u^{kn} u_{ik}\right) = det\left(\delta_i^n\right) = 1 .$$

Thus,

$$det\ u^{kn} = \frac{1}{U} .$$

The molecular chaos hypothesis

Related to distribution function $g(\boldsymbol{r}, \boldsymbol{v}, t)$ and in the Transport Theory frame, one shall make the following assumptions:

- **The gas is sufficiently diluted so that only two-particle collisions need to be taken into account.**
- **Any possible effects of external force on the magnitude of the collision cross section can be ignored.**
- **The distribution function $g(r, v, t)$ does not vary appreciably during a time interval of the order of the duration of a molecular collision. It is supposed too that it does not significantly vary over a spatial distance of the order of the range of intermolecular forces.**

- **When considering a collision between two molecules one can neglect possible correlations between their velocities prior to the collision.**

This fundamental approach in the theory is called the assumption of molecular chaos. It is justified when the gas density is sufficiently low. It means that the mean free path is much greater than the range of intermolecular forces. Thus, two molecules that before collision are at a relative separation of the order of the mean free path, are supposed to be sufficiently far from each other so that the correlation between their initial velocities is irrelevant.

References

1. Fierros Palacios, Angel. "La ecuación de campo para la densidad de masa". Rev. del IMP, Vol. XXIV, Num 1. (Enero-Junio 1994).
2. Halmos R. Paul. "Measure Theory". Springer-Verlag, New York Heildelberg Berlin (1974).
3. Landau, L.D. and Lifshitz, E.M. "Theory of Elasticity". Addison-Wesley Publishing Co. (1959).
4. Landau, L.D. and Lifshitz, E.M. "The Classical Theory of Fields". Addison-Wesley Publishing Co. 2th edition (1979).
5. Landau, L. D. and. Lifshitz, E.M. "Fluid Mechanics" Addison-Wesley Publishing Co. (1959).
6. Reif Federik. "Statistical and Thermal Physics". International Student Edition, McGraw- Hill Book Company (1965).

Chapter III

Analytical Mechanics

§13. Analytical treatment of mechanics

Ever since Newton laid the solid foundation of dynamics by formulating the laws of motion, the science of mechanics developed along two main lines. One branch, which we shall call Vectorial Mechanics, arises directly from Newton´s laws basing its theoretical frame on the knowledge of all the forces acting on any particle or any given physical system. The analysis and synthesis of forces and momenta is thus, the basic concern of Vectorial Mechanics.

While in Newton´s mechanics the action of a force is measured by the momentum produced by that force, the great philosopher and scientist Leibniz, a contemporary of Newton, advocated another quantity, the *vis viva* (living force) as the proper measure for the dynamic action of a force. This *vis viva* of Leibniz coincides† with the quantity we call today kinetic energy. Thus, Leibniz replaced the momentum of Newton by the latter quantity. At the same time he replaced the concept of force by the work of the force, concept that was later replaced by a still more basic quantity known by the name of work function. Leibniz is thus the originator of that second branch of mechanics usually called Analytical Mechanics whose theoretical scheme is based on the entire study of motion on two fundamental scalar quantities: the kinetic energy and the potential energy or the work function.

Since motion is by its very nature a phenomenon usually derived from the action of forces, it seems puzzling that two scalar quantities be sufficient to completely determine it. The energy theorem, which states that the sum of the kinetic and potential energies remains unchanged during the motion, yields only one equation, while to describe the motion of a

† *Apart from the unessential factor 2.*

single particle in space three equations are required. In the case of mechanical systems composed of two or more particles and even more complicated, the discrepancy becomes even greater. However, those two fundamental scalar quantities contain the complete dynamic information of even the most complicated physical system. In consequence, these concepts must not be considered as a simple equation but as the basis of a fundamental principle. In Analytical Mechanics the complete set of motion equations for any system can be determined from a single principle including all these equations. This principle takes its form from the minimization process of a certain fundamental quantity known by the name of the action integral. Due to the fact that this principle of least action is independent of any special reference system, the equations of Analytical Mechanics are valid for any set of coordinates. This allows us to adjust the coordinates employed to the specific nature of each problem. For all that has been said before, it can be stated that the Analytical Mechanics is much more than a powerful tool for the solution of dynamic problems in physics and engineering; it must rather be considered as a completely mathematical science.

§14. The Hamilton-Type variational principle

From a formal view point the problem of minimizing a definite integral is considered as the proper domain of the calculus of variations, while the problem of minimizing a function belongs to ordinary calculus. Historically, the two problems arose simultaneously and a clear-cut distinction was not made till the time of Lagrange who developed the technique of the calculus of variations. However, the basic differential equation of variational problems was discovered by Euler and Lagrange, whereas a general method for the solution of physical problems with the aid of the calculus of variations was introduced by Lagrange in 1788 in his book *Mecanique Analytique*.

The problem of establishing Fluid Dynamics as a branch of Lagrange´s Analytical Mechanics can be approached from a lagrangian formalism as the one used in Classical Theory of Fields and with the aid of a Hamilton-Type Variational Principle.

In general, the route to follow should be: an action functional is proposed as a space-time integral of a lagrangian density smoothly dependent of the coordinates and time, and of certain field variables which describe the dynamical state of the continuous system. So a unique Hamilton-Type Extremal Action Principle with adequate boundary conditions is postulated, and the field differential equations are obtained. Next, a lagrangian density (*T-V*) type is proposed; with *T* and *V* the kinetic and potential energies of the continuous medium, respectively. At substituting that function in the field differential equations before obtained, the motion equation for any newtonian fluid is recovered. In the theoretical frame of Lagrange´s Analytical Mechanics and for any fluid, a general lagrangian density $\circ$ is proposed as a continuous function and with continuous derivatives as far as third order in their arguments.

The action integral is defined as usual; that is to say

$$W = \int_{t_1}^{t_2} \int_R \circ \, dV dt ; \tag{14.1}$$

where *R* is some region in the three-dimensional Euclidean Space, *dV* is the volume element, and *dt* the time differential with *t* considered as the evolution parameter.

According to Hamilton-Type Extremal Action Principle, the action functional must be invariant under a continuous and infinitesimal geometric variation with respect to the set of geometrical parameters $\{\alpha\}$ introduced in paragraph 5, that is

$$\delta W = \frac{\partial W}{\partial \alpha} \delta \alpha = 0 . \tag{14.2}$$

The mathematical scheme is completed when the following boundary condition of general character is imposed over the coordinates

$$\delta x^i(t_1) = \delta x^i(t_2) = 0 . \tag{14.3}$$

Evidently and according to the usual definition of geometrical variations, $\delta t = 0$ for every *t*, due to the fact that the time and the set of geometrical parameters used are independent between them. As a result of the invariance of the action, the field differential equations are obtained.

§15. Temporary variations

In order to complete the theoretical scheme is necessary to explore the resulting consequences when a new type of variations are introduced in the dynamic description, which are the temporary. With their help, it can be demonstrated that the action functional is invariant under transformations with respect to the evolution parameter. As a consequence of that invariance, and due to the fact of time uniformity it will be possible to obtain the energy balance equation for any fluid. To obtain such objective it is necessary to demonstrate the following theorem:

The action integral for any fluid is invariant under continuous and infinitesimal temporary transformations.

Let us consider the following infinitesimal transformation of the evolution parameter t

$$t \to t' = t + \delta^+ t\,, \tag{15.1}$$

where $\delta^+ t$ is the infinitesimal part of the transformation. Regarding coordinates the following result is obtained

$$x^{i\prime}\left(t + \delta^+ t\right) = x^i\left(t\right) + \delta^+ x^i\left(t\right), \tag{15.2}$$

with

$$\delta^+ x^i\left(t\right) = v^i \delta^+ t\,; \tag{15.3}$$

where $v^i(t)$ is the velocity whose temporary variation is

$$v^{i\prime}\left(t + \delta^+ t\right) = v^i\left(t\right) + \delta^+ v^i\left(t\right), \tag{15.4}$$

and

$$\delta^+ v^i\left(t\right) = \frac{d}{dt} v^i\left(t\right) \delta^+ t\,. \tag{15.5}$$

The temporary variation of the action integral (14.1) can be expressed as follows

$$\delta^+ \int_{t_1}^{t_2} \int_R \circ\, dV\, dt\,. \tag{15.6}$$

The former operation can be made if Reynolds′ transport theorem is used. Then and according to (4.4) we have that

$$\delta^+ \int_{t_1}^{t_2}\int_R \circ dVdt = \delta^+ \int_{t_1}^{t_2}\int_{Ro} \circ J dV_o dt$$

$$= \int_{t_1}^{t_2}\int_{Ro} \left[\delta^+ \circ J dV_o\, dt + \circ \delta^+ J dV_o\, dt + \circ J dV_o\, \delta^+ dt\right]$$

$$= \int_{t_1}^{t_2}\int_{Ro} \left[\delta^+ \circ J dV_o\, dt + \circ J\, div\left(\delta^+ \boldsymbol{x}\right) dV_o dt + \circ J dV_o \frac{d}{dt}\left(\delta^+ t\right) dt\right]$$

$$= \int_{t_1}^{t_2}\int_{R} \left\{\delta^+ \circ + \circ\left[div\left(\delta^+ \boldsymbol{x}\right) + \frac{d}{dt}\left(\delta^+ t\right)\right]\right\} dV dt\,;$$

where the relation (5.5) was used. Now and according to (15.3) the following is obtained

$$\delta^+ \int_{t_1}^{t_2}\int_R \circ dV dt = \int_{t_1}^{t_2}\int_R \left[\delta^+ \circ + \circ\left(\frac{\partial v^i}{\partial x^i}\delta^+ t + \frac{d}{dt}\delta^+ t\right)\right] dV dt\,. \quad (15.7)$$

If in the former relation the third term is integrated, the result (5.5), and Reynolds′ transport theorem are used again, it can be demonstrated that

$$\int_{t_1}^{t_2}\int_R \circ \frac{d}{dt}\left(\delta^+ t\right) dV dt = \left[\int_R \circ dV \delta^+ t\right]_{t_1}^{t_2} - \int_{t_1}^{t_2}\int_R (\mathcal{D}\circ) dV dt \delta^+ t. \quad (15.7')$$

The first term of the right hand side of (15.7′) is zero according to the following boundary condition

$$\delta^+ t_1 = \delta^+ t_2 = 0. \quad (15.8)$$

In the remaining term it is clear that $\mathcal{D}$ is Reynolds′ differential operator whose general form is given in (7.1). Moreover, the two first terms in the integral (15.7) are

$$\delta^{+}\circ + \circ\, div\, \mathbf{v}\, \delta^{+}t = \left(\frac{d\circ}{dt} + \circ\, div\, \mathbf{v}\right)\delta^{+}t\,; \qquad (15.9)$$

which is no other thing but the product of the derivative $\mathcal{D}\circ$ and $\delta^{+}t$.

In this case it is fulfilled that

$$\delta^{+}W = 0\,. \qquad (15.10)$$

In other words, the action integral is invariant under temporary variations. With this, the demonstration of the theorem is complete. Since this result is valid for any lagrangian density, it is of a general character.

§16. The field equation for the mass density

The theoretical frame of Hamilton-Type Variational Principle gives the methodology needed to obtain the field differential equation for the mass density. According to what was said in paragraph 14, be

$$\circ = \circ\left(J; \boldsymbol{grad}\ J\right) \qquad (16.1)$$

a lagrangian density which is a function of the Jacobian J, a scalar field, and its first gradient.

On the other hand it is useful to define a lagrangian per unit mass, as follows

$$\circ = \rho\,\lambda; \qquad (16.2)$$

where clearly

$$\lambda = \lambda\left(J; \boldsymbol{grad}\ J\right) \qquad (16.3)$$

is a function that we will call the specific lagrangian. Then and taking into account the condition (14.2), in (14.1) the following is obtained

$$\int_{t_1}^{t_2}\int_{R} \rho\,\delta\lambda\, dV\, dt = 0\,, \qquad (16.4)$$

where the definition (16.2) was used and the fact that

$$\delta\rho + \rho \, div(\delta \boldsymbol{x}) = 0. \tag{16.5}$$

The results (16.4) and (16.5) are valid in general so they will be systematically used in all cases studied with the Hamilton-Type Variational formalism.

Now and according to the functional form of the specific lagrangian, it is easy to see that

$$\rho\delta\lambda = \rho\frac{\partial\lambda}{\partial J}\delta J + \rho\frac{\partial\lambda}{\partial\nabla J}\delta\nabla J; \tag{16.6}$$

where ∇ is a symbol known by the name of nabla used to abbreviate the word gradient. Applying the methods of the calculus of variations in (16.6) the following result for the variation of ***grad*** J is obtained

$$\delta\frac{\partial J}{\partial x^n} = \frac{\partial J}{\partial x^n}\frac{\partial}{\partial x^i}\left(\delta x^i\right); \tag{16.7}$$

where the relationship (5.5) was used. According to that relation and with the former result, in (16.6) the following is obtained

$$\rho\delta\lambda = \frac{\partial}{\partial x^i}\left[\rho\left(J\frac{\partial\lambda}{\partial J} + \nabla J\frac{\partial\lambda}{\partial\nabla J}\right)\delta x^i\right]$$
$$-\frac{\partial}{\partial x^i}\left[\rho\left(J\frac{\partial\lambda}{\partial J} + \nabla J\frac{\partial\lambda}{\partial\nabla J}\right)\right]\delta x^i; \tag{16.8}$$

where an integration by parts was made. Now, let us consider that the first term of the right hand side of the former equation can be integrated using Green's theorem; that is to say

$$\int_{t_1}^{t_2}\int_R\left\{\frac{\partial}{\partial x^i}\left[\rho\left(J\frac{\partial\lambda}{\partial J} + \nabla J\frac{\partial\lambda}{\partial\nabla J}\right)\delta x^i\right]\right\}dVdt$$
$$= \int_{t_1}^{t_2}\oint_S\rho\left[J\frac{\partial\lambda}{\partial J} + \nabla J\frac{\partial\lambda}{\partial\nabla J}\right]\delta x^i da_i \, dt;$$

with S the limit surface of the region R and $d\boldsymbol{a}$ the differential of area. The surface integral is zero for the following reason. Be a continuous medium contained in a region R that is not deformed in the infinity and the surface of integration extended to infinity. In this case J=1 and $\textbf{grad}\, J = 0$ in such a way that the integral becomes null.

If the remainder of (16.8) is replaced in (16.4) the following is obtained

$$\int_{t_1}^{t_2}\int_R \left\{ \frac{\partial}{\partial x^i}\left[\rho\left(J\frac{\partial \lambda}{\partial J} + \nabla J \frac{\partial \lambda}{\partial \nabla J} \right)\right]\right\} \delta x^i dV dt = 0\,. \quad (16.9)$$

Since the local variations of $\boldsymbol{x}$ are arbitrary and linearly independent between them and due to the fact that dV as well as dt are totally arbitrary increments and in consequence different from zero, the former equation is only fulfilled if the integrand is nullified. This last condition is also a general result which arises from the theoretical formalism utilized so consequently they will be systematically used in subsequent chapters where relations similar to the result (16.9) should appear. Thus, it is clear that

$$\rho\, \textbf{grad}\left[J\frac{\partial \lambda}{\partial J} + \nabla J \frac{\partial \lambda}{\partial \nabla J} \right] + \left[J\frac{\partial \lambda}{\partial J} + \nabla J \frac{\partial \lambda}{\partial \nabla J} \right] \textbf{grad}\, \rho = 0; \quad (16.10)$$

where the derivative has been calculated. However, in the first member of the previous equation we have that

$$\rho\left[\frac{\partial J}{\partial x^i}\frac{\partial \lambda}{\partial J} + \frac{\partial \nabla J}{\partial x^i}\frac{\partial \lambda}{\partial \nabla J} \right] = 2\rho \frac{\partial \lambda}{\partial x^i} = 0.$$

To obtain the former result the chain rule has been used. The term is zero because the quantities λ, $\partial\lambda/\partial J$ and $\partial\lambda/\partial\nabla J$ are not explicit functions of $\boldsymbol{x}$. In consequence and due to the fact that $\textbf{grad}\, \rho \neq 0$, in (16.10) it is finally obtained that

$$\frac{\partial \lambda}{\partial J} + \frac{1}{J}\frac{\partial \lambda}{\partial \nabla J}\, \textbf{grad}\, J = 0. \quad (16.11)$$

This is the field differential equation for the mass density in terms of a specific lagrangian λ whose functional form is given by the relationship (16.3). Now, be

$$\lambda = \square_o \left[J\Gamma^i_{ik} u_k + u_k \frac{\partial J}{\partial x^k} \right] \tag{16.12}$$

the explicit form of specific lagrangian. In this relationship $\boldsymbol{u}(\boldsymbol{x})$ is the displacement vector and $\square_o$ a constant referred to the equilibrium value of the specific internal energy of continuous medium under consideration. As it can be easily seen from (16.12), λ is only a function of quantities related to the geometry of the region R. Then if (16.12) is introduced in the field equation (16.11) we have that

$$\frac{\partial \lambda}{\partial J} = \square_o \Gamma^i_{ik} \; u_k \; = -\frac{\square_o \rho}{\rho_o} \; ;$$

where the relationship (11.8) and the result (11.7) were used. Moreover,

$$\frac{1}{J}\frac{\partial \lambda}{\partial \nabla J} \, \boldsymbol{grad} \; J = \frac{\square_o}{J} \boldsymbol{u} \cdot \boldsymbol{grad} \; J .$$

With these results substituted in the field equation (16.11) it is obvious that the scalar equation for mass density is recovered; that is to say, the equation (11.9).

Finally, the temporary variation of lagrangian density (16.1) is such that $\overset{+}{\delta} \circ = 0$, because it precisely deals with the field equation (16.11) multiplied by $J\rho \; div \; \boldsymbol{v} \; \overset{+}{\delta} t$. Moreover, from (15.7) and (15.7′) it can be shown that the invariance of the action integral (1.41) under continuous temporary transformation results in

$$\frac{d\circ}{dt} = 0 ; \tag{16.13}$$

which is no other thing but the mass balance equation. In fact, if the definitions (16.2), and (6.1) are used, the mass balance equation is obtained.

Selected Topics

The Hamilton-Type least action principle

Hamilton-Type Variational Principle is an analytical scheme proposed for the mathematical treatment of the dynamical state of any fluid. It arises as a consequence of demanding that the action integral be invariant under a continuous and infinitesimal geometrical variation†. As a result of the action invariance, the momentum balance equation for any continuous medium is obtained.

Related to that principle it is necessary to clear up two very important aspects.

1. **It is not like the typical Hamilton Principle of Classical Theory of Fields.**
2. **It is not required that as a result of action integral invariance under coordinate transformations, conservation laws be obtained either, as it occurs in the case of Noether´s theorem related to classical fields.**

Regarding the first point, it is known that Hamilton´s Principle of Classical Theory of Fields, demands that the action integral is an extremal or at least remains steady under local variations; that is, changes of the coordinates or field functions which do not imply scale variations in the coordinates or of field functions themselves, or temporary transformations either. In other words, the parameters used are independent of position and time.

The proposed variation in Hamilton´s Principle must be attributable to parameters which are not associated to the scenario where the physical events occur but, they are entities of non-geometrical nature by means of which the dynamic conditions are varied keeping the scenario where natural phenomena occur, fixed; in such a way that the action invariance is established as a dynamic quality of the system.

Within the context of Hamilton-Type Principle it is assumed that the contrary occurs. The variations of the action are understood as coming from changes suffered by the geometrical parameters or in the evolution parameter which is the time, in such a way that the invariance of the action is established not as a dynamic quality but as a symmetry of the system; symmetry that is intimately connected to the deepest properties of time and space. The former looks more like that which is demanded in the theoretical frame of Noether´s theorem than the conditions imposed on Hamilton´s Principle. It is for these reasons that the proposed theoretical scheme is described as a Hamilton-Type Variational Principle.

Noether´s theorem of the Classical Theory of Fields upon demanding the invariance of the action under geometrical transformations; that is, changes that only affect the scenario of physical events, states that what is

† *According to Classical Theory of Fields, in order for the action to be an invariant it is required that the integrand be a scalar density.*

obtained are conservation laws as for example, the law of momentum conservation as a consequence of the space homogeneity or the law of energy conservation as a result of time uniformity. On the other hand, in order to demonstrate Noether´s theorem it is necessary to have previously found the field equations as a result of the application of Hamilton´s Principle. In other words, in the first place the dynamic conditions are varied for a fixed scenario and later, the scenario of the physical events is varied keeping the dynamic entities unchanged. For all that has been said before, field equations are obtained from Hamilton´s Principle, whereas from Noether´s theorem conservation laws arise.

On the other hand, in the theoretical frame of Hamilton-Type Principle and as a consequence of certain symmetries of the system, only balance equations are obtained: the momentum balance equation or the energy balance equation as a consequence of the homogeneity of space and the uniformity of time, respectively. For the case of any fluid, the unique valid principle is Hamilton-Type Principle, in such a way that it will not be possible to obtain conservation laws, because the expressions resulting as field equations which arise from this principle are known as generalized momentum balance equation or generalized energy balance equation. In other words, from Hamilton-Type Principle certain relationships between coordinates and field functions are obtained, that are neither totally field equations nor totally conservation laws; that is they are balance equations.

Noether´s theorem

The importance of this theorem is due to the fact that it allows to construct quantities which are constant along any extremal, that is a curve which satisfies Euler-Lagrange´s equations: $E_j(L) = 0$, that is to say

$$\frac{d}{dt}\left(\frac{\partial L}{\partial \dot{x}^j}\right) - \frac{\partial L}{\partial x^j} = 0 .$$

In the case of physical applications, relations which may be interpreted as conservation laws can be obtained.

Noether´s theorem can be stated as follows:

If the action integral of a fundamental problem in the calculus of variations is invariant under the r-parametrical group of transformations;

the r distinct quantities θs known as Noether's invariants, are constant along any extremal.

In theoretical physics only two types of transformations totally different between themselves, must be considered:

1. Coordinate transformations of the type

$$x^{j} = x^{j}\left(x^{h}\right);$$

which do not affect the parameter t.

2. Parametric transformations of the type

$$\tau = \tau(t),$$

which do not affect the coordinates x^{h}.

Description of classical fields

The Hamilton-Type Variational Principle can be used to obtain the equations describing a field even when a definite mechanical system does not exist. Due to the fact that neither the lagrangian density and of course nor the specific lagrangian of a field are associated to any specific mechanical system, it is not necessary that those functions are given explicitly as the difference between kinetic and potential energy densities or between the specific kinetic and potential energies. Instead of them, any expression for ∘ or λ that leads to the wanted field equations, can be used.

As it is easy to see, in any of the terms of (16.12) the specific kinetical and potential energies respectively, can be identified. Nevertheless, such specific lagrangian recovers the right scalar equation for mass density and this is all that is required; in complete accordance to what is demanded in the theoretical frame of Classical Theory of Fields.

The calculus of variations

Let us consider the following variations of scalar equation for the mass density

$$\rho(\boldsymbol{x},t) = \frac{\rho_o}{J}\,\boldsymbol{u}\cdot\boldsymbol{grad}\,J\,.$$

a) Local variations. This variation is given by the following expression

$$\delta\rho = -\rho\, div(\delta\boldsymbol{x}) + \mathbf{O}$$

where

$$\mathbf{O} = \frac{\rho_o}{J}\left[\tilde{u}\,\delta\boldsymbol{x}\cdot\boldsymbol{grad}\,J + \boldsymbol{u}\cdot\boldsymbol{grad}\,J\, div\,\boldsymbol{v}\right]$$

is a term of higher order. In this case, in a first approach it is clear that

$$\delta\rho + \rho\, div(\delta\boldsymbol{x}) = 0\,.$$

b) Temporary variations. The variation of mass density is the following

$$\delta^+\rho = -\rho\, div\,\boldsymbol{v}\,\delta^+ t + \frac{\rho_o}{J}\delta^+\boldsymbol{u}\cdot\boldsymbol{grad}\,J + \frac{\rho_o}{J}\boldsymbol{u}\cdot\delta^+\boldsymbol{grad}\,J$$

$$= -\rho\, div\,\boldsymbol{v}\,\delta^+ t + \frac{\rho_o}{J}\delta^+\boldsymbol{u}\cdot\boldsymbol{grad}\,J + \frac{\rho_o}{J}\boldsymbol{u}\cdot\boldsymbol{grad}\,J\, div\,\boldsymbol{v}\,\delta^+ t.$$

The last term of the former result can be neglected because it is a higher order term, whereas

$$\delta^+\boldsymbol{u} = \frac{d\boldsymbol{u}}{dt}\delta^+ t = 0\,;$$

in such a way that

$$\delta^+\rho + \rho\, div\,\boldsymbol{v}\,\delta^+ t = 0\,.$$

The fundamental processes of the calculus of variations

Consider the function $f(x)$ which by hypothesis gives a stationary value to the next definite integral

$$I = \int_a^b F(y, y', x)\,dx\,; \tag{1}$$

where $y' \equiv dy/dx$. In order to prove that a stationary value is obtained, it is necessary to evaluate the same integral for a slightly modified function $\bar{f}(x)$ and show that the rate of change of the integral (1) due to the change in the modified function becomes zero. According to the following diagram,

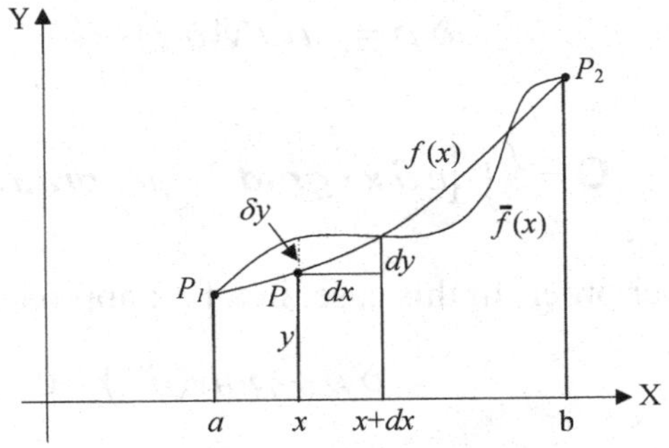

it is clear that the modified function $\bar{f}(x)$ can be written as follows

$$\bar{f}(x) = f(x) + \square\phi(x). \tag{2}$$

Here, □is an arbitrary parameter whose value becomes zero, by means of which the original function $f(x)$ can be modified using small increments. Moreover, $\phi(x)$ is some arbitrary function which satisfies the same general continuity conditions as $\bar{f}(x)$. In other words, $\phi(x)$ has to be a function continuous and differentiable.

In the book´s language it is evident that $\square\phi(x) \equiv \delta f(x)$.

Now, the values of the modified function $\bar{f}(x)$ with the values of the original function $f(x)$ at certain definite point x of the independent variable must be compared by forming the difference between $\bar{f}(x)$ and $f(x)$. Such difference is called the variation of the function $f(x)$ and is denoted by $\delta f(x)$; that is to say

$$\delta f(x) = \bar{f}(x) - f(x) \equiv \square\phi(x). \tag{3}$$

The difference between the process $\delta f(x)$ and $df(x)$ is the following: in spite of both being infinitesimal changes of the function $f(x)$, the process $df(x)/dx$ refers to the infinitesimal change of the given function $f(x)$ caused by the infinitesimal change dx of the independent variable x.

On the other hand, $\delta f(x)$ is an infinitesimal change of $f(x)$ which produces a new function $f(x)+\delta f(x)$. It is in the nature of the process of variation that only the dependent function $f(x)$ should be varied while the variation of x serves no useful purposes. Moreover, if the two limiting values $f(a)$ and $f(b)$ of the function $f(x)$ are prescribed, these two values cannot be varied, which means that

$$[\delta f(a)]_{x=a} = [\delta f(b)]_{x=b} = 0 . \tag{4}$$

In this case, it deals with a variation between definite limits.

The commutative properties of the δ-process

The δ-process displays two characteristic properties: one of them is that variation and differentiation are a commuting process and the other is that variation and integration are a commuting process too. In fact, let us investigate in the first place the variation of the derivative. It is clear that the derivative of the new function $\bar{f}(x)$ minus the derivative of the original function $f(x)$ is what is called the variation of the derivative; that is to say

$$\delta \frac{d}{dx}[f(x)] = \delta[f'(x)] = \bar{f}'(x) - f'(x)$$

$$= f'(x) + \frac{d}{dx}[\delta f(x)] - f'(x) = \frac{d}{dx}[\delta f(x)].$$

On the other hand, the derivative of the variation is

$$\frac{d}{dx}[\delta f(x)] = \frac{d}{dx}[\bar{f}(x) - f(x)] = \bar{f}'(x) - f'(x)$$

$$= \delta \frac{d}{dx}[f(x)] .$$

Consequently it is fulfilled that

$$\delta \frac{d}{dx} = \frac{d}{dx} \delta .$$

Let us consider the variation of a definite integral, that is

$$\delta \int_a^b F(x,...)dx = \int_a^b \bar{F}(x,...)dx - \int_a^b F(x,...)dx$$

$$= \int_a^b [\bar{F}(x,...) - F(x,...)]dx = \int_a^b \delta F(x,...)dx.$$

Therefore it is fulfilled that

$$\delta \int_a^b \equiv \int_a^b \delta .$$

Finally, let us consider the process of variation of the action integral for Fluid Dynamics; that is to say

$$\delta \int_{t_1}^{t_2} \int_R \circ dV dt .$$

As on the other hand

$$L = \int_R \circ dV$$

it is fulfilled that

$$\delta \int_{t_1}^{t_2} \int_R \circ dV dt = \delta \int_{t_1}^{t_2} L dt = \int_{t_1}^{t_2} \delta L dt .$$

Now,

$$\delta L = \delta \int_R \circ dV .$$

According to Reynolds′ transport theorem, the symbols of variation and integral can be interchanged if $dV = J dV_o$, in such a way that

$$\begin{aligned} \delta \int_{R_o} \circ J dV_o &= \int_{R_o} [J \delta \circ + \circ \delta J] dV_o \\ &= \int_{R_o} (\delta \circ + \circ \, div \mathbf{v}) dV ; \end{aligned}$$

where R_o is a fixed region independent of the geometrical parameters.

On the other hand and due to the fact that $\circ = \rho \lambda$

$$\delta \circ = \lambda \delta \rho + \rho \delta \lambda ;$$

in such a way that

$$\delta \int_R \circ dV = \int_R \rho \delta \lambda dV .$$

Consequently it is fulfilled that

$$\delta\int_R \equiv \int_R \delta\,;$$

so that, the symbol of variation can be interchanged with the volume integral, too.

Temporary variation of the action

From invariance condition under temporary transformations of the action integral, $\delta^+W=0$, the following result is obtained

$$\delta^+\int_{t_1}^{t_2}\int_R \circ\, dVdt = 0\,.$$

In this case, at realizing the variation process and using Reynolds′ transport theorem, we have that

$$\int_{t_1}^{t_2}\int_R\left[\delta^+\circ + \circ\left(div\,\boldsymbol{v} + \frac{d}{dt}\delta^+t\right)\right]dVdt = 0\,. \tag{1}$$

In the integrand of (1) the integration of the third term using the relationship $dV=JdV_o$, has the following result

$$\int_{t_1}^{t_2}\int_R\left[\circ\frac{d}{dt}\left(\delta^+t\right)dV\right]dt = \int_{t_1}^{t_2}\int_{Ro}\frac{d}{dt}\left(\circ dV\delta^+t\right)dt - \int_{t_1}^{t_2}\int_{Ro}\left[J\frac{d\circ}{dt} + \circ\frac{dJ}{dt}\right]dV_o dt\delta^+t$$

$$= \left[\int_R \circ dV\delta^+t\right]_{t_1}^{t_2} - \int_{t_1}^{t_2}\int_R\left[\frac{d\circ}{dt} + \circ\, div\,\boldsymbol{v}\right]dVdt\delta^+t\,;$$

where an integration by parts was made. The term enclosed between square parenthesis is zero according to boundary condition (15.8). Substituting in (1) what remains of this calculus, it can be seen that

$$\int_{t_1}^{t_2}\int_R\left[\delta^+\circ + \left(\circ\, \cancel{div}\,\boldsymbol{v} - \circ\, \cancel{div}\,\boldsymbol{v} - \frac{d\circ}{dt}\delta^+t\right)\right]dVdt = 0\,;$$

in such a way that finally the following result is obtained

$$\int_{t_1}^{t_2}\int_R \left[\delta^+ \circ - \frac{d\circ}{dt}\delta^+ t\right] dV dt = 0 \, .$$

That is the resulting expression from the invariance condition under a continuous and infinitesimal temporary transformation of the action.

The variational method

The application of calculus of variations to dynamics showed that it is a powerful theoretical methodology that should not be considered as a simple alternative formulation of Newton´s laws. The establishment of the *relativity principle* confirming that the variational fundaments of mechanics have a significant depth and amplitude that distinguish the analytical scheme as a mathematical tool more powerful than Newton´s Vectorial Mechanics. This supremacy is based on the following observations.

- The *Principle of Relativity* requires that the laws of nature shall be formulated in an invariant fashion, that is to say, independently of the election of a special frame of reference. The methods of the calculus of variations automatically satisfy this principle due to the fact that the minimum of a scalar density does not depend on the coordinates in which that quantity is specified. While the newtonian equations of motion do not satisfy the principle of relativity, the principle of least action remains valid. The only modification refers to the action integral which had to be brought into harmony with the requirement of invariance demanded.
- The *Theory of General Relativity* has shown that matter cannot be separated from field idea. The basic equations of physics must be formulated as partial rather than ordinary differential equations. While Newton´s mechanics of particles scheme can hardly be brought into harmony with the field concept, the variational methods are not restricted to the mechanics of particles but can be extended to the dynamics of continuous systems.
- The *Principle of General Relativity* is automatically satisfied if the fundamental action integral of the variational principle is chosen as an invariant under any coordinate transformation. Since the *differential geometry of Riemann* furnishes one such invariants,

one has no difficulty in setting up the required field equations. However, the present knowledge of mathematics does not give any clue to the formulation of a co-variant[†], and at the same time consistent system of field equations.

Green´s theorem

One of the most valuable transformations within *tensor analysis* is that of *Green´s theorem* known also as the *divergence theorem*. It refers to the relation existing between a volume integral and an integral over its bounding surface. The surface integral occurring is established for a certain quantity $\boldsymbol{f}\cdot\boldsymbol{n}$; where $\boldsymbol{n}$ is the outward unit vector normal to the surface and $\boldsymbol{f}$ is the flux of some physical property. In this case, the integral of $\boldsymbol{f}\cdot\boldsymbol{n}$ over the whole surface of that studied physical property is therefore the total flux out of a closed surface bounding this volume. Green´s theorem says that this total flux equals the integral of *div* $\boldsymbol{f}$ throughout the enclosed volume.

Let $\boldsymbol{f}(\boldsymbol{x},t)$ be any vectorial field, as a function of position and time. Its divergence is defined as follows

$$div\boldsymbol{f} = \underset{\Delta V\to 0}{\circ im}\frac{1}{\Delta V}\int_S \boldsymbol{f}\cdot\boldsymbol{n}\,dS\,;$$

where S is the surface which encloses the volume V considered, dS is the surface differential, and ΔV a small volume element. From the definition of divergence, the net outward flux $\boldsymbol{f}$ from a small volume element ΔV, can be written as follows

$$\Phi = \underset{\Delta V\to 0}{\circ im}\frac{1}{\Delta V}div\boldsymbol{f}\,\Delta V\,.$$

On the other hand, if the total volume V can be divided into N small elements of volume each one is equal to ΔV, the total flux $\boldsymbol{f}$ leaving this volume is

$$\Phi = \oint_S \boldsymbol{f}\cdot\boldsymbol{n}\,dS\,;$$

[†] *Co-variance means that the equations can be written so that both sides have the same, well-defined, transformation properties under Lorentz transformations. They must be relations between four-vectors, or Lorentz scalars, or in general four-tensors of the same rank.*

and clearly it is equal to the sum of the flux leaving all of the small volumes considered; that is to say,

$$\Phi = \sum_{\circ=1}^{N} \Phi_{\Delta V} ;$$

or also,

$$\oint_S \boldsymbol{f} \cdot \boldsymbol{n}\, dS = \underset{\substack{\Delta V \to 0 \\ N \to \infty}}{\circ im} \sum_{\circ=1}^{N} div \boldsymbol{f}\, \Delta V .$$

But the right hand term of the former expression is just the following

$$\underset{\substack{\Delta V \to 0 \\ N \to \infty}}{\circ im} \sum_{\circ=1}^{N} div \boldsymbol{f}\, \Delta V = \int_V div \boldsymbol{f}\, dV ,$$

in such a way that in components it has the following final result

$$\int_V \frac{\partial f_i}{\partial x^i} dV = \oint_S f_i \, da_i ;$$

where $\boldsymbol{da} = \boldsymbol{n} dS$ is the area element. This is one of Green´s theorem forms. It is valid for any physical property that can be represented by means of a tensor or a scalar. In fact, if the physical property of interest is a tensorial function of position and time, the divergence theorem has the following form

$$\int_V \frac{\partial T_{ij}}{\partial x^j} dV = \oint_S T_{ij} \, da_j ;$$

where $\tilde{T}(\boldsymbol{x},t)$ is a second rank tensor. Finally, if we have any physical entity which is a scalar function of position and time, Green´s theorem has the following form

$$\int_V \frac{\partial \phi}{\partial x^j} dV = \oint_S \phi \, da_j ,$$

with $\phi\,(\boldsymbol{x},t)$ any scalar function.

The mass balance

The invariance of action under a continuous and infinitesimal transformation with respect to the evolution parameter, has as a consequence that

$$\frac{d\circ}{dt} = 0 .$$

According to definition (16.2) from the former result we have that

$$\rho \frac{d\lambda}{dt} + \lambda \frac{d\rho}{dt} = 0 ; \tag{1}$$

so, using the hydrodynamics derivative definition (6.1), and by direct calculus it is easy to demonstrate the following result

$$\lambda\left[\frac{\partial \rho}{\partial t} + div(\rho \boldsymbol{v})\right] - \lambda \rho \, div \, \boldsymbol{v} + \rho \frac{\partial \lambda}{\partial t} = \frac{d\circ}{dt} , \tag{2}$$

because λ is not an explicit function of $\boldsymbol{x}$. Let us consider now that the lagrangian density has trhe following explicit form: $\circ = -\rho\, \varepsilon^{2}$, with ρ again the mass density, and ε^{2} a constant with units of velocity squared; in such a way that $\lambda = -\varepsilon^{2}$, and

$$\frac{d\circ}{dt} = \varepsilon^{2} \rho \, div \, \boldsymbol{v} ; \tag{3}$$

where the continuity equation (9.2) was used. Finally, from (2) and (3) the following relationship is obtained

$$\lambda\left[\frac{\partial \rho}{\partial t} + div(\rho \boldsymbol{v})\right] + \rho \frac{\partial \lambda}{\partial t} = 0 .$$

Time uniformity has as a consecuence that the specific lagrangian does not become and explicit function of time; so that $\partial\lambda/\partial t = 0$, and it is fulfilled that

$$\frac{\partial \rho}{\partial t} + div(\rho \boldsymbol{v}) = 0 ;$$

because $\lambda \neq 0$. The former relationship is again the mass balance equation.

$$\frac{\partial \rho}{\partial t} + div(\rho \mathbf{v}) = 0;$$

where the definition for hydrodynamics derivative (6.1) was used. The former relationship is again the mass balance equation.

References

1. Aris, Rutherford. "Vectors, Tensors, and the Basic Equations of Fluid Mechanics". Prentice-Hall, Inc. (1962).
2. Fierros Palacios, Angel. "La ecuación de campo para la densidad de masa". Rev. del IMP. Vol. XXIV, No. 1 (enero-junio, 1994).
3. Viniegra H., F. Salcido G, A. y Fierros Palacios, A. "Las ecuaciones de balance de un fluido perfecto a partir de un principio variacional tipo Hamilton". Rev. del IMP. Vol. XVI, No.1 (enero, 1984).
4. Lovelock D. & Rund H. "Tensors, Differential Forms, & Variational Principles" John Wiley & Sons New York - London. Sydney. Toronto. (1975).

Chapter IV

Ideal Fluids

§17. Field functions for an ideal fluid

An ideal fluid sometimes also called perfect is a continuous system whose dynamical evolution can be completely characterized by means of a set of field functions like the velocity field $\boldsymbol{v}(\boldsymbol{x},t)$ and two thermodynamic variables concerning the fluid, such as the mass density $\rho(\boldsymbol{x},t)$ and the hydrostatic pressure $p(\boldsymbol{x},t)$. When a perfect fluid moves what is happening is that a momentum transference is being produced in the system. This momentum transference is totally reversible due to the mechanical transport of a volume element of fluid from a place of the system where the velocity is high to another where its value is low. The gradient of pressure over the system is the hydrodynamic force responsible of motion. For the case of an ideal fluid the pressure is defined as follows

$$p(\boldsymbol{x},t) = \rho^2 \frac{\partial \square}{\partial \rho}; \tag{17.1}$$

where $\square(\rho)$ is a thermodynamic function known as the specific internal energy of the fluid. It is only a function of mass density. In fact, when the deformation suffered by the region limiting the system is the result of a process of hydrostatic or volumetric compression or expansion, it can be demonstrated that the small deformations strain tensor is reduced to the sum of the elements of its principal diagonal, that is its trace. This demonstration will be postponed until the viscous fluid is studied. This trace is likewise in this case, the fractional change of a volume element of fluid in the region occupied by the system. It can be proved that, when the deformations are small the trace of the small deformations strain tensor is nearly equal to the reciprocal of the mass density of the fluid. Consequently and as it was previously stated, the specific internal energy of the system only depends on this thermodynamical function and in this case it is said that the fluid is ideal or perfect.

According to everything that has been mentioned before, the hydrodynamic force acting over the fluid is the gradient of the pressure in such a way that

$$\boldsymbol{F}_H = -\boldsymbol{grad}\left(\rho^2 \frac{\partial \square}{\partial \rho}\right). \tag{17.2}$$

During the motion of a perfect fluid in general terms what we have is that the system is subjected to a volumetric or hydrostatic compression or expansion process that clearly originates pressure changes in different parts of the fluid. As a consequence of that, the mass density of the continuous medium as well as its internal energy suffer certain variations that propitiate a hydrodynamic regime to be established in the fluid characterized by the appearance in it of gradients of the velocity field.

§18. Hamilton-Type variational principle and field differential equations

To make the dynamic of ideal fluids a branch of Lagrange´s Analytical Mechanics, a lagrangian function which contains all the relevant information of the system is required. Consequently be

$$\circ = \circ(\boldsymbol{x}, \boldsymbol{v}, \rho, t) \tag{18.1}$$

the proper lagrangian density. According to (16.2)

$$\lambda = \lambda(\boldsymbol{x}, \boldsymbol{v}, \rho, t) \tag{18.2}$$

is the corresponding specific lagrangian.

According to Hamilton-Type Variational Principle, the action integral (14.1) must be held down to the invariance condition (14.2) in such a way that the specific lagrangian which appears there is the one corresponding to the ideal fluid. If the functional dependence of λ given in the relation (18.2) is considered, it is evident that

$$\rho\delta\lambda = \rho\left[\frac{\partial\lambda}{\partial x^i}\delta x^i + \frac{\partial\lambda}{\partial v^i}\delta v^i + \frac{\partial\lambda}{\partial\rho}\delta\rho\right]. \tag{18.3}$$

With the aid of the general result (16.5) and due to the fact that

$$\delta v^i = \frac{d}{dt}\left(\delta x^i\right) \tag{18.4}$$

in (18.3) the following is obtained

$$\begin{aligned}\rho\delta\lambda &= \left[\rho\frac{\partial\lambda}{\partial x^i}\delta x^i + \rho\frac{\partial\lambda}{\partial v^i}\frac{d}{dt}\left(\delta x^i\right) - \left(\rho^2\frac{\partial\lambda}{\partial\rho}\right)\frac{\partial}{\partial x^i}\left(\delta x^i\right)\right] \\ &= \left[\rho\frac{\partial\lambda}{\partial x^i} - \frac{d}{dt}\left(\rho\frac{\partial\lambda}{\partial v^i}\right) + \frac{\partial}{\partial x^i}\left(\rho^2\frac{\partial\lambda}{\partial\rho}\right)\right]\delta x^i \\ &\quad + \frac{d}{dt}\left[\rho\frac{\partial\lambda}{\partial v^i}\delta x^i\right] - \frac{\partial}{\partial x^i}\left[\rho^2\frac{\partial\lambda}{\partial\rho}\delta x^i\right];\end{aligned} \tag{18.5}$$

where two integrations by parts were made. The second term of the right hand side of (18.5) can be integrated to obtain

$$\int_{t_1}^{t_2}\int_R \frac{d}{dt}\left[\rho\frac{\partial\lambda}{\partial v^i}\delta x^i\right]dVdt$$

$$= \int_{t_1}^{t_2}\left[\int_R \frac{d}{dt}\left[\rho\frac{\partial\lambda}{\partial v^i}dV\delta x^i\right]dt - \int_{t_1}^{t_2}\int_R\left(\rho\frac{\partial\lambda}{\partial v^i}\right)\mathrm{div}\,\boldsymbol{v}\,\delta x^i dVdt\right.$$

$$= \left[\int_R\left(\rho\frac{\partial\lambda}{\partial v^i}\right)dV\delta x^i\right]_{t_1}^{t_2} - \int_{t_1}^{t_2}\int_R\left(\rho\frac{\partial\lambda}{\partial v^i}\right)\mathrm{div}\,\boldsymbol{v}\,\delta x^i dVdt \quad ;$$

where an integration by parts was made and the relations (4.4) and (5.6) were used. The term enclosed in the square parenthesis is zero according to the boundary conditions (14.3), in such a way that only the following expression survives

$$-\int_{t_1}^{t_2}\int_R \left(\rho\frac{\partial\lambda}{\partial v^i}\right) div\, \boldsymbol{v}\, \delta x^i dVdt\,. \qquad (18.6)$$

The last term in (18.5) can be integrated using Green's theorem to obtain that

$$\int_{t_1}^{t_2}\int_R \frac{\partial}{\partial x^i}\left[\rho^2\frac{\partial\lambda}{\partial\rho}\delta x^i\right]dVdt = \int_{t_1}^{t_2}\oint_S \rho^2\frac{\partial\lambda}{\partial\rho}\delta x^i da_i dt\,.$$

The surface integral is zero due to an argument similar to the one used in paragraph 16. In fact, if the continuous medium considered is contained in an infinity region which is not deformed in the infinity, the integration surface can be made to extend to infinity where the pressure $\rho^2\,\partial\lambda/\partial\rho$ is zero over the surface.

Finally with what remains of the expression (18.5) and with the result (18.6) replaced in the general expression (16.4) we reach the following

$$\int_{t_1}^{t_2}\int_R \left[\rho\frac{\partial\lambda}{\partial x^i} - \mathcal{D}\left(\rho\frac{\partial\lambda}{\partial v^i}\right) + \frac{\partial}{\partial x^i}\left(\rho^2\frac{\partial\lambda}{\partial\rho}\right)\right]\delta x^i dVdt = 0\,. \quad (18.7)$$

Evidently and taking into account the conditions imposed to this type of integrals the former result is only valid if the integrand becomes null; that is to say, if

$$\mathcal{D}\left(\rho\frac{\partial\lambda}{\partial v^i}\right) - \rho\frac{\partial\lambda}{\partial x^i} - \frac{\partial}{\partial x^i}\left(\rho^2\frac{\partial\lambda}{\partial\rho}\right) = 0\,. \qquad (18.8)$$

These are the field differential equations looked for. In fact, we are dealing with a vectorial equation in partial derivative for the specific lagrangian λ known as generalized momentum balance equation for the ideal fluid.

If the definition (7.1) for Reynolds′ differential operator is used it can be demonstrated that

$$\mathcal{D}\left(\rho\frac{\partial\lambda}{\partial v^i}\right)=\rho\frac{d}{dt}\left(\frac{\partial\lambda}{\partial v^i}\right), \tag{18.9}$$

only because the other term of the development becomes zero. In fact, $\mathcal{D}\rho$ is nothing else but the continuity equation (9.2). In this case in (18.8) the following is obtained

$$\frac{d}{dt}\left(\frac{\partial\lambda}{\partial v^i}\right)-\frac{\partial\lambda}{\partial x^i}-\frac{1}{\rho}\frac{\partial}{\partial x^i}\left(\rho^2\frac{\partial\lambda}{\partial\rho}\right)=0\ ; \tag{18.10}$$

and these are the corresponding Euler-Lagrange′s motion differential equations for the perfect fluid.

§19. Generalized energy balance equation

When the action integral (14.1), with $\circ$ the lagrangian density (18.1) of the ideal fluid is subjected to temporary variations, the kinematic entities as well as the field functions contained in it, suffer infinitesimal changes. According to the results obtained in the section *Selected Topics* of the former chapter, it is easy to see that the invariance condition of action under continuous temporary transformations can be written as follows

$$\int_{t_1}^{t_2}\int_R\left[\delta^+\circ-\frac{d\circ}{dt}\delta^+t\right]dVdt=0; \tag{19.1}$$

and again, this result becomes a general relationship with which we have to operate in all cases studied with the proposed theoretical frame. If the functionality for the lagrangian density given in the relation (18.1) is taken into account, the calculus of variations directly lead to the following formula

$$\delta^{+}\circ=\frac{\partial\circ}{\partial x^{i}}\delta^{+}x^{i}+\frac{\partial\circ}{\partial v^{i}}\delta^{+}v^{i}+\frac{\partial\circ}{\partial\rho}\delta^{+}\rho+\frac{\partial\circ}{\partial t}\delta^{+}t\,. \tag{19.2}$$

From *Selected Topics* of the former chapter it can be seen that the temporary variation of mass density is given by the following formula

$$\delta^{+}\rho=-\rho\, div\mathbf{v}\delta^{+}t\,. \tag{19.3}$$

Then, in (19.2) and with the aid of this result and the relations (15.3) and (15.5) the following is obtained

$$\delta^{+}\circ=\left[\frac{\partial\circ}{\partial x^{i}}v^{i}+\frac{\partial\circ}{\partial v^{i}}\frac{dv^{i}}{dt}-\rho\frac{\partial\circ}{\partial\rho}div\,\mathbf{v}+\frac{\partial\circ}{\partial t}\right]\delta^{+}t\,. \tag{19.4}$$

Now, the terms appearing in (19.4) can be written as follows

$$\frac{\partial\circ}{\partial x^{i}}v^{i}=\lambda v^{i}\frac{\partial\rho}{\partial x^{i}}+\rho v^{i}\frac{\partial\lambda}{\partial x^{i}} \tag{19.5}$$

$$\frac{\partial\circ}{\partial v^{i}}\frac{dv^{i}}{dt}=\frac{d}{dt}\left[\frac{\partial\circ}{\partial v^{i}}v^{i}\right]-\frac{d}{dt}\left(\rho\frac{\partial\lambda}{\partial v^{i}}\right)v^{i} \tag{19.6}$$

$$-\rho\frac{\partial\circ}{\partial\rho}div\,\mathbf{v}=-\frac{\partial}{\partial x^{i}}\left[\rho^{2}\frac{\partial\lambda}{\partial\rho}v^{i}\right]+\frac{\partial}{\partial x^{i}}\left[\rho^{2}\frac{\partial\lambda}{\partial\rho}\right]v^{i} \tag{19.7}$$

$$\frac{\partial\circ}{\partial t}=\rho\frac{\partial\lambda}{\partial t}+\lambda\frac{\partial\rho}{\partial t}\,; \tag{19.8}$$

where systematically the definition (16.2) was used and some integrations by parts were made. Moreover, it is obvious that

$$\lambda\left(\frac{\partial\rho}{\partial t}+v^{i}\frac{\partial\rho}{\partial x^{i}}\right)=-\circ\, div\mathbf{v} \tag{19.9}$$

and

$$-\frac{d}{dt}\left(\rho\frac{\partial\lambda}{\partial v^i}\right) = -\mathcal{D}\left(\rho\frac{\partial\lambda}{\partial v^i}\right) + \frac{\partial\circ}{\partial v^i}\, div\,\mathbf{v} \tag{19.10}$$

where the continuity equation (19.2) and the definition of Reynolds′ differential operator were used. With all the results previously obtained substituted in (19.4) the following is obtained

$$\delta^+\circ = \left[\rho\frac{\partial\lambda}{\partial x^i} - \mathcal{D}\left(\rho\frac{\partial\lambda}{\partial v^i}\right) + \frac{\partial}{\partial x^i}\left(\rho^2\frac{\partial\lambda}{\partial\rho}\right)\right] v^i\delta^+ t$$

(19.11)

$$+\left[\mathcal{D}\left(\frac{\partial\circ}{\partial v^i}v^i\right) - \circ\, div\,\mathbf{v} - \frac{\partial}{\partial x^i}\left(\rho^2\frac{\partial\lambda}{\partial\rho}v^i\right) + \rho\frac{\partial\lambda}{\partial t}\right]\delta^+ t.$$

However, the first square parenthesis is zero because it deals with the field differential equations (18.8). In consequence, when the remains of (19.11) are substituted in (19.1) we reach the following result

$$\int_{t_1}^{t_2}\int_R\left[\mathcal{D}\mathcal{H} - \frac{\partial}{\partial x^i}\left(\rho^2\frac{\partial\lambda}{\partial\rho}v^i\right) + \rho\frac{\partial\lambda}{\partial t}\right]\delta^+ t\, dV\, dt = 0\,. \tag{19.12}$$

Evidently and according to the very nature of temporary variations and of the increments dV and dt, the former relation is only fulfilled if again the integrand of (19.10) becomes null in such a way that finally the following fundamental relationship is obtained

$$\mathcal{D}\mathcal{H} - \frac{\partial}{\partial x^i}\left(\rho^2\frac{\partial\lambda}{\partial\rho}v^i\right) + \rho\frac{\partial\lambda}{\partial t} = 0\,; \tag{19.13}$$

and this is the generalized energy balance equation for the perfect fluid. It is a partial derivative scalar equation for the specific lagrangian of ideal fluid. In (19.13),

$$\mathcal{H} \equiv \frac{\partial \circ}{\partial v^i} v^i - \circ \tag{19.14}$$

is the general form for the hamiltonian density in terms of the lagrangian density of the perfect fluid.

§20. Euler´s motion equation

In order to obtain the motion equation that any ideal fluid must fulfil, the following analytical expression for lagrangian density is proposed

$$\circ = \frac{1}{2}\rho v^2 - \rho \square - \rho\phi(\boldsymbol{x})\,; \tag{20.1}$$

where $1/2\rho v^2$ is the kinetic energy density, $\rho\square(\rho)$ is the potential energy density, and $\phi(\boldsymbol{x})$ some conservative potential which only depends on the position in such a way that

$$\boldsymbol{f} = -\boldsymbol{grad}\,\phi(\boldsymbol{x})\,, \tag{20.2}$$

with $\boldsymbol{f}(\boldsymbol{x}, t)$ the external force per unit mass. For convenience it is assumed that $\boldsymbol{f}=0$ so that the lagrangian density (20.1) only contains hydrodynamic terms and is of the form (t-u), with t and u the kinetic and potential energy densities respectively. When it is required to include the body force in some specific situation, the complete expression (20.1) and also the definition (20.2) must be used. For the present case

$$\circ = \rho\left[\frac{1}{2}v^2 - \square(\rho)\right], \tag{20.3}$$

and according to definition (16.2), the corresponding specific lagrangian is

$$\lambda = \frac{1}{2}v^2 - \square(\rho)\,. \tag{20.4}$$

Finally, if in Euler-Lagrange's differential motion equations (18.10) the relationship (20.4) is used, the following is obtained

$$\frac{\partial \boldsymbol{v}}{\partial t}+(\boldsymbol{v}\cdot\boldsymbol{grad})\boldsymbol{v}+\frac{1}{\rho}\boldsymbol{grad}\,p=0; \tag{20.5}$$

where the definitions for the hydrodynamics derivative (6.1) and for the hydrostatic pressure (17.1) respectively, have been used. The result (20.5) is a partial derivative vectorial equation for the velocity field and pressure known as Euler's motion equation for the ideal fluid.

Upon deriving the different forms of the equations of motion for this case, the processes of energy dissipation, which may occur when a real fluid moves have not been taken into account. These are processes intimately related to internal frictions occurring between different parts of the system and which are the origin of what is known by the name of viscosity of the continuous medium. Heat exchanges between different parts of the system are not considered even between the system and its environment either. The former discussion is valid only for real fluids in which the thermal conductivity and the viscosity are irrelevant. In those cases it is said that such fluids are ideal or perfect.

§21. Energy conservation law

The energy conservation law for any perfect fluid can be determined from generalized energy balance equation (19.13) and with the aid of the relations (20.3) and (20.4) for the corresponding lagrangians. According to (20.3) and with the general definition (19.14) the explicit form of hamiltonian density is

$$\mathcal{H}=\frac{1}{2}\rho v^2+\rho\square. \tag{21.1}$$

On the other hand and with the aid of Reynolds' differential operator,

$$\mathcal{D}\mathcal{H}=\frac{\partial\mathcal{H}}{\partial t}+div(\mathcal{H}\boldsymbol{v}); \tag{21.2}$$

where an integration by parts was made. Besides,

$$\frac{\partial}{\partial x^i}\left(\rho^2 \frac{\partial \lambda}{\partial \rho} v^i\right) = -\frac{\partial}{\partial x^i}\left(p v^i\right); \tag{21.3}$$

according to the definition (17.1) for the hydrostatic pressure. Time uniformity has as a consequence that the specific lagrangian for the continuous medium does not become an explicit function of time. In this case and due to the fact that $\partial\lambda/\partial t{=}0$ in generalized energy balance equation (19.13) the following is obtained

$$\frac{\partial \mathcal{H}}{\partial t} + div\left[\mathbf{v}\left(\mathcal{H} + p\right)\right] = 0\,. \tag{21.4}$$

Using in this formula the result (21.1) the energy conservation law for any perfect fluid is finally obtained; that is

$$\frac{\partial}{\partial t}\left(\frac{1}{2}\rho v^2 + \rho\square\right) = -div\left[\rho\mathbf{v}\left(\frac{1}{2}v^2 + w\right)\right]. \tag{21.5}$$

In this relationship

$$w = \square + \frac{p}{\rho} \tag{21.6}$$

is the specific enthalpy or the heat function. The result (21.5) is another of the fundamental equations of fluid dynamics for perfect fluids. It has a very simple physical interpretation. In fact, on their left hand side we have the change on time of the total energy of the system. Then, the right hand side is an expression of the quantity of energy which flows out of the volume occupied by the fluid in unit time. This energy flux is due to the simple mass transfer by effect of the motion of the system. The expression

$$\rho\mathbf{v}\left(\frac{1}{2}v^2 + w\right); \tag{21.7}$$

is known as the energy flux density vector. On the other hand, the absence of heat exchanges between different parts of the fluid and between the fluid and its environment means that the motion is adiabatic throughout the fluid. In adiabatic motion the specific entropy remains constant in such a way it fulfils the following relationship: $ds/dt=0$, that is to say

$$\frac{\partial s}{\partial t}+\mathbf{v}\cdot\mathbf{grad}\, s=0\,, \qquad (21.8)$$

where again the definition (6.1) was used. The last equation is the analytical form of the general formula describing the adiabatic motion on the ideal fluid.

On the other hand, and if as it usually happens, the specific entropy remains constant along the flux and across all the region that contains the fluid in a given initial instant, it will retain everywhere the same constant value at all times and for any subsequent motion of the continuous medium. In this case the adiabatic equation is simply

$$s = constant\,; \qquad (21.9)$$

and therefore the motion is said to be isentropic. This former equation is totally equivalent to the energy conservation law (21.5).

With all that has been previously mentioned and within the theoretical scheme of Hamilton-Type Variational Principle, it was possible to make from the Fluid Dynamics a branch of Lagrange´s Analytical Mechanics as a consistent theory. Therefore, we now have the same number of field differential equations for an identical number of field variables. Namely, the generalized momentum balance equation (18.8), the generalized energy balance equation (19.13), and the field equation for the mass density (16.11).

Within the theoretical frame of Classical Hydrodynamics, we also have a consistent theory since we now have the vectorial Euler´s motion equation (20.5) for the three components of the velocity field, the energy conservation law (21.5), and the scalar equation for mass density (11.9). So that, the theoretical scheme presented is consistent in both ways: within the scope of Lagrange´s Analytical Mechanics as well as within the frame of Newton´s Vectorial Mechanics.

Besides, we have the continuity equation (9.2) as an additional condition for any fluid, and for the case of the ideal fluid and within the Vectorial Mechanics scheme we also have the isentropic equation (21.9). In the Analytical Mechanics language both equations can be considered as non-holonomic constraint conditions.

Finally, from (16.13) the validity of the mass-energy equivalence principle can be extended to the Fluid Dynamics scope. In fact, the invariance of the action under transformations with respect to the evolution parameter and due to time uniformity, the mass density balance equation instead of the energy balance equation was obtained. This result is indicative of the existence of *mass-energy relation* for continuous mechanical system. To prove the former statement let us consider that for any fluid confined in the given region R of the three-dimensional Euclidean Space and from (19.14) we have that $\mathcal{H}=-\circ$, because and according to (16.1), $\circ$ is independent from the velocity field. In this case, from (16.13) the following is obtained

$$\mathcal{H} = constant \,. \tag{21.10}$$

Let us suppose that

$$\mathcal{H} = \rho\varepsilon^2 \,, \tag{21.11}$$

where ε^2 is a constant with unities of square velocity that have the mission to dimensionally balance the relation (21.11). On the other hand from Relativistic Dynamics the following result is obtained

$$L = -mc^2\sqrt{1-\frac{v^2}{c^2}}\;;$$

where L is the lagrangian for a free particle of mass m that moves with constant velocity equal to v and c is the light velocity in the empty space. However, the former relation is also valid for extended bodies of total mass m that moves as a whole with constant rapidity equal to v. In particular it must be true for the continuous medium free of forces under consideration. On the other hand, for continuous mechanical systems, the mass m is given by the relation (9.1) so that from the last equation, from (9.1) and (8.2) the following is obtained

$$\circ = -\rho c^2\sqrt{1-\frac{v^2}{c^2}}\,,$$

with ∘ the lagrangian density for the fluid studied and v the flux velocity. In the newtonian limit of relativistic mechanics, that is to say when $v/c \to 0$ we have that $\circ \to -\rho c^2$. Therefore, in (21.11) it is fulfilled that $\varepsilon^2 = c^2$. Moreover and according to *Lorentz contraction*

$$V = V_o \sqrt{1 - \frac{v^2}{c^2}}.$$

The former result is known as the proper volume of region R of the three-dimensional Euclidean Space containing the continuous medium under study. This volume moves with the fluid with a velocity v equal to the flux rapidity, so that

$$\rho = \frac{\rho_o}{\sqrt{1 - \frac{v^2}{c^2}}}; \tag{21.12}$$

where ρ_o is the mass density at rest. Therefore,

$$\mathcal{H} = \frac{\rho_o c^2}{\sqrt{1 - \frac{v^2}{c^2}}}; \tag{21.13}$$

so that the condition (21.10) is satisfied because the flux rapidity is a constant. From the former relation it is easy to see that in the scope of Relativisvistic Mechanics the kinetic energy of a fluid does not becomes null when its flux rapidity is zero. In this situation $\mathcal{H}$ has the following finite value

$$\mathcal{H} = \rho_o c^2, \tag{21.14}$$

and this is the formula for the rest energy of the continuous medium. Stating the equivalence between the densities of energy and mass in the scope of Fluid Dynamics. It is evident that since $\mathcal{H}$ is the energy per unit volume and ρ_o the mass of the fluid per unit volume too, the equation (21.14) is reduced to the well known Einstein´s relationship of Relativistic Mechanics: $E=m_oc^2$; except that m_o is not referred to the rest mass of a free particle but to the total mass at rest of the continuous medium free of forces.

On the other hand, and for small values of the flux rapidity, that is to say, when $v/c \ll 1$, in (21.13) an expansion in power series of v/c can be made to obtain the following result

$$\mathcal{H} \approx \rho c^2 + \frac{1}{2}\rho v^2 . \tag{21.15}$$

This is so because clearly, when $v/c \to 0$ we have that $\rho \to \rho_o$ which it is the newtonian limit of the general formula for the relativistic mass density ρ. With the exception of the rest energy, the former relationship is the expression for the kinetic energy density of the continuous medium. From (21.12) it is evident that the relativistic relationship between the total mass of the fluid and its flux velocity is fulfilled; that is to say

$$m = \frac{m_o}{\sqrt{1 - \frac{v^2}{c^2}}} .$$

Finally, from total derivative with respect to time of (21.11) and since the product $\rho\varepsilon^2 \neq 0$, it is directly obtained that

$$div\, \boldsymbol{v} = 0 . \tag{21.16}$$

This is the continuity equation when the mass density is referred to a mechanical system at rest. The former relationship is satisfied identically because the system under consideration does not move so that the flux velocity is zero. Since the type of fluid considered in R, at any moment was specified, the mass-energy relation is valid for any continuous medium.

Selected Topics

Euler-Lagrange's equations

Let us consider any mechanical system which occupies, at instants t_1 and t_2, positions defined by the set of values of the coordinates given by $q^{(1)}$ and $q^{(2)}$, respectively. In Lagrange's Analytical Mechanics theoretical frame, the condition imposed to the system is that it moves between these positions in such a way that the integral

$$W = \int_{t_1}^{t_2} L(q, \dot{q}, t)\,dt$$

takes the least possible value.

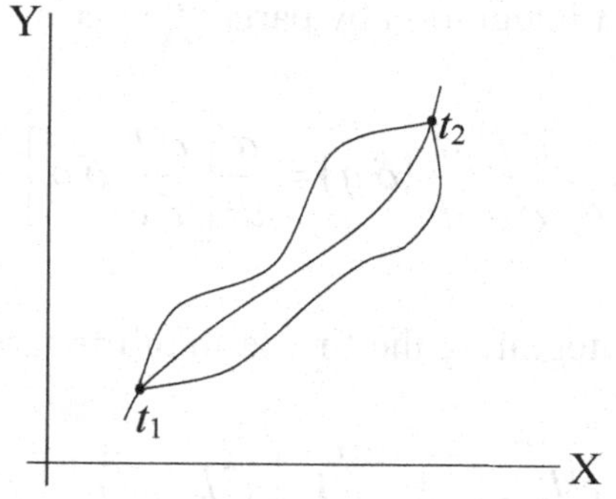

The function L is the lagrangian of the system concerned, and the integral is called the action. Let $q(t)$ be a function of time known as the generalized coordinate for which W reaches a minimum value. This means that the action integral is increased when $q(t)$ is replaced by any function of the form $q(t)+\delta q(t)$. Here $\delta q(t)$ is a function which is small everywhere in the interval of time from t_1 to t_2; $\delta(q)$ is called a variation of the function $q(t)$. Since, for $t=t_1$ and for $t=t_2$, the modified function must take the values $q^{(1)}$ and $q^{(2)}$ respectively, it follows that

$$\delta q(t_1) = \delta q(t_2) = 0\,.$$

This is the boundary condition that the set of coordinates must satisfy. According to Lagrange´s scheme within the least action principle, the necessary condition for W to have a minimum is that the variation of the integral becomes zero; that is to say that

$$\delta \int_{t_1}^{t_2} L(q,\dot{q},t)dt = 0 .$$

The process of variation produces the following result

$$\int_{t_1}^{t_2} \left[\frac{\partial L}{\partial q} \delta q + \frac{\partial L}{\partial \dot{q}} \delta \dot{q} \right] dt = 0 . \qquad (1)$$

Since $\delta\dot{q} = d(\delta q)/dt$, the second term of the integrand can be modified using the method of integration by parts†; that is

$$\frac{\partial L}{\partial \dot{q}} \delta \dot{q} = \frac{\partial L}{\partial \dot{q}} \frac{d}{dt} (\delta q) = \frac{d}{dt} \left[\frac{\partial L}{\partial \dot{q}} \delta q \right] - \frac{d}{dt} \left(\frac{\partial L}{\partial \dot{q}} \right) \delta q .$$

In this case at integrating the first term of the former result we see that

$$\int_{t_1}^{t_2} \frac{d}{dt} \left[\frac{\partial L}{\partial \dot{q}} \delta q \right] dt = \int_{t_1}^{t_2} d \left[\frac{\partial L}{\partial \dot{q}} \delta q \right] = \left[\frac{\partial L}{\partial \dot{q}} \delta q \right]_{t_1}^{t_2} = 0 ,$$

by the boundary conditions. In consequence, in (1) we have that

$$\int_{t_1}^{t_2} \left[\frac{\partial L}{\partial q} - \frac{d}{dt} \left(\frac{\partial L}{\partial \dot{q}} \right) \right] \delta q \, dt = 0 .$$

† *The method of integration by parts is based on the formula: d(AB)=AdB+BdA.*

Since δq as well as dt are increments totally arbitrary and therefore different from zero, the former relation is only fulfilled if the integrand becomes null; that is to say

$$\frac{d}{dt}\left(\frac{\partial L}{\partial \dot{q}}\right)-\frac{\partial L}{\partial q}=0\,. \tag{2}$$

These are Euler-Lagrange´s motion differential equations. The necessary and sufficient condition for the action integral W to have a stationary value, with the generalized coordinates subjected to the given boundary conditions is that Euler-Lagrange´s equations must be satisfied.

Isentropic flux

In order to demonstrate the fact that the flux of an ideal fluid occurs at a constant specific entropy, let us remember that for this continuous dynamical system the specific internal energy is only a function of mass density, that is to say $\square=\square(\rho)$. The total derivative with respect to time of this function is

$$\frac{d\square}{dt}=\frac{\partial \square}{\partial \rho}\frac{d\rho}{dt}=-\rho\frac{\partial \square}{\partial \rho}div\boldsymbol{v}\,;$$

where the continuity equation was used. On the other hand

$$\rho\frac{\partial \square}{\partial \rho}=\frac{p}{\rho}\,,$$

with $p(\boldsymbol{x},t)$ the hydrostatic pressure. Therefore it is fulfilled that

$$\rho\frac{d\square}{dt}=-p\,div\boldsymbol{v}\,. \tag{1}$$

From $\square=Ts-p/\rho$ which is Euler´s form for the first and second thermodynamic laws; with s the specific entropy and T the temperature, the following is obtained

$$\rho \frac{d\mathcal{L}}{dt} = \rho T \frac{ds}{dt} - p \, div\, \mathbf{v}; \tag{2}$$

where Gibbs-Duhem's relationship and the continuity equation were taken into account. So that the equations (1) and (2) be consistent to each other it is necessary that $ds/dt=0$; that is to say

$$s = constant;$$

which means that the flux of an ideal fluid, occurs in such a way that the specific entropy of the system remains constant. In other words, the flux is isentropic.

Equivalence between the isentropic flux and the energy balance equation

The equation (2) of the former problem can be written as follows

$$\rho \frac{d\mathcal{L}}{dt} = \rho T \frac{ds}{dt} + \mathbf{v} \cdot \mathbf{grad}\, p - div(p\mathbf{v});$$

where an integration by parts was made. Now,

$$\rho \frac{d\mathcal{L}}{dt} = \frac{\partial}{\partial t}(\rho\Box) + div(\rho\mathbf{v}\Box)$$

and

$$\mathbf{v} \cdot \mathbf{grad}\, p = -\frac{\partial}{\partial t}\left(\frac{1}{2}\rho v^2\right) - div\left[\rho\mathbf{v}\left(\frac{1}{2}v^2\right)\right].$$

Here two integrations by parts were made and Euler's motion equation was used. In this case it is fulfilled that

$$\frac{\partial}{\partial t}\left(\frac{1}{2}\rho v^2 + \rho\Box\right) + div\left[\rho\mathbf{v}\left(\frac{1}{2}v^2 + w\right)\right] = \rho T \frac{ds}{dt};$$

where

$$w = \square + \frac{p}{\rho} \qquad (2)$$

is the heat function or specific enthalpy. In consequence, the condition

$$s = constant$$

implies that ds/dt=0, so that

$$\frac{\partial}{\partial t}\left(\frac{1}{2}\rho v^2 + \rho\square\right) = -div\left[\rho \boldsymbol{v}\left(\frac{1}{2}v^2 + w\right)\right];$$

and this is the energy balance equation for the perfect fluid.

Bernoulli's equation

If the flow is considered steady, the velocity is constant in time at any point occupied by the ideal fluid under study. In other words, $\boldsymbol{v}(\boldsymbol{x})$ is a function of the coordinates only, so that $\partial \boldsymbol{v}/\partial t=0$.

In this case, from the well known thermodynamic relation

$$dw = T ds + V dp\,, \qquad (1)$$

where V=1/ρ is the specific volume and T the temperature, we have that

$$dw = Vdp = \frac{dp}{\rho}\,; \qquad (2)$$

due to the fact that for the perfect fluid s =*constant*. Therefore,

$$\boldsymbol{grad}\, w = \frac{1}{\rho}\boldsymbol{grad}\, p\,. \qquad (3)$$

From Euler's equation for steady flow

$$(\boldsymbol{v} \cdot \mathbf{grad})\boldsymbol{v} = -\frac{1}{\rho}\,\mathbf{grad}\, p \tag{4}$$

and since

$$(\boldsymbol{v} \cdot \mathbf{grad})\boldsymbol{v} = \frac{1}{2}\,\mathbf{grad}\, v^2 - \boldsymbol{v} \times \mathbf{rot}\,\boldsymbol{v}\,,$$

it is fulfilled that

$$\frac{1}{2}\,\mathbf{grad}\, v^2 - \boldsymbol{v} \times \mathbf{rot}\,\boldsymbol{v} = -\mathbf{grad}\, w\,; \tag{5}$$

where the result (3) was used.

At this moment it is necessary to introduce the concept of streamlines. These are lines such that the tangent to a streamline at any point gives the direction of the velocity at that point. They are determined from the following system of differential equations

$$\frac{dx}{v_x} = \frac{dy}{v_y} = \frac{dz}{v_z}\,.$$

Next we form the scalar product of equation (5) with the unit vector tangent to the streamlines at each point. Let $\boldsymbol{m}$ be this unit vector. The projection of the gradient on any direction is as is well known, the derivative in that direction. Hence, the projection of $\mathbf{grad}\, w$ in the direction of $\boldsymbol{m}$ is $\partial w / \partial m$ †. On the other hand, since the vector $\boldsymbol{v} \times \mathbf{rot}\,\boldsymbol{v}$ is perpendicular to $\boldsymbol{v}$, its projection on the direction of $\boldsymbol{m}$ is the same that $\boldsymbol{v}$ has so it becomes null because the angle formed by these vectors is equal to 90° and the cosine of 90° is zero. In this case in (5) the following is obtained

$$\frac{\partial}{\partial m}\left(\frac{1}{2} v^2 + w\right) = 0\,. \tag{6}$$

† *In fact, we have that* $/\mathbf{grad}\, w\, / \, /\boldsymbol{m}/ \cos(\mathbf{grad}\, w,\ \boldsymbol{m})$. *But the angle between* $\mathbf{grad}$ w *and* $\boldsymbol{m}$ *is equal to zero and cosine of* 0° *is equal to 1. Besides* $/\boldsymbol{m}/ = 1$ *because it is a unit vector and then:* $/\mathbf{grad}\, w\,/ = \partial w/\partial m$.

From here it follows that along a streamline

$$\frac{1}{2}v^2 + w = constant\,. \tag{7}$$

In general, the constant takes different values for different streamlines. Relationship (7) is known as Bernoulli's equation. If the flow takes place in a gravitational field, it becomes necessary to add the acceleration $\boldsymbol{g}$ due to the gravity to the right hand side of (6). If the direction of gravity is the same as the Z-axis, with z increasing upwards, then† the cosine of the angle between the direction of $\boldsymbol{g}$ and $\boldsymbol{m}$ is equal to $-dz/dm$; therefore, the projection of $\boldsymbol{g}$ over $\boldsymbol{m}$ is $-gdz/dm$ and in (6) we have that

$$\frac{\partial}{\partial m}\left(\frac{1}{2}v^2 + w + gz\right) = 0\,.$$

Thus Bernoulli's equation establishes that along a streamline and when the flow takes place in a gravitational field

$$\frac{1}{2}v^2 + w + gz = constant\,. \tag{8}$$

Proper time

Suppose that in a certain inertial reference system we have clocks which are moving with respect to the observer in an arbitrary manner. However, at each different moment of time this motion can be considered as uniform. Thus, at each moment of time we can introduce a coordinate system rigidly linked to the moving clocks, which with the clocks constitutes an inertial reference system. In the course of an infinitesimal time interval dt as measured by a clock linked to the rest frame, the other moving clocks with respect to the last one move one distance equal to $\sqrt{dx^2+dy^2+dz^2}$. The ques-

† 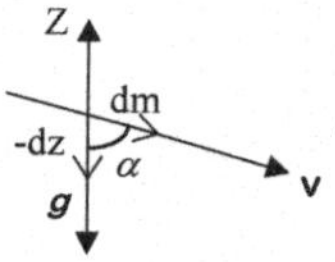

cos α = −dz/dm. Then the projection of $\boldsymbol{g}$ *over* $\boldsymbol{m}$ *is equal to:* $|\boldsymbol{g}||\boldsymbol{m}|\cos\alpha = -gdz/dm$.

tion is: which is the time interval dt' that is measured by the moving clock along the time interval dt?

Since in each coordinate system linked to the moving clocks, the latter are at rest, we have that $dx'=dy'=dz'=0$. On the other hand and according to the relativity principle the intervals ds and ds' are invariant; thus

$$ds^2 = c^2dt^2 - dx^2 - dy^2 - dz^2$$

and

$$ds'^2 = c^2dt'^2.$$

In consequence and given that $ds^2 = ds'^2$ it is fulfilled that

$$dt' = \frac{ds}{c} = \frac{1}{c}\sqrt{c^2dt^2 - dx^2 - dy^2 - dz^2}\ ;$$

where c is the velocity of light in the empty space. Thus,

$$dt' = dt\sqrt{1 - \frac{dx^2 + dy^2 + dz^2}{c^2dt^2}}\ .$$

However,

$$v^2 = \frac{dx^2 + dy^2 + dz^2}{dt^2}$$

is the square velocity of the moving clocks, therefore

$$dt' = \frac{ds}{c} = dt\sqrt{1 - \frac{v^2}{c^2}}$$

is the time interval measured by the moving clocks. Integrating this expression we have that

$$t_2' - t_1' = \int_{t_1}^{t_2} dt\sqrt{1 - \frac{v^2}{c^2}}\ .$$

The time read by a clock moving with a given object is called the proper time for this object. Finally,

$$ds = cdt\sqrt{1-\frac{v^2}{c^2}} . \qquad (1)$$

The action integral and the lagrangian for a free particle

To determine the form of the action integral for a free material particle it is important to recognize that this integral must not depend on the reference system elected to make the description of the mechanical system under study. In other words, the action must be invariant under Lorentz transformations. Therefore it is necessary that it becomes the integral of a scalar in such a way that the integrand becomes a differential of the first order. With these restrictions, it is clear that the only scalar of this kind that one can construct for a free particle is the interval *ds* multiplied by some constant α, so that the action must have the form

$$W = -\alpha\int_a^b ds ; \qquad (1)$$

where the definite integral is calculated along the work line of the particle between the two particular events of the arrival of the particle at the initial position and at the final position at definite times t_1 and t_2 i.e. between two given world points.

It can be assumed that α corresponds to some characteristic of the particle and must be a positive quantity for all and each of the particles considered. To demonstrate the former statement it is important to note that the action integral (1) has its maximum value along a straight world line.

However it is also true that it can be integrated along a curved world line in such a way that it will always be possible to make the action integral arbitrarily small. Then, it is easy to see that this integral with the positive sign cannot have a minimum value, but if one considers the opposite sign it clearly has a minimum along the straight world line. On the other hand, the action integral can be represented as an integral with respect to time as follows

$$W = \int_{t_1}^{t_2} L dt \,,$$

where L is Lagrange's function of the mechanical system under study. According to the result (1) of the former problem,

$$W = -\int_{t_1}^{t_2} \alpha c \sqrt{1 - \frac{v^2}{c^2}} dt \,;$$

where now v is the velocity of the material particle. If this result is compared to the last definition we see that

$$L = -\alpha c \sqrt{1 - \frac{v^2}{c^2}} \,.$$

In Classical Mechanics each particle is characterized by its mass m in such a way that it is interesting to find the relation that exists between α and m. It is expected that when $c \to \infty$ the lagrangian becomes reduced to the classical expression $L = mv^2/2$. Therefore and to carry out this transition, the square root in the former lagrangian can be expanded in powers of v/c. Then, neglecting terms of higher order, we find that

$$L \approx -\alpha c + \frac{\alpha v^2}{2c} \,.$$

As it is well known, terms which are exact differentials with respect to time are of no importance in the lagrangian. They can be omitted from it due to the fact that they give quantities whose value does not contribute in anything to the variation of the action. On the other hand, any constant is an exact differential of the product of the constant with t. In this case and omitting the constant αc, we get that $L = \alpha v^2/2c$; while in Classical Mechanics we have that $L = mv^2/2$. Consequently, $\alpha = mc$ and of course,

$$W = -mc \int_a^b ds$$

is the action integral for the free particle while its lagrangian has the following form

$$L = -mc^2\sqrt{1 - \frac{v^2}{c^2}}\,.$$

The relativistic form of lagrangian density

To obtain the form that the relativistic lagrangian density for any fluid must have, let us consider a region R of three-dimensional Euclidean Space which contains any fluid. Note that the classical lagrangian is such that

$$L = \int_R \circ\, dV\,; \tag{1}$$

with $\circ$ the lagrangian density and dV the volume element in region R.

On the other hand, the mass m of the fluid contained in this region is given by the following expression

$$m = \int_R \rho(\boldsymbol{x}, t)\, dV\,, \tag{2}$$

where $\rho\,(\boldsymbol{x}, t)$ is the mass density of the dynamical system. If the relativistic lagrangian for a free particle or for an extended body free of forces is substituted in (1) and the relation (2) is used, we have that

$$\int_R \left(\circ - \rho c^2 \sqrt{1 - \frac{v^2}{c^2}} \right) dV = 0\,. \tag{3}$$

Since dV was arbitrarily selected in such a way that it is different from zero, the former equation is only satisfied if the integrand becomes zero; that is to say, if

$$\circ = -\rho c^2 \sqrt{1 - \frac{v^2}{c^2}}\,. \tag{4}$$

This is the relativistic form for the lagrangian density for the fluid free of forces contained in the region R considered.

If we expand $\circ$ in powers of v/c the following is obtained

$$\circ \approx \frac{1}{2}\rho v^2 - \rho c^2; \tag{5}$$

where the flux velocity v is a constant. In the general case of a perfect fluid subjected to internal stresses and also to body forces, the lagrangian density takes the following form

$$\circ = \frac{1}{2}\rho v^2 - \rho\Box - \rho\phi(\boldsymbol{x}) - \rho c^2; \tag{6}$$

where $\Box(\rho)$ is the specific internal energy of ideal fluid, $\boldsymbol{v}(\boldsymbol{x},t)$ is now the velocity field, and $\phi(\boldsymbol{x})$ a conservative potential which only depends on the position in such a way that

$$\boldsymbol{f} = -\boldsymbol{grad}\,\phi(\boldsymbol{x}), \tag{7}$$

with $\boldsymbol{f}(\boldsymbol{x},t)$ the external force per unit mass. For convenience it is always assumed that $\boldsymbol{f} = 0$ in order so that (6) only contains hydrodynamic terms. Thus, for the case of the perfect fluid it is fulfilled that

$$\circ = \rho\left(\frac{1}{2}v^2 - \Box(\rho) - c^2\right). \tag{8}$$

As a consequence, the corresponding specific lagrangian is

$$\lambda = \frac{1}{2}v^2 - \Box(\rho) - c^2. \tag{9}$$

When in Euler-Lagrange´s equations the relationship (9) is substituted, the motion equation for ideal fluid is obtained. On the other hand, it can be demonstrated by direct calculation that the explicit form of the hamiltonian density for the perfect fluid is

$$\mathcal{H} = \frac{1}{2}\rho v^2 + \rho\Box + \rho c^2. \tag{10}$$

In this case, in the energy balance equation

$$\frac{\partial \mathcal{H}}{\partial t} + div[\boldsymbol{v}(\mathcal{H} + p)] = 0,$$

we have that

$$\frac{\partial}{\partial t}\left(\frac{1}{2}\rho v^2 + \rho\square\right) + div\left[\rho\boldsymbol{v}\left(\frac{1}{2}v^2 + w\right)\right] = -c^2\left[\frac{\partial \rho}{\partial t} + div(\rho\boldsymbol{v})\right] \quad (11)$$

so that, the energy conservation law is only satisfied if the mass of the fluid is preserved. That is to say if

$$\frac{\partial \rho}{\partial t} + div(\rho\boldsymbol{v}) = 0,$$

it is fulfilled that

$$\frac{\partial}{\partial t}\left(\frac{1}{2}\rho v^2 + \rho\square\right) = -div\left[\rho\boldsymbol{v}\left(\frac{1}{2}v^2 + w\right)\right]$$

because c^2 is a constant different from zero. Thus, in Fluid Dynamics the condition which the mass is preserving along the flux process, has as a consequence the energy conservation law; and this is a general result whose validity can be extended to any fluid.

The geometrical form of relativistic lagrangian density

In order to demonstrate that the relativistic lagrangian density for the case of mass density for any fluid free of forces

$$\circ = -\rho c^2\sqrt{1 - \frac{v^2}{c^2}},$$

can be written in terms of purely geometric quantities, it is necessary to take into account the scalar equation for the mass density

$$\rho = \frac{\rho_o}{J} \boldsymbol{u} \cdot \boldsymbol{grad}\, J \,.$$

In consequence it is possible to write that

$$\circ = \frac{A}{J} \boldsymbol{u} \cdot \boldsymbol{grad}\, J \,,$$

with

$$A \equiv -\rho_o c^2 \sqrt{1 - \frac{v^2}{c^2}}$$

a constant. Therefore, the functional form of relativistic lagrangian density is simply

$$\circ = \circ\left(J; \boldsymbol{grad}\, J\right);$$

that is to say, this function only depends on geometric entities as the Jacobian and its first gradient.

The mass-energy relation

The following elemental derivation of the equivalence principle between the mass and the energy is due to Albert Einstein†. Although it makes use of the principle of relativity, it does not consider the formal machinery of the theory but uses only the following three well known results:

1. The momentum conservation law.
2. The formula for the pulse of radiation, that is the momentum of a pulse of radiation moving in a fixed direction.
3. The well known expression for the aberration of light; that is the influence of the motion of the Earth on the apparent location of the fixed stars (Bradley).

† *A. Einstein. Out of My Later Yars. Philosophical Library, New York, 1950(116-119)*

Let us consider the following system

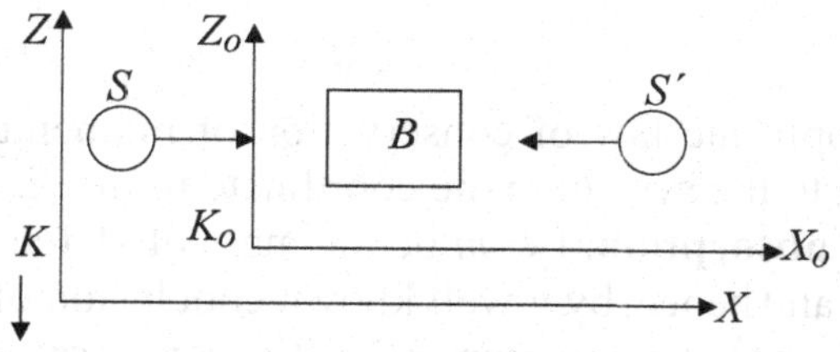

Let the body B rest freely in space with respect to the system K_o. Two pulses of radiation S and S' each of energy $E/2$ move in the positive and negative X_o direction respectively and are eventually absorbed by B. With this absorption the energy of B increases by the quantity E. The body B stays at rest with respect to K_o by reasons of symmetry.

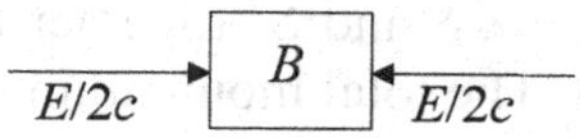

Now let us consider the same process but with respect to the system K which moves with respect to K_o with the constant velocity v in the negative Z_o direction.
With respect to K the description of the process is as follows

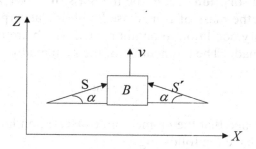

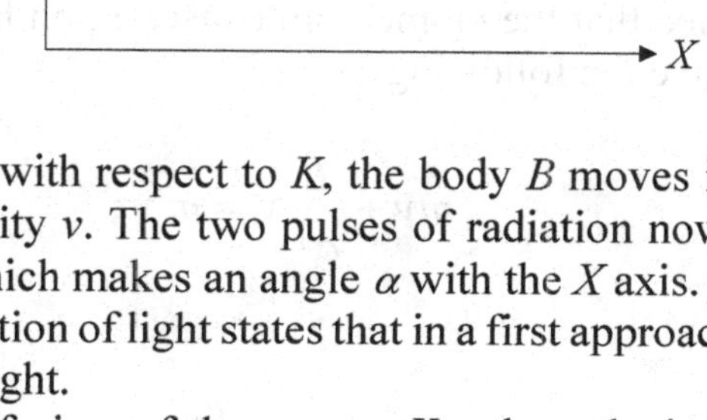

It is evident that with respect to K, the body B moves in the positive Z direction with velocity v. The two pulses of radiation now have direction with respect to K which makes an angle α with the X axis.

The law of aberration of light states that in a first approach $\alpha = v/c$, where c is the velocity of light.

From the point of view of the system K_o, the velocity v of B remains unchanged by the absorption of S and S'.

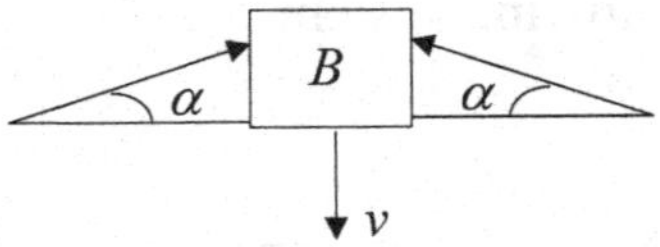

Now we apply the law of conservation of momentum with respect to the Z direction to the system in the coordinate frame K.

Before the absorption. Let m be the mass of B. Each of the pulses has the energy $E/2$ and hence, by a well known conclusion of Classical Electrodynamics, each has the momentum $E/2c$. Rigorously speaking this is the momentum of S with respect to K_o. However, when v is small with respect to c, the momentum from the view point of K is the same, except for a quantity of second order of magnitude. That is v^2/c^2 is a quantity whose magnitude is of second order compared to unity. The z-component of this momentum is $(E/2c)$ *sen* α with sufficient accuracy, except for quantities of higher order. In a first approach, it can be written that *sen* $\alpha \approx \alpha$ so that the momentum is $(E/2c)\alpha$. S and S' together therefore have a momentum Ev/c^2 in the Z direction. The total momentum of the system before absorption is therefore

$$mv + \frac{E}{c^2} v .$$

After the absorption. Let m' be the mass of B. We anticipate here the possibility that the mass of B increased with the absorption of the energy E, as a necessary condition to obtain results which are consistent with the consideration made. The momentum of the system after absorption is then

$$m'v .$$

As it is assumed that the momentum conservation law is fulfilled in the Z direction we have the following

$$mv + \frac{E}{c^2} v = m'v ;$$

or also

$$m' - m = \frac{E}{c^2} .$$

This equation expresses the law of the equivalence between the mass and the energy. The energy increase E is connected to the mass increase E/c^2. Since the energy according to the usual definition leaves an additive constant free, it is possible to choose such constant in such a way that†

$$E = mc^2 .$$

Momentum and energy

In the scope of Analytical Mechanics of particles the momentum is defined in terms of the classical lagrangian and the velocity as follows $\boldsymbol{p} = \partial L / \partial \boldsymbol{v}$. Since in Relativistic Mechanics

$$L = -mc^2 \sqrt{1 - \frac{v^2}{c^2}} \tag{1}$$

is the lagrange´s function for a material particle or an extended body of mass m moving as a whole with velocity $\boldsymbol{v}$, it is easy to prove by direct calculus that

$$\frac{\partial L}{\partial \boldsymbol{v}} = \frac{m\boldsymbol{v}}{\sqrt{1 - \frac{v^2}{c^2}}} \equiv \boldsymbol{p} \tag{2}$$

is the relativistic momentum. For small velocities of the particle; that is to say when $v/c \ll 1$or also when $c \to \infty$, the former expression turns into the classical momentum $\boldsymbol{p} = m\boldsymbol{v}$. In the former formulae, c is the velocity of light in the empty space.

On the other hand, the total energy of the particle or extended body is defined as follows

$$\mathsf{E} \equiv \frac{\partial L}{\partial \boldsymbol{v}} \cdot \boldsymbol{v} - L = \boldsymbol{p} \cdot \boldsymbol{v} - L .$$

---

† *So that the relationship is fulfilled it is required that:* $E = c^2(m' - m) + \text{constant}$. *In this case,* $\text{constant} = c^2(2m - m')$.

In this case

$$\mathsf{E} = \frac{mv^2}{\sqrt{1-\dfrac{v^2}{c^2}}} + mc^2\sqrt{1-\frac{v^2}{c^2}}$$

$$= \frac{\cancel{mv^2} + mc^2 - \cancel{mv^2}}{\sqrt{1-\dfrac{v^2}{c^2}}}$$

so that

$$\mathsf{E} = \frac{mc^2}{\sqrt{1-\dfrac{v^2}{c^2}}}. \tag{3}$$

According to this important result of relativistic mechanics, the energy of a particle or an extended body does not become null when $\boldsymbol{v}=0$, but has the following finite value

$$\mathsf{E} = mc^2. \tag{4}$$

This quantity is known as the *mass-energy relation* or the *rest energy* of the physical system under study. For small values of the velocity; that is to say when $v/c \ll 1$, we can expand (3) in powers of v/c to obtain that

$$\mathsf{E} \approx mc^2 + \frac{1}{2}mv^2. \tag{5}$$

With the exception of the rest energy, the former result is the classical expression for the kinetic energy of the system.

Momentum and energy in fluid dynamics

Within the theoretical scheme of Hamilton-Type Variational Principle, the momentum per unit volume for any fluid can be defined as the one of Analytical Mechanics of particles; that is in terms of the velocity field and from the relativistic form of the lagrangian density given by the equation (4) of page 85. In this case, by direct calculus it can be demonstrated that

$$\frac{\partial \mathcal{L}}{\partial \boldsymbol{v}} = \frac{\rho \boldsymbol{v}}{\sqrt{1 - \frac{v^2}{c^2}}} \equiv \boldsymbol{\pi}\,; \tag{1}$$

where π is the relativistic momentum per unit volume. The total energy per unit volume in the fluid is defined in the same fashion as in Analytical Mechanics; that is to say,

$$\mathcal{E} = \boldsymbol{\pi} \cdot \boldsymbol{v} - \mathcal{L} \equiv \frac{\partial \mathcal{L}}{\partial \boldsymbol{v}} \cdot \boldsymbol{v} - \mathcal{L}\,. \tag{2}$$

In consequence, it is fulfilled that

$$\mathcal{E} = \frac{\rho c^2}{\sqrt{1 - \frac{v^2}{c^2}}}\,. \tag{3}$$

When the flux velocity is zero, the energy density of the fluid has the following finite value

$$\mathcal{E} = \rho c^2\,. \tag{4}$$

This is the rest energy per unit volume of the fluid. Since $\mathcal{E}$ is the energy per unit volume and ρ is the mass per unit volume too, the *mass-energy relation* for any fluid takes the following familiar form

$$E = mc^2\,; \tag{5}$$

where m is the total mass of the fluid contained in the given volume. Again, if the flux velocity is very small compared to the light velocity in the empty space; that is to say, when $v/c << 1$, we have that

$$\mathcal{E} \approx \frac{1}{2}\rho v^2 + \rho c^2 ; \tag{6}$$

where we make an expansion in powers of v/c.

Aside of the rest energy density, in the former relationship we have the kinetic energy density characteristic of Fluid Dynamics.

References

1. Fierros Palacios, Angel. "El principio variacional tipo Hamilton y las ecuaciones de balance de la dinámica de los fluidos". Rev. del IMP. Vol. XXIV, No. 3 (julio, 1992).
2. Fierros Palacios Angel. "The mass-energy equivalence principle in fluid dynamics". To be publishing.
3. Landau, L. D. and Lifshitz, E. M. "Fluid Mechanics". Addison-Wesley Publishing Co. (1959).
4. Landau, L.D. and Lifshitz, E.M. "The Classical Theory of Fields". Addison-Wesley Publishing Co. (1962).
5. Morse, P. and Feshbach, H. "Methods of Theoretical Physics". Part I McGraw-Hill Book Co. (1953).
6. Viniegra H., F. Salcido G., A. y Fierros P., A. "Las ecuaciones de balance de un fluido perfecto a partir de un principio variacional tipo Hamilton". Rev. del IMP. Vol. XI, No. 1 (enero, 1979).

Chapter V

Potential Flow

§22. Wave equation

In order to obtain the differential equation that governs the potential flow it is necessary to take into account the physical conditions which characterize it. If the vorticity is zero in the whole space occupied by the fluid, the velocity field is irrotational and the flow is potential; in such a way that the description of dynamical behaviour of the system can be made in terms of a scalar function of position and time $\phi(\boldsymbol{x},t)$ such that

$$\boldsymbol{v}(\boldsymbol{x},t) = \boldsymbol{grad}\ \phi\,(\boldsymbol{x},t)\,. \tag{22.1}$$

This function is known as the velocity potential. When the flow of compressible fluids is investigated one starts studying the case of small oscillations. An oscillatory motion of small amplitude is known by the name of a sound wave. This phenomenon is studied within the scope of gas dynamics where the effect of external force is usually neglected. The pass of a sound wave through an ideal compressible fluid can be theoretically treated from some completely equivalent view points.

The problem can be faced considering that it deals of the propagation of pressure waves moving through the continuous medium with finite velocity. Also the geometrical view point can be adopted assuming that what is propagated through the system is a deformation wave; or also solving the problem in terms of the velocity potential and imagining that in the fluid alternate compression and rarefaction are produced.

Let R be a region of the three-dimensional Euclidean Space occupied by a perfect compressible fluid. If through the system an oscillatory motion of small amplitude is generated, at each point of the fluid alternate compressions and rarefactions are produced at the passing of the disturbance. Since the oscillations are small, the perturbation in the velocity $\boldsymbol{v}$ is small too, so that the term $(\boldsymbol{v}_o \cdot \boldsymbol{grad})\boldsymbol{v}'$ in Euler´s equation (20.5) may be neglected. To write the last term in this form we assume that the velocity field is equal to: $\boldsymbol{v}(\boldsymbol{x},t) = \boldsymbol{v}_o + \boldsymbol{v}'$ with $\boldsymbol{v}_o$ a constant vector such that $\boldsymbol{v}_o >> \boldsymbol{v}'$. It is important to make clear that for the present case the definition (22.1) corresponds to the small perturbative term; so that in the following we write $\boldsymbol{v}(\boldsymbol{x},t)$ instead of $\boldsymbol{v}'(\boldsymbol{x},t)$ in order not to drag along the prime throughout the chapter.

For the same reason as in the case of velocity field, the relative changes in the fluid density and pressure are small in such a way that they can be written as

$$p(\boldsymbol{x},t) = p_o + p' \tag{22.2}$$

and

$$\rho(\boldsymbol{x},t) = \rho_o + \rho' \tag{22.3}$$

where p_o and ρ_o are the equilibrium constant referred to pressure and density; whereas $p' << p_o$ and $\rho' << \rho_o$ are their variations at a constant specific entropy at the pass of the sound wave. Another information of general character concerning the phenomenon studied that we have is that a sound wave in a perfect compressible fluid is an adiabatic motion for which the specific entropy is a constant along the region which is occupied by it; in such a way that the adiabatic equation (21.8) is fulfilled. In this case it is said that the flow is barotropic because pressure and density are directly related; that is to say, we may have that $\rho = \rho(p)$ or also that

$$p = p(\rho). \tag{22.4}$$

In consequence, to a small change in p corresponds a small variation in ρ, in such a way that

$$p' = c_o^2 \rho' \tag{22.5}$$

where c_o is the velocity of sound in the medium defined as

$$c_o = \sqrt{(\partial p / \partial \rho)_s}; \tag{22.6}$$

and it is such that $|\,\boldsymbol{v}\,| << c_o$. If the effect of the external force is neglected and the relations (22.2) and (22.3) are taken into account in Euler's motion equation the following is obtained

$$\rho_o \frac{\partial}{\partial t}(grad_i\,\phi) + grad_i\,p' = 0.$$

Since the operators $\partial/\partial t$ and ***grad*** are independent between them, they can be interchanged so that in the former relation we have the following result

$$grad_i\left[\rho_o \frac{\partial \phi}{\partial t} + p'\right] = 0. \tag{22.7}$$

Integrating this equation we have that

$$p' = -\rho_o\left(\frac{\partial \phi}{\partial t}\right); \tag{22.8}$$

where the constant of integration is considered as zero without generality lose. In such case and according to relation (22.5),

$$\rho' = -\frac{\rho_o}{c_o^2}\left(\frac{\partial \phi}{\partial t}\right). \tag{22.9}$$

Now, if the continuity equation (9.2) is linearised it can be written as follows

$$\frac{\partial \rho'}{\partial t} + \rho_o\, div\, \boldsymbol{v} = 0. \tag{22.10}$$

If in this equation the definitions (22.1) and (22.9) are substituted the following result is reached

$$\nabla^2\phi - \frac{1}{c_o^2}\frac{\partial^2\phi}{\partial t^2} = 0\,. \qquad (22.11)$$

This is the wave equation which the velocity potential satisfies; in it $\nabla^2 = div\ \boldsymbol{grad}$ is the laplacian operator. This is the motion equation that satisfies a sound disturbance of small amplitude which is propagated through a compressible perfect fluid with constant velocity equal to c_o.

§23. Field differential equations

The problem of potential flow can be solved with all the theoretical rigidity with the aid of the proposed lagrangian formalism and the application of Hamilton-Type Variational Principle. In fact, let

$$\circ = \circ\left(\boldsymbol{grad}\,\phi\,;\frac{\partial\phi}{\partial t}\right) \qquad (23.1)$$

be the lagrangian density proposed to deal with the case of small oscillations in a compressible ideal fluid. Clearly,

$$\lambda = \lambda\left(\boldsymbol{grad}\,\phi;\frac{\partial\phi}{\partial t}\right) \qquad (23.2)$$

is the corresponding specific lagrangian.

The application of Hamilton-Type Variational Principle to the action integral (14.1) subjected to the invariance condition (14.2) conducting again to the general result (16.4), where the specific lagrangian (23.2) now appears.

The local variation of velocity potential takes the following form

$$\delta\phi(\boldsymbol{x},t) = grad_i\,\phi\,\delta x^i(t) \qquad (23.3)$$

so that

$$\delta\lambda = \frac{\partial\lambda}{\partial(\nabla\phi)}\left[\frac{\partial}{\partial x^i}\{grad_i\,\phi\,\delta\boldsymbol{x}(t)\}\right] + \frac{\partial\lambda}{\partial\left(\frac{\partial\phi}{\partial t}\right)}\left[\frac{\partial}{\partial t}\{grad_i\,\phi\,\delta\boldsymbol{x}(t)\}\right], \tag{23.4}$$

where the former result was used and the differential operators $\delta, \partial/\partial\boldsymbol{x}$, and $\partial/\partial t$ were interchanged due to the fact that they are independent between them. The terms of right hand side of the last relation can be expressed as follows

$$\frac{\partial\lambda}{\partial(\nabla\phi)}\left[\frac{\partial}{\partial x^i}\{grad_i\,\phi\,\delta\boldsymbol{x}\}\right] = \frac{\partial}{\partial x^i}\left[\left(\frac{\partial\lambda}{\partial(\nabla\phi)}\right)\{grad_i\,\phi\,\delta\boldsymbol{x}\}\right] - \frac{\partial}{\partial x^i}\left[\frac{\partial\lambda}{\partial(\nabla\phi)}\right]grad_i\,\phi\delta\boldsymbol{x} \approx -\frac{\partial}{\partial x^i}\left[\frac{\partial\lambda}{\partial(\nabla\phi)}\right]grad_i\,\phi\delta\boldsymbol{x};$$

where an integration by parts was made and the first term of the right hand side of the former result was neglected because it is of higher order in the approximations. In the same way,

$$\frac{\partial\lambda}{\partial\left(\frac{\partial\phi}{\partial t}\right)}\left[\frac{\partial}{\partial t}\{grad_i\,\phi\delta\boldsymbol{x}\}\right] = \frac{\partial}{\partial t}\left[\left(\frac{\partial\lambda}{\partial\left(\frac{\partial\phi}{\partial t}\right)}\right)\{grad_i\,\phi\delta\boldsymbol{x}\}\right] - \frac{\partial}{\partial t}\left[\frac{\partial\lambda}{\partial\left(\frac{\partial\phi}{\partial t}\right)}\right]grad_i\,\phi\delta\boldsymbol{x} \approx -\frac{\partial}{\partial t}\left[\frac{\partial\lambda}{\partial\left(\frac{\partial\phi}{\partial t}\right)}\right]grad_i\,\phi\delta\boldsymbol{x}.$$

In this case in (23.4) only the following is obtained

$$\delta\lambda = -\left\{\frac{\partial}{\partial x^i}\left[\frac{\partial\lambda}{\partial(\nabla\phi)}\right] + \frac{\partial}{\partial t}\left[\frac{\partial\lambda}{\partial\left(\frac{\partial\phi}{\partial t}\right)}\right]\right\} grad_i\, \phi\, \delta\, \boldsymbol{x}(t).$$

Finally, as far as terms of first order, in the general formula (16.4) we have that

$$\int_{t_1}^{t_2}\int_R \left\{\left[\frac{\partial}{\partial x^i}\left\{\frac{\partial\lambda}{\partial(\nabla\phi)}\right\} + \frac{\partial}{\partial t}\left\{\frac{\partial\lambda}{\partial\left(\frac{\partial\phi}{\partial t}\right)}\right\}\right]\rho_o grad_i\phi\right\}\delta\boldsymbol{x} dV dt = 0\,. \quad (23.5)$$

Again, the former result is only valid if the integrand becomes null. In this case and due to the fact that ρ_o ***grad*** $\phi \neq 0$; it is fulfilled that

$$\frac{\partial}{\partial x^i}\left[\frac{\partial\lambda}{\partial(\nabla_i\,\phi)}\right] + \frac{\partial}{\partial t}\left[\frac{\partial\lambda}{\partial\left(\frac{\partial\phi}{\partial t}\right)}\right] = 0\,. \quad (23.6)$$

These are the field differential equations which satisfy the specific lagrangian for potential flow.

§24. Potential energy density

Let m be the mass of the ideal fluid which is placed in the considered region R of the three-dimensional Euclidean Space; and suppose that it is about a sufficiently small quantity so that the potential energy density u is a constant. If this condition is fulfilled, it is clear that uV_o is the potential energy of the given mass; where $V_o = m/\rho_o$ is the initial volume and ρ_o the mass density before the compression occurs. Thus, as a result of the sound disturbance, the volume changes from V_o to $V_o+\Delta V$; as is it easy to see in the following diagram pV

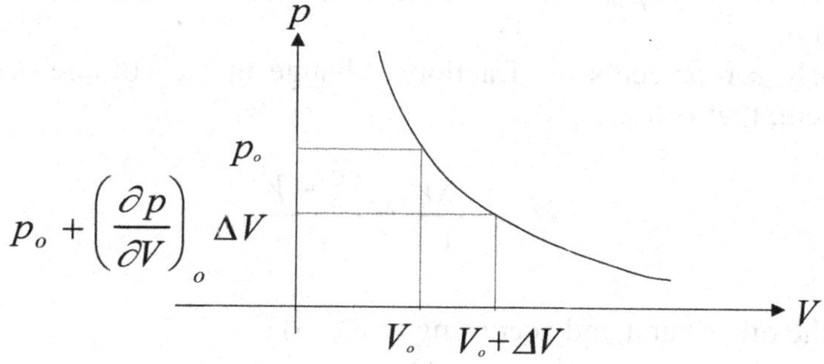

In a change in volume dV, the work required to compress the system is $-pdV$ and this is equal to the increase in the potential energy. In consequence, the potential energy corresponding to a volume change from V_o to $V_o+\Delta V$ is

$$uV_o = -\int_{V_o}^{V_o+\Delta V} pdV \,. \tag{24.1}$$

On the other hand and according to the approximation (22.2)

$$p_o \int_{V_o}^{V_o+\Delta V} dV + \int_{V_o}^{V_o+\Delta V} p'dV = p_o\Delta V + \int_{V_o}^{V_o+\Delta V} p'dV \,. \tag{24.2}$$

It is convenient to express the change of volume in terms of the associated change in density.

As in general $V=m/\rho$ its variation is

$$\Delta\left(\frac{m}{\rho}\right) = -\frac{m\rho'}{\rho^2} = -\frac{m\rho'}{\rho_o^2 + 2\rho_o\rho' + \rho'^2} \approx -\frac{m}{\rho_o}\frac{\rho'}{\rho_o}$$

$$= -V_o\left(\frac{\rho'}{\rho_o}\right) = -V_o\left(\frac{\rho-\rho_o}{\rho_o}\right) = -V_o\sigma\,; \tag{24.3}$$

where $\sigma = (\rho - \rho_o)/\rho_o$. If $\sigma > 0$ we have compression and if $\sigma < 0$ we have expansion.

Clearly σ represents the fractional change in the volume occupied by the system, that is to say

$$\sigma = -\frac{\Delta V}{V_o} = \frac{V_o - V}{V_o}. \tag{24.4}$$

On the other hand and according to (22.5)

$$p' = \rho_o c_o^2 \sigma; \tag{24.5}$$

so that

$$p'dV = -V_o \rho_o c_o^2 \sigma d\sigma. \tag{24.6}$$

In order to change the limits of the integral in the expression (24.2) we must take into account the following argumentation. The lower limit corresponds to the fact that the region has not yet been deformed because the sound wave has not passed there yet in such a way that $\sigma = 0$. In the upper limit one has that for a change from V_o to $V_o + \Delta V$ the system has been disturbed by sound the wave in which case σ must substitute $V_o + \Delta V$. Thus, in (24.2) the following is obtained

$$\int_{V_o}^{V_o+\Delta V} p'dV = -\int_0^{\sigma} V_o \rho_o c_o^2 \sigma d\sigma = -\frac{V_o \rho_o c_o^2 \sigma^2}{2}; \tag{24.7}$$

where the result (24.6) was used. In this case

$$uV_o = -p_o \Delta V + \frac{V_o \rho_o c_o^2 \sigma^2}{2}. \tag{24.8}$$

However, from (24.4) it is easy to see that $\Delta V = -V_o \sigma$ so that in the last relation it is obtained that

$$u = p_o \sigma + \frac{\rho_o c_o^2}{2} \sigma^2. \tag{24.9}$$

On the other hand, it is also obvious that $\sigma = (\rho'/\rho_o)$ in such a way that

$$\sigma^2 = \frac{\rho'^2}{\rho_o^2}. \tag{24.10}$$

Then in (24.9) the following is obtained

$$u = \frac{p_o \rho'}{\rho_o} + \frac{p'\rho'}{2\rho_o} \approx \frac{p'\rho'}{2\rho_o}; \tag{24.11}$$

because this is the dominant term. In this case in a first approach be

$$u \equiv \frac{1}{2}\rho_o \square , \tag{24.12}$$

where

$$\square = \frac{p'\rho'}{\rho_o^2} \tag{24.13}$$

is the variation of internal specific energy of the fluid at a constant specific entropy at the passing of the sound wave.

§25. The lagrangian density and the specific lagrangian

The explicit form of these functions can be determined from the expression (20.3) for the hydrodynamic lagrangian density of perfect fluid. As it occurs with the other quantities involved in the potential flow problem, at the pass of the sound disturbance this function changes, that is to say

$$\circ = \circ_o + \circ', \tag{25.1}$$

with $\circ_o$ the equilibrium lagrangian density and $\circ'$ its variation at a constant specific entropy. Now and according to (22.3), in the definition (20.3) the following is obtained

$$\circ = \frac{1}{2}\rho_o v^2 + \frac{1}{2}\rho' v^2 - \rho_o \square_o - \rho_o \square - \rho' \square_o - \rho' \square . \tag{25.2}$$

However, ½ $\rho_o v$ >> ½ $\rho' v$. On the other hand, $\rho' \Box_o$ is practically constant and almost not contributing to the potential energy change, as well as the term constant $\rho_o \Box_o$. Finally, the term $\rho' \Box'$ is a quantity of second order in such a way that its contribution is neglected. Thus, in the first approximation the following is obtained

$$\circ_o = \rho_o \left(\frac{1}{2} v^2 - \Box_o \right) \tag{25.3}$$

and

$$\circ' = -\rho_o \Box' . \tag{25.4}$$

In these former results the following approximation was used

$$\Box' = \Box - \Box_o . \tag{25.5}$$

$\Box'$ is the variation of $\Box$ at a constant specific entropy and $\Box_o$ its constant equilibrium value. According to the results (24.12) and (25.4), it is clear that the potential energy density can be written as follows

$$u = -\frac{1}{2} \circ' . \tag{25.6}$$

From Euler´s form of thermodynamic first and second laws

$$\Box = Ts - \frac{p}{\rho} \tag{25.7}$$

and from Gibbs-Duhem´s relationship

$$sT' - \frac{p'}{\rho} = 0 \tag{25.8}$$

the following result is obtained

$$\square = \frac{p_o \rho'}{\rho_o^2} + \frac{p' \rho'}{\rho_o^2}, \tag{25.9}$$

where the calculus was made at a first approximation. On the other hand and according to (22.8) and (22.9),

$$\square = -\frac{p_o}{\rho_o c_o^2}\left(\frac{\partial \phi}{\partial t}\right) + \frac{1}{c_o^2}\left(\frac{\partial \phi}{\partial t}\right)^2 \approx \frac{1}{c_o^2}\left(\frac{\partial \phi}{\partial t}\right)^2 \tag{25.10}$$

because it is the dominant term. Then in (25.6) we have that

$$u = \frac{\rho_o}{2c_o^2}\left(\frac{\partial \phi}{\partial t}\right)^2. \tag{25.11}$$

According to (22.1) the kinetic energy density is

$$t = \frac{1}{2}\rho_o \left|\mathbf{grad}\,\phi\right|^2. \tag{25.12}$$

Consequently and due to the fact that $\circ = t - u$, the explicit form for lagrangian density for potential flow calculated as far as first order terms is the following

$$\circ = \frac{1}{2}\rho_o \left[\left|\mathbf{grad}\,\phi\right|^2 - \frac{1}{c_o^2}\left(\frac{\partial \phi}{\partial t}\right)^2\right]; \tag{25.13}$$

and evidently

$$\lambda = \frac{1}{2}\left[\left|\mathbf{grad}\,\phi\right|^2 - \frac{1}{c_o^2}\left(\frac{\partial \phi}{\partial t}\right)^2\right] \tag{25.14}$$

is the appropriate explicit form of the specific lagrangian to study the small vibrations in a compressible ideal fluid. In fact, when each term of (23.6) is calculated using the relationship (25.14) the following is obtained

$$\frac{\partial}{\partial x^i}\left[\frac{\partial \lambda}{\partial(\nabla_i \phi)}\right] = \nabla^2 \phi$$

and

$$\frac{\partial}{\partial t}\left[\frac{\partial \lambda}{\partial\left(\frac{\partial \phi}{\partial t}\right)}\right] = -\frac{1}{c_o^2}\frac{\partial^2 \phi}{\partial t^2};$$

in such a way that adding these results the wave equation (22.11) is recovered.

When the fluid under study turns to be incompressible its density is a constant and obviously $\rho'=0$. In this case, Π as well as u become null so that in (25.14) only the term of kinetic energy density is kept, that is to say

$$\tilde{\lambda} = \frac{1}{2}\left|\mathbf{grad}\ \phi\right|^2 . \tag{25.15}$$

For the same reason, the field equations (23.6) are reduced as follows

$$\frac{\partial}{\partial x^i}\left[\frac{\partial \tilde{\lambda}}{\partial(\nabla_i \phi)}\right] = 0; \tag{25.16}$$

so that the wave equation (22.1) is transformed into that of Laplace

$$\nabla^2 \phi = 0 , \tag{25.17}$$

as it is easy to see if equation (25.15) is substituted in the relation (25.16).

§26. The linearised continuity equation

In paragraph 22 the wave equation was calculated from Euler´s motion and continuity equations conveniently linearised, by means of which it was demonstrated that these relations are totally equivalents. This result is only true for the case of potential flow and obeys to the fact that all the calculus were made as far as terms of the first order in the approximations. That also means that from Hamilton-Type Principle the field equations can be obtained in terms of a specific lagrangian that instead of depending on the spatial and time derivatives of velocity potential becomes a function of the variations in field velocity and mass density at the pass of sonic disturbance, that is to say of $\boldsymbol{v}(\boldsymbol{x},t)$ and $\rho'(\boldsymbol{x},t)$. On the other hand and in order to complete the theoretical scheme it is necessary to calculate the explicit form which can be adopted by the specific lagrangian in terms of the field variables before mentioned, so that with their aid in the field equations previously obtained the linearised continuity equation (22.10) may be recovered. Hence, be

$$\circ_{\rho} = \circ_{\rho}\left(v^{i}, \rho'\right) \tag{26.1}$$

the proper lagrangian density, so that

$$\lambda_{\rho} = \lambda_{\rho}\left(v^{i}, \rho'\right) \tag{26.2}$$

is the corresponding functional form of the specific lagrangian. In this case

$$\delta\lambda_{\rho} = \frac{\partial\lambda_{\rho}}{\partial v^{i}}\delta v^{i} + \frac{\partial\lambda_{\rho}}{\partial\rho'}\delta\rho' . \tag{26.3}$$

From (22.1) it is clear that

$$\begin{aligned} \delta v^{i} &= \delta\left(grad_{i}\phi\right) = \frac{\partial}{\partial x^{i}}\left(\delta\phi\right) = \frac{\partial}{\partial x^{i}}\left(\frac{\partial\phi}{\partial x^{j}}\delta x^{j}\right) \\ &= \frac{\partial}{\partial x^{i}}\left(v^{j}\delta x^{j}\right), \end{aligned} \tag{26.4}$$

where the result (23.3) was used. Moreover and according to (22.9) it is easy to see that

$$\delta\rho' = -\frac{\rho_o}{c_o^2}\frac{\partial}{\partial t}(\delta\phi) = -\frac{\rho_o}{c_o^2}\frac{\partial}{\partial t}\left[\frac{\partial\phi}{\partial x^j}\delta x^j\right]$$
$$= -\frac{\rho_o}{c_o^2}\frac{\partial}{\partial t}\left[v^j\delta x^j\right]. \tag{26.5}$$

Then, in (26.3) at a first approach only the following is obtained

$$\begin{aligned}\delta\lambda_\rho &= \left(\frac{\partial\lambda_\rho}{\partial v^i}\right)\left[\frac{\partial}{\partial x^i}\left(v^j\delta x^j\right)\right] - \left(\frac{\partial\lambda_\rho}{\partial\rho'}\right)\left[\frac{\rho_o}{c_o^2}\frac{\partial}{\partial t}\left(v^j\delta x^j\right)\right]\\ &= \frac{\partial}{\partial x^i}\left[\left(\frac{\partial\lambda_\rho}{\partial v^i}\right)v^j\delta x^j\right] - \frac{\partial}{\partial x^i}\left[\frac{\partial\lambda_\rho}{\partial v^i}\right]v^j\delta x^j\\ &\quad - \frac{\partial}{\partial t}\left[\frac{\rho_o}{c_o^2}\left(\frac{\partial\lambda_\rho}{\partial\rho'}\right)v^j\delta x^j\right] + \frac{\rho_o}{c_o^2}\frac{\partial}{\partial t}\left[\frac{\partial\lambda_\rho}{\partial\rho'}\right]v^j\delta x^j\\ &\approx \left[\frac{\rho_o}{c_o^2}\frac{\partial}{\partial t}\left(\frac{\partial\lambda_\rho}{\partial\rho'}\right) - \frac{\partial}{\partial x^i}\left(\frac{\partial\lambda_\rho}{\partial v^i}\right)\right]v^j\delta x^j.\end{aligned} \tag{26.6}$$

To reach the former result two integrations by parts were made. Substituting the expression (26.6) in the general formula (16.4) the following is obtained

$$\int_{t_1}^{t_2}\int_R\left\{\left[\frac{\rho_o}{c_o^2}\frac{\partial}{\partial t}\left(\frac{\partial\lambda_\rho}{\partial\rho'}\right) - \frac{\partial}{\partial x^i}\left(\frac{\partial\lambda_\rho}{\partial v^i}\right)\right]v^j\right\}\delta x^j\,dV dt = 0. \tag{26.7}$$

Again, so that the former equation is valid it is necessary that the integrand becomes null. Since on the other hand $\boldsymbol{v} \neq 0$, it is clear that

$$\frac{\partial}{\partial x^i}\left[\frac{\partial\lambda_\rho}{\partial v^i}\right] - \frac{\rho_o}{c_o^2}\frac{\partial}{\partial t}\left[\frac{\partial\lambda_\rho}{\partial\rho'}\right] = 0; \tag{26.8}$$

and these are the required field differential equations. The explicit form of the proper specific lagrangian for this case is directly obtained substituting in (25.14) the definitions (22.1) and (22.9); that is

$$\lambda_\rho = \frac{1}{2}\left[v^2 - \frac{c_o^2}{\rho_o^2}\rho'^2\right]. \tag{26.9}$$

With this formula introduced in (26.8) the following results are obtained

$$\frac{\partial}{\partial x^i}\left[\frac{\partial \lambda_\rho}{\partial v^i}\right] = div\,\boldsymbol{v}$$

whereas

$$-\frac{\rho_o}{c_o^2}\frac{\partial}{\partial t}\left[\frac{\partial \lambda_\rho}{\partial \rho'}\right] = \frac{1}{\rho_o}\frac{\partial \rho'}{\partial t};$$

so that adding these equations the linearised continuity equation (22.10) is recovered.

It is evident that if the ideal fluid studied becomes incompressible we have that $\rho' = 0$; so that in the field equation (26.8) we only have the following term

$$\frac{\partial}{\partial x^i}\left[\frac{\partial \widetilde{\lambda}_\rho}{\partial v^i}\right] = 0. \tag{26.10}$$

For the same reason, the specific lagrangian (26.9) only contains one term, that is to say

$$\widetilde{\lambda}_\rho = \frac{1}{2}v^2. \tag{26.11}$$

In this case the corresponding Laplace equation is simply

$$div\,\boldsymbol{v} = 0. \tag{26.12}$$

§27. The energy conservation law of sound waves

To obtain the generalized energy sound balance equation it is necessary to subject the action integral (14.1) to a temporary variation process to obtain again the general result (19.1), where clearly the lagrangian density $\circ_\rho$ whose functional form is given in the definition (26.1) must appear. In this case it is easy to see that

$$\delta^+ \circ_\rho = \frac{\partial \circ_\rho}{\partial v^i}\delta^+ v^i + \frac{\partial \circ_\rho}{\partial \rho'}\delta^+ \rho' . \tag{27.1}$$

If the definitions (15.5) and (19.3) are used here and the calculus is made as far as terms of the first order in the approximations the following will be obtained

$$\delta^+ \circ_\rho = \left[\frac{d}{dt}\left(\frac{\partial \circ_\rho}{\partial v^i} v^i \right) - \frac{\partial}{\partial x^i}\left(\rho_o^2 \frac{\partial \lambda_\rho}{\partial \rho'} v^i \right) \right] \delta^+ t . \tag{27.2}$$

In this case in (19.1) the following result is reached

$$\int_{t_1}^{t_2}\int_R \left\{ \frac{d}{dt}\left[\frac{\partial \circ_\rho}{\partial v^i} v^i - \circ_\rho \right] - \frac{\partial}{\partial x^i}\left[\rho_o^2 \frac{\partial \lambda_\rho}{\partial \rho'} v^i \right] \right\} \delta^+ t \, dV dt = 0 ; \tag{27.3}$$

which is only valid if the usual condition imposed on this type of integrals is fulfilled; that is to say if

$$\frac{\partial \mathcal{H}_\rho}{\partial t} - \frac{\partial}{\partial x^i}\left[\rho_o^2 \frac{\partial \lambda_\rho}{\partial \rho'} v^i \right] = 0 ; \tag{27.4}$$

where and according to the general formula (19.14), $\mathcal{H}_\rho$ is the corresponding hamiltonian density. Moreover, in (27.4) the definition of hydrodynamics derivative (6.1) was used and the term $\boldsymbol{v}\cdot\mathbf{grad}\mathcal{H}_\rho$ has been ignored because it is of a higher order in the approximations. Evidently, the former relationship is the generalized sonic energy balance equation.

Now, from (26.9) and (16.2) it is obvious that

$$\circ_{\rho} = \frac{1}{2}\rho_o v^2 - \frac{c_o^2 \rho'^2}{2\rho_o} \tag{27.5}$$

is the explicit form of lagrangian density. In this case,

$$\mathcal{H}_\rho = \frac{1}{2}\rho_o v^2 + \frac{c_o^2 \rho'^2}{2\rho_o} \tag{27.6}$$

is the corresponding hamiltonian density. Moreover,

$$\frac{\partial}{\partial x^i}\left[\rho_o^2 \frac{\partial \lambda_\rho}{\partial \rho'} v^i\right] = -div(p'\boldsymbol{v}); \tag{27.7}$$

where the results (26.9) and (22.5) were used. With all the former substituted in (27.4) it is finally obtained that

$$\frac{\partial}{\partial t}\left[\frac{1}{2}\rho_o v^2 + \frac{c_o^2 \rho'^2}{2\rho_o}\right] + div(p'\boldsymbol{v}) = 0\,. \tag{27.8}$$

This is the energy conservation law of sound waves propagated through a compressible perfect fluid with constant velocity equal to c_o. This equation is valid for the flux at any instant.

§28. Pressure and density variations

The generalized energy sonic balance equation (27.4) was calculated as far as terms of the first order in the approximations. Next the higher order terms that were ignored will be examined in order to obtain some interesting results.

If in the equation (19.1) all the terms resulting of temporary variation which give as a result the relation (27.2) are considered, the following is obtained

$$\int_{t_1}^{t_2}\int_R \left\{ \frac{\partial \mathcal{H}_\rho}{\partial t} - \frac{\partial}{\partial x^i}\left(\rho_o^2 \frac{\partial \lambda_\rho}{\partial \rho'} v^i \right) \right. \tag{28.1}$$

$$\left. - v^i \left[\frac{d}{dt}\left(\frac{\partial \circ_\rho}{\partial v^i} \right) - \frac{\partial}{\partial x^i}\left(\mathcal{H}_\rho + \rho_o^2 \frac{\partial \lambda_\rho}{\partial \rho'} \right) \right] \right\} \delta^+ t dV dt = 0 .$$

The first two terms of the former relation are zero because they refer to the generalized energy sonic balance equation. Then, in (28.1) only the higher order terms before ignored remain, that is to say

$$\int_{t_1}^{t_2}\int_R \left\{ v^i \left[\frac{d}{dt}\left(\frac{\partial \circ_\rho}{\partial v^i} \right) - \frac{\partial}{\partial x^i}\left(\mathcal{H}_\rho + \rho_o^2 \frac{\partial \lambda_\rho}{\partial \rho'} \right) \right] \right\} \delta^+ t \, dV dt = 0 . \tag{28.2}$$

Again, since $\delta^+ t$, dV, and dt are arbitrary and therefore different from zero, the former equation is only fulfilled if

$$\frac{\partial}{\partial x^i}\left(\mathcal{H}_\rho + \rho_o^2 \frac{\partial \lambda_\rho}{\partial \rho'} \right) = \frac{\partial}{\partial v^i}\left[\frac{\partial \circ_\rho}{\partial t} + v^j \frac{\partial \circ_\rho}{\partial x^j} \right] - \frac{\partial \circ_\rho}{\partial x^i} , \tag{28.3}$$

where the total derivative with respect to time being developed and the partial derivatives $\partial / \partial t, \partial / \partial \mathbf{v}$, and $\partial / \partial \mathbf{x}$ were interchanged.

The right hand side of (28.2) is zero because $\circ_\rho$ is not an explicit function of either t or x. Then if what remains of this expression is integrated, the following is obtained

$$\mathcal{H}_\rho + \rho_o^2 \frac{\partial \lambda_\rho}{\partial \rho'} = f(t) ; \tag{28.4}$$

where $f(t)$ is some function of time. Let

$$f(t) = \frac{c_o^2 \rho'^2}{\rho_o} - 2p' \tag{28.5}$$

be, so that in (28.4) and after substituting here the value of $\mathcal{H}_\rho$ given in (27.6) and the definition (28.5), the following result is obtained

$$p' = -\frac{1}{2}\rho_o v^2 + \frac{1}{2}\frac{c_o^2}{\rho_o}\rho'^2 . \tag{28.6}$$

This is the value for the change in pressure due to the pass of sound wave. On the other hand and according to (22.5), the corresponding value of the change in density calculated as far as terms of the first order is equal to

$$\rho' = -\frac{1}{2}\frac{\rho_o}{c_o^2}v^2 ; \tag{28.7}$$

and clearly it is such that $\rho' < 0$.

Now it is evident that

$$\frac{d}{dt}\left(\frac{\partial \circ_\rho}{\partial v^i}\right) = \frac{d}{dt}\left(\rho_o v^i\right) = 0 \tag{28.8}$$

as it is easy to see from (27.5). This result is particularly important in relation to a wave packet which it is not anything else but sound disturbance that at any given instant occupies a finite region of space. For this case the total momentum of the fluid in the wave packet can be determined.

If the relations (22.3) and (22.5) are taken into account it is easy to be convinced that

$$\rho \boldsymbol{v} - \frac{\boldsymbol{q}}{c_o^2} = \boldsymbol{a} \tag{28.9}$$

where

$$\boldsymbol{q} \equiv p'\boldsymbol{v} \tag{28.10}$$

is a vector representing the sound energy flux and $\boldsymbol{a} = \rho_o \boldsymbol{v}$ some vectorial quantity with units of mass flux. When the relation (28.9) is integrated over all the volume occupied by the wave packet the total momentum of the fluid in the sound disturbance is obtained; that is to say

$$\int_V \rho \boldsymbol{v} dV = \frac{1}{c_o^2} \int_V \boldsymbol{q} dV + \boldsymbol{A}; \tag{28.11}$$

where clearly, $\boldsymbol{A}$ is the value of $\boldsymbol{a}$ over the whole volume occupied by the wave packet. The existence of this term means that the propagation of a sound wave packet in a compressible perfect fluid is accompanied by matter transfer.

§29. Deformation waves

In order to offer another point of view from the studied phenomenon, let us consider that by the effect of small disturbance the geometry of region R considered suffers a deformation process. If each volume element of the fluid at the pass of disturbance is deformed, the deformation process is transferred through the system with the sound velocity; in such a way that the motion pattern is periodically repeated in the whole region occupied by the fluid as the wave goes forward. In order to obtain the wave equation which describes the phenomenon, the scalar equation for mass density (11.9) is used. Thus, at a first approximation we have that

$$\rho' = -\frac{\rho_o}{J_o} J'; \tag{29.1}$$

where it has been considered that $J = J_o + J'$, with $J_o >> J'$ and the approximation (22.3) was used. Now and according to the result (22.5), in the last relation we have that

$$p' = -\frac{\rho_o c_o^2}{J_o} J' . \tag{29.2}$$

In order to obtain the motion equation that governs the phenomenon it is necessary to eliminate the field velocity of the scheme with the aid of linearised Euler's and continuity equations, but properly written in terms of J'. From equation (29.2) the gradient of p' can be obtained so that

$$-\frac{1}{\rho_o}\,\textit{\textbf{grad}}\; p' = \frac{c_o^2}{J_o}\,\textit{\textbf{grad}}\; J' ;$$

in such a way that in the linearised Euler's equation

$$\frac{\partial \mathbf{v}}{\partial t} + \frac{1}{\rho_o}\,\textit{\textbf{grad}}\; p' = 0, \tag{29.3}$$

the following is obtained

$$\frac{\partial \mathbf{v}}{\partial t} - \frac{c_o^2}{J_o}\,\textit{\textbf{grad}}\; J' = 0; \tag{29.3$'$}$$

which is the form of this equation in terms of J'. Now, deriving the equation (29.1) with respect to time we have that

$$\frac{\partial \rho'}{\partial t} = -\frac{\rho_o}{J_o}\frac{\partial J'}{\partial t}; \tag{29.4}$$

so that in the linearised continuity equation (22.10) the form of this equation is obtained in terms of J'; that is to say

$$\frac{\partial J'}{\partial t} = J_o\, div\, \mathbf{v}. \tag{29.5}$$

If this last equation is derived again with respect to time and the operators $\partial/\partial t$ and *div* are interchanged because they are independent between them; the following is obtained

$$\frac{\partial^2 J'}{\partial t^2} = J_o\, div\left(\frac{\partial \mathbf{v}}{\partial t}\right). \tag{29.6}$$

However and according to (29.3′)

$$J_o\, div\left(\frac{\partial \boldsymbol{v}}{\partial t}\right) = c_o^2 \nabla^2 J';$$

in which case in (29.6) it is finally obtained that

$$\nabla^2 J' - \frac{1}{c_o^2}\frac{\partial^2 J'}{\partial t^2} = 0\,. \qquad (29.7)$$

This is the wave equation that governs the small geometrical local changes occurring in a compressible perfect fluid at the pass of a sound disturbance. Evidently, these disturbances are propagated through the fluid with the same velocity by which sound travels in the medium.

§30. Pressure waves

If we think that the sonic disturbance is a pressure wave, the linearised Euler´s equation is the relationship (29.3), whereas the linearised continuity equation in terms of p' and $\boldsymbol{v}$ can be obtained from the relations (29.2) and (29.5), that is to say

$$\frac{\partial p'}{\partial t} + \rho_o c_o^2\, div\, \boldsymbol{v} = 0\,. \qquad (30.1)$$

From linarised Euler´s equation (29.3), when its divergence is calculated and the operators $\partial/\partial t$ and *div* are interchanged, the following is obtained

$$\frac{\partial}{\partial t}(div\, \boldsymbol{v}) = -\frac{1}{\rho_o}\nabla^2 p'\,. \qquad (30.2)$$

If now the equation (30.1) is derived again with respect to time, the following result is obtained

$$-\frac{1}{\rho_o c_o^2}\frac{\partial^2 p'}{\partial t^2} = \frac{\partial}{\partial t}(div\, \boldsymbol{v})\,. \qquad (30.3)$$

Comparing these two equations it is immediately obtained that

$$\nabla^2 p' - \frac{1}{c_o^2}\frac{\partial^2 p'}{\partial t^2} = 0 . \qquad (30.4)$$

Again, a wave equation is obtained for pressure change; change that obviously is propagated through the fluid with the same velocity the sound has in the medium.

Selected Topics

The sound velocity

When the wave equation that satisfies a sound disturbance of small amplitude is obtained, the sound velocity in the medium is defined in terms of the change of pressure with density at a constant specific entropy, according to the following relationship

$$c_o = \sqrt{(\partial p/\partial \rho)_s} .$$

In order to evaluate this rate of change it is essential to know how the temperature varies along the dynamic process. In a sound wave one hopes that in the compression region the temperature raises and in the decompression region its value decreases. Newton was the first one who calculated the rate of change of pressure with density, and assumed that the temperature remains invariable. His argument was that the heat was conducted from one region to another so rapidly that the temperature was neither raised nor diminished. This wrong argument implies that the sound velocity is isothermal. The right deduction was given by Laplace who anticipated the opposite idea, that is that pressure and temperature in a sound wave changed adiabatically†. The real variation of pressure with density in a small amplitude sonic disturbance is such that it does not permit any heat flow; so that is an adiabatic variation for which

---

† *The heat flow from compressed region to rarified region is negligible in the measure that the wavelength is large compared to the mean free path. Under this condition the small amount in the heat flow in a sound wave does not affect the rapidity of propagation; even when it produces a small absorption in the sound energy. It can be correctly stated, that this absorption increases when the value of the mean free path becomes approximately equal to the wavelength . However these wavelengths are smaller by factor of around millions than those of the wavelengths of audible sounds.*

$$pV^{\gamma} = constant\,.$$

Here V is the volume. Since the density varies inversely with volume, the adiabatic relationship between pressure and density is equal to

$$p = constant\;\rho^{\gamma}\,;$$

in such a way that from this relation it is obtained that

$$\left(\frac{\partial p}{\partial \rho}\right)_s = \frac{\gamma p}{\rho} = c_o^2\,.$$

Let us consider the sound velocity in a perfect gas. For this case the thermic equation of state is

$$pV = \frac{p}{\rho} = \frac{RT}{\mu}\,,$$

where R is the gas constant and μ the molecular weight.

It is easy to demonstrate that for this situation

$$c_o = \sqrt{(\gamma RT/\mu)}\,;$$

so, it is evident the fact that the sound velocity depends only on the gas temperature and not on pressure or density. In all the former formulae

$$\gamma = c_p/c_v\,,$$

with c_p and c_v the values of specific heat at a pressure and volume constant respectively.

In general, the relation between the velocity v and the temperature oscillation in a sound wave can be obtained. It can be demonstrated that if the density remains fixed, to a small change in T corresponds to a small variation in p, that is to say

$$T' = \left(\frac{\partial T}{\partial p}\right)_s p'.$$

From the thermodynamic relationship

$$\left(\frac{\partial T}{\partial p}\right)_s = \left(\frac{T}{c_p}\right)\left(\frac{\partial V}{\partial T}\right)_p$$

and given that† $p' = \rho c_o v$ the following is obtained

$$T' = \frac{c_o \varepsilon T v}{c_p},$$

where

$$\varepsilon = \frac{1}{V}\left(\frac{\partial V}{\partial T}\right)_p$$

is the volume coefficient of expansion.

Plane waves

Let us consider a sound wave in which all quantities depend on the coordinate x only. In this case the flow is completely homogenous in the y-z plane, in such a way that it deals with the case of a plane wave. The corresponding form of the wave equation is

$$\frac{\partial^2 \phi}{\partial t^2} - c_o^2 \frac{\partial^2 \phi}{\partial x^2} = 0.$$

To solve this equation, it is re-written in the following manner

---

† *See the next paragraph*

$$\left(\frac{\partial}{\partial t}-c_o\frac{\partial}{\partial x}\right)\left(\frac{\partial}{\partial t}+c_o\frac{\partial}{\partial x}\right)\phi=0\,.$$

Next, new variables are introduced

$$\xi=t-\frac{x}{c_o}, \quad \text{and} \quad \eta=t+\frac{x}{c_o},$$

so that

$$t=\frac{1}{2}(\eta+\xi), \quad \text{and} \quad x=\frac{c_o}{2}(\eta-\xi).$$

If the chain rule is used we have that

$$\frac{\partial}{\partial t}=\frac{\partial\xi}{\partial t}\frac{\partial}{\partial\xi}\equiv\frac{\partial}{\partial\xi}; \text{ because } \frac{\partial\xi}{\partial t}=1,$$

and

$$\frac{\partial}{\partial x}=\frac{\partial\xi}{\partial x}\frac{\partial}{\partial\xi}\equiv-\frac{1}{c_o}\frac{\partial}{\partial\xi}; \text{ since } \frac{\partial\xi}{\partial x}=-\frac{1}{c_o}.$$

Adding the former results we see that

$$\frac{\partial}{\partial\xi}=\frac{1}{2}\left(\frac{\partial}{\partial t}-c_o\frac{\partial}{\partial x}\right).$$

In the same way it is demonstrated that

$$\frac{\partial}{\partial\eta}=\frac{1}{2}\left(\frac{\partial}{\partial t}+c_o\frac{\partial}{\partial x}\right)$$

in such a way that the original wave equation becomes

$$\frac{\partial^2 \phi}{\partial \xi \partial \eta} = 0.$$

Integrating this equation with respect to ξ, we find that

$$\frac{\partial \phi}{\partial \eta} = F(\eta),$$

where $F(\eta)$ is some arbitrary function of η. Integrating again it is obtained that

$$\phi = \phi_1(\xi) + \phi_2(\eta),$$

where ϕ_1 and ϕ_2 are arbitrary functions of their arguments. Thus,

$$\phi = \phi_1\left(t - \frac{x}{c_o}\right) + \phi_2\left(t + \frac{x}{c_o}\right).$$

It is evident that the other quantities (p', ρ', v) in a plane wave are given by functions of the same form. For example, for the change of density we would have that

$$\rho' = \phi_1\left(t - \frac{x}{c_o}\right) + \phi_2\left(t + \frac{x}{c_o}\right).$$

Let us suppose that $\phi_2 = 0$ in such a way that

$$\rho' = \phi_1\left(t - \frac{x}{c_o}\right),$$

whose meaning is evident. In fact in any plane where it is fulfilled that $x = constant$, the density varies with time and at any given instant it is different for different values of x. However, it has the same value for coordinates x and times t such that the following relationship is satisfied

$$t - \frac{x}{c_o} = constant\,,$$

or too

$$x = constant + c_o t\,.$$

The former means that if at some instant $t = 0$ and at some point the fluid density has a certain value, then after a time t the same value of the density is found at a distance $c_o t$ along the X-axis measured from the original point. The same is true of all other quantities in the wave. Therefore, the pattern of motion is propagated through the medium in the x direction with a velocity c_o known by the name of the velocity of sound.

Thus $\phi_1(t-x/c_o)$ represents what is called a traveling plane wave propagated in the positive direction of the X-axis. It is clear that $\phi_2(t+x/c_o)$ represents a wave propagated in the opposite direction.

On the other hand, of the three components of the velocity $\boldsymbol{v} = \boldsymbol{grad}\ \phi$ in a plane wave, only the x component is not zero; that is to say

$$v_x = \frac{\partial \phi}{\partial x}.$$

In consequence, the fluid velocity in a sound wave is directed in the direction of propagation. For this reason sound waves in a fluid are said to be longitudinal.

In a traveling plane wave, the velocity $v_x = v$ is related to the pressure p' and the density ρ' in a simple manner.

In fact, let $\phi = f(x-c_o t)$ be so that

$$v = \frac{\partial \phi}{\partial x} = f'(x - c_o t). \tag{1}$$

On the other hand and given that

$$p' = -\rho_o \frac{\partial \phi}{\partial t},$$

it is clear that

$$p' = \rho_o c_o f'(x - c_o t). \tag{2}$$

Comparing the expressions (1) and (2) we find that

$$v = \frac{p'}{\rho_o c_o}.$$

Finally and given that $p' = c_o^2 \rho'$ we have that

$$v = \frac{c_o \rho'}{\rho_o};$$

that is no other thing but the relation between the velocity and the variation in the density.

The wave equation in terms of the change in the density

In order to obtain this relationship let us consider the linearised continuity equation

$$\frac{\partial \rho'}{\partial t} + \rho_o \, div \, \mathbf{v} = 0.$$

Deriving this equation with respect to time it is obtained that

$$\frac{\partial^2 \rho'}{\partial t^2} + \rho_o div\left(\frac{\partial \mathbf{v}}{\partial t}\right) = 0; \tag{1}$$

where the differential operators $\partial/\partial t$ and *div* were interchanged because they are independent between them. On the other hand, from the linearised Euler´s equation

$$\frac{\partial \mathbf{v}}{\partial t} + \frac{1}{\rho_o} \textbf{grad}\ p' = 0$$

and given that $p' = c_o^2 \rho'$ we have that

$$\frac{\partial \mathbf{v}}{\partial t} = -\frac{c_o^2}{\rho_o} \textbf{grad}\ \rho' .$$

Substituting this relation in (1) the next wave equation is obtained

$$\nabla^2 \rho' - \frac{1}{c_o^2} \frac{\partial^2 \rho'}{\partial t^2} = 0 .$$

This is the motion equation that governs the density change; change that is propagated through the considered continuous medium with the same velocity the sound has in the fluid.

Variations in pressure and density

These changes are given by the next relations

$$p' = -\frac{1}{2} \rho_o v^2 + \frac{c_o^2}{2\rho_o} \rho'^2 \tag{1}$$

and

$$\rho' = -\frac{\rho_o}{2c_o^2} v^2 . \tag{2}$$

Eliminating the term v^2 between these two formulae a second degree equation for ρ' in terms of p' is obtained; that is to say

$$\rho'^2 + 2\rho_o \rho' - \frac{2\rho_o}{c_o^2} p' = 0 . \tag{3}$$

In this case, it is clear that

$$\rho' = \rho_o \left[-1 \pm \sqrt{1 + \frac{2p'}{\rho_o c_o^2}} \right] . \tag{4}$$

If we expand in power series the following is obtained

$$\rho' \approx \rho_o\left[-1+1+\frac{p'}{\rho_o c_o^2}\right].$$

Evidently

$$p' = c_o^2 \rho',$$

where only the positive root is considered.

Geometrical acoustics

A plane wave has the distinctive property that its direction of propagation and its amplitude are the same in all the space. An arbitrary sound wave does not possess this quality. However, cases may occur where a sound wave that is not plane may still be regarded as plane in any small region of space. For this to be so it is evidently necessary that the amplitude and the direction of propagation of the disturbance should vary only slightly over distances of the order of the wavelength. If this condition holds, it is possible to talk about rays, as these lines where the tangent to them at any point is in the same direction as the direction of propagation; so that we can say that the sound is propagated in a rectilinear way, and ignore its wave nature. The study of the laws of propagation of sound in such cases is what is known as geometrical acoustics and corresponds to the limit of small wavelengths, that is to say, when $\lambda_s \to 0$. In order to derive the basic equation that governs the phenomenon, it is assumed that the wave velocity potential $\phi(\boldsymbol{x},t)$ can be written as follows

$$\phi(\boldsymbol{x},t) = ae^{i\psi}. \tag{1}$$

In the case where the sound wave is not plane but geometrical acoustics can be applied, the amplitude a is a slowly varying function of coordinates and time, while the wave phase $\psi = \boldsymbol{k}\cdot\boldsymbol{r} - \omega t + \alpha$ is almost linear, with α a constant as the wave vector $\boldsymbol{k}$ and the frequency ω are as well. Over small regions of space and short intervals of time, the phase ψ may be expanded in series up to terms of the first order in such a way that

$$\psi = \psi_o + \boldsymbol{r}\cdot \mathbf{grad}\,\psi + t\frac{\partial \psi}{\partial t}. \tag{2}$$

Since in any small region of space and during short intervals of time, the sonic disturbance may be regarded as a plane wave, it is possible at any point of the space to define the wave vector and the frequency as follows

$$\boldsymbol{k} = \frac{\partial \psi}{\partial \boldsymbol{r}} \equiv \textbf{\textit{grad}}\, \psi$$
$$\omega = -\frac{\partial \psi}{\partial t}. \tag{3}$$

The function ψ is known by the name of the *eikonal*.

In a sound wave we have that

$$\frac{\omega^2}{c_o^2} = k^2 \tag{4}$$

with

$$k^2 = k_x^2 + k_y^2 + k_z^2 \tag{5}$$

and c_o the sound velocity in the medium. Substituting (3) in (4) and taking into account (5) it is obtained that

$$\left(\frac{\partial \psi}{\partial x}\right)^2 + \left(\frac{\partial \psi}{\partial y}\right)^2 + \left(\frac{\partial \psi}{\partial z}\right)^2 = \frac{1}{c_o^2}\left(\frac{\partial \psi}{\partial t}\right)^2. \tag{6}$$

This is the basic relationship of geometrical acoustics. It is called the *Eikonal Equation*.

References

1. Fierros Palacios, Angel. "Las ecuaciones diferenciales de campo para flujo potencial a partir de un principio variacional tipo Hamilton". Rev. Mex. de Fís. 38, No. 5 (1992) 764-777.
2. Fierros Palacios, Angel. "La transmisión de un movimiento oscilatorio de amplitud pequeña en un fluido perfecto compresible". Rev. del IMP. Vol. XXIV, No. 4 (octubre, 1992).
3. Goldstein, H. "Classical Mechanics". Addison-Wesley Publishing Co. Inc. (1959).
4. Landau, L.D. and Lifshitz, E. M. "Fluid Mechanics". Addison-Wesley Publishing Co. (1959).
5. Morse, P. and Feshbach, H. "Methods of Theoretical Physics". Part I. McGraw-Hill Book Co. (1953).

Chapter VI

Viscous Fluids

§31. Dissipative processes in a viscous fluid

To obtain the field differential equations describing the dynamical state of any viscous fluid it is necessary to take into account the effect of energy dissipation occurring during the motion of a fluid and affecting the motion itself. In particular, this dissipative process is the result of the thermodynamic irreversibility of the motion that always occurs when the real fluids move, and it is due to internal friction mechanisms known by the generical name of viscosity. The viscosity of the fluid is the result of an irreversible transfer of momentum from points in the continuous medium where the velocity is large to those where it is small. There is another mechanism of energy dissipation arising when the temperature of the fluid is not uniform throughout all the region occupied by the fluid. The difference in temperature gives origin to a heat transfer between different parts of the system and between the system and its environs. This phenomenon is known as thermal conduction. This other mechanism generates a thermal energy transfer from points where the temperature is high to those where it is low. In general, it does not involve any macroscopic motion, and it can occur even when the fluid remains at rest.

In a viscous fluid the physical mechanism generating the motion are the variations of density in the system; changes that can originate a complex hydrodynamic regime as in that of the ideal fluid case, characterized by the presence of velocity gradients in the continuous medium. On the other hand, the variations of density occur when the geometry of the region occupied by the fluid is deformed because of the action of some external agent. When the deformation is produced, the system is removed from its equilibrium state that is not only mechanical but thermal too, propitiating the appearance of internal stresses that are trying to return it to its original situation of equilibrium.

As it has already been said, the internal stresses are produced by short range molecular forces acting from any point in the fluid affecting only those nearby. In general, these are forces of limited range exerted over the surface limit of the region occupied by the system.

§32. The thermodynamics of deformation

When in some deformed region occupied by any viscous fluid a change is produced in the original deformation, the position of whichever two points changes in such a way that the displacement vector joining those points, experiments a variation. If the changes in the initial deformation are small, the variations in the displacement vector are infinitesimal. As a consequence of such process the internal stresses generated inside the fluid realize an infinitesimal quantity of mechanical work.

In order to make an adequate description of the process let us consider the total force over some portion of the given continuous medium. Firstly, this total force is equal to the sum of all the forces on all the volume elements in that portion of the continuous medium, in such a way that it is possible that the volume integral is written as $\int \boldsymbol{F} dV$ where $\boldsymbol{F}$ is the force per unit volume and $\boldsymbol{F} dV$ the force on the volume element dV. Secondly, the forces by which various parts of the portion considered act one over the other can not give anything but zero in the total resultant force, since they are canceled by Newton´s third law. The required total force can therefore be regarded as the sum of the forces exerted on the given portion of the continuous medium by the other volume elements surrounding it. These forces act on the surface of that volume element, so that the resultant force can be represented as the sum of forces acting on all the surface elements; that is to say, it can be written as an integral over the surface.

Thus and for any portion of the continuous medium, each of the three components $\int F_i \, dV$ of the resultant of all internal stresses can be transformed into an integral over the surface. As it is well known from vector analysis, the integral of a scalar over an arbitrary volume can be transformed into an integral over the surface if the scalar is the divergence of a vector. In the present case we have the integral of a vector and not of a scalar. This means that the vector F_i must be the divergence of a tensor of rank two; that is to say

$$F_i = \frac{\partial \sigma_{ij}}{\partial x^j}; \tag{32.1}$$

where

$$\sigma_{ij} = \sigma_{ji} \tag{32.2}$$

are the components of a symmetrical tensor known as the mechanical stress tensor. The force on any elemental volume can be written as an integral of a tensor of rank two over the closed surface boundaring that volume, that is to say

$$\int_V F_i \, dV = \int_V \frac{\partial \sigma_{ij}}{\partial x^j} dV = \oint_S \sigma_{ij} da_j \,, \tag{32.3}$$

where ***da*** is the surface element vector. Therefore and under a deformation process the work done by the internal stresses is

$$\int_V \Delta \mathcal{R} dV = \int_V \left(\frac{\partial \sigma_{ij}}{\partial x^j} \right) \Delta u_i \, dV \,. \tag{32.4}$$

Integrating by parts the former relation and using Greeen´s theorem the following is obtained

$$\int_V \Delta \mathcal{R} \, dV = \oint_S \sigma_{ij} \, \Delta u_i \, da_j \; - \; \int_V \sigma_{ij} \frac{\partial \Delta u_i}{\partial x^j} dV \,. \tag{32.5}$$

By considering an infinite medium which is not deformed at infinity and the surface of integration extended to infinity it can be obtained that $\sigma_{ij} = 0$ on the surface, so that the surface integral in (32.5) becomes null. Since the stress tensor is symmetrical, what remains of the last expression can be written as follows

$$\begin{aligned} \int_V \Delta \mathcal{R} \, dV &= -\frac{1}{2} \int_V \sigma_{ij} \left(\frac{\partial \Delta u_i}{\partial x^j} + \frac{\partial \Delta u_j}{\partial x^i} \right) dV \\ &= -\frac{1}{2} \int_V \sigma_{ij} \Delta \left(\frac{\partial u_i}{\partial x^j} + \frac{\partial u_j}{\partial x^i} \right) dV = - \int_V \sigma_{ij} \, \Delta u_{ij} \, dV ; \end{aligned} \tag{32.6}$$

where u_{ij} are the components of small deformations strain tensor, as it is easy to see from (10.9). Since dV is a volume element which was arbitrarily elected and consequently can not be equal to zero, from the former relation the following result is obtained

$$\Delta \mathcal{R} = -\sigma_{ij}\,\Delta u_{ij}\,. \tag{32.7}$$

This is the work done by the internal stresses when the deformations suffered by the region R occupied the continuous medium under consideration are small. When the last calculus is made, we also suppose that the process of deformation occurs so slowly that thermodynamic equilibrium is established in the system at every instant in accordance to the external conditions; in such a way that the process is thermodynamically reversible, assumption that is almost always justified in practice. On the other hand and due to the deformation, the specific internal energy of the fluid changes. Thus, to an infinitesimal variation of the specific internal energy corresponds a difference between the quantity of heat acquired and the work done by the internal stresses, that is to say $d\square = Tds - d\mathcal{R}$, so that and according to (32.7)

$$d\square = Tds + \sigma_{ij}du_{ij}\,, \tag{32.8}$$

where T is the temperature. The former relationship is the thermodynamic identity for the deformation. When the deformation process is a hydrostatic compression the stress tensor is simply

$$\sigma_{ij} = -p\delta_{ij}\,, \tag{32.9}$$

and clearly

$$\sigma_{ij}du_{ij} = -pdu_{nn}\,. \tag{32.10}$$

Let us consider an infinitesimal volume element of fluid equal to dV that after the deformation is transformed into dV'. In order to be able to calculate this last value it is assumed that dV' is centered at the point where the principal axes of the strain tensor are coincident with the coordinate axes of a cartesian reference system. Then, the elements of length dx_1, dx_2, and dx_3 along these axes will become, after the deformation, $dx'_1=(1+u^{(1)})dx_1$, $dx'_2=(1+u^{(2)})dx_2$, and $dx'_3=(1+u^{(3)})dx_3$.

Evidently, the corresponding volume element of fluid after and before the deformation are $dV=dx_1dx_2dx_3$, and $dV'=dx'_1dx'_2dx'_3$, respectively. In this case it is possible to write that $dV'=dV(1+u^{(1)})(1+u^{(2)})(1+u^{(3)})$. If the former expression is developed and neglecting higher order terms it is clear that $dV'=dV(1+u^{(1)}+u^{(2)}+u^{(3)})$. However, the sum $u^{(1)}+u^{(2)}+u^{(3)}$ of the principal values of a small deformation stress tensor is an invariant that in any coordinate system is equal to the sum of the principal diagonal components; that is to say $u_{nn}=u_{11}+u_{22}+u_{33}$. In this case it is finally obtained that

$$dV' = dV\left(1+u_{nn}\right). \qquad (32.11)$$

If unitary volumes are considered u_{nn} is simply the change in volume, that is to say

$$u_{nn} = \Delta V = V - V_o \qquad (32.12)$$

so that $du_{nn}=dV$; in such a way that the thermodynamic identity for deformation (32.8) takes the usual form, that is

$$d\square = Tds - pdV\,. \qquad (32.13)$$

On the other hand and due to the fact that $V=1/\rho$ is the specific volume, in (32.12) we have that

$$u_{nn} = \frac{1}{\rho} - \frac{1}{\rho_o}\,. \qquad (32.14)$$

Finally and according to the relationship (32.8) it is easy to see that for any viscous fluid it is fulfilled that

$$\square = \square\left(s, u_{ij}\right); \qquad (32.15)$$

and then

$$\left(\frac{\partial \square}{\partial s}\right)_{u_{ij}} = T \qquad (32.16)$$

and

$$\left(\frac{\partial \square}{\partial u_{ij}}\right)_{s} = \sigma_{ij}\,. \qquad (32.17)$$

It is clear that if the deformations are the result of processes of expantion or compression hydrostatic or volumetric, in (32.15) and as it is easy to see from relation (32.14) we only have that

$$\square = \square(s,\rho) = \square(\rho) \qquad (32.18)$$

because the specific entropy is a constant. In this case it is said that the fluid is ideal or perfect.

§33. Field differential equations

A viscous fluid is a continuous system whose dynamical evolution can be completely characterized with the aid of a set of field functions that as in the case of the ideal fluid are the velocity field $\boldsymbol{v}(\boldsymbol{x},t)$, the hydrostatic pressure $p(\boldsymbol{x},t)$, and the mass density $\rho(\boldsymbol{x},t)$. Also, the stress tensor $\tilde{\sigma}(\boldsymbol{x},t)$ and the small deformations strain tensor $\tilde{u}(\boldsymbol{x})$ must be taken into account.

However and as it was before seen in the last paragraph, the pressure is consider within the structure of stress tensor and the mass density in some way is contained in the small deformations strain tensor. In fact, in the so called constitutive equations which are no other thing than additional postulates to the theory proposed in order to take into account the answer of the mechanical system under specific stresses, the pressure is the spherically symmetrical part of the stress tensor for newtonian fluid. In general, they are tensorial equations. For all that has been previously said and within the theoretical frame of Lagrange´s Analytical Mechanics, a lagrangian density is proposed containing all the dynamical information concerning the viscous fluid and whose functional form is

$$\circ = \circ\left(\boldsymbol{x},\boldsymbol{v},u_{ij},t\right); \tag{33.1}$$

so that the corresponding specific lagrangian has the following form

$$\lambda = \lambda\left(\boldsymbol{x},\boldsymbol{v},u_{ij},t\right). \tag{33.2}$$

The invariance condition (14.2) to which the action integral (14.1) is subjected gives again as a result the general formula (16.4); where now the functional relationship (33.2) for the proper specific lagrangian for viscous fluid must appear. According to it,

$$\rho\delta\lambda = \rho\left[\frac{\partial\lambda}{\partial x^i}\delta x^i + \frac{\partial\lambda}{\partial v^i}\delta v^i + \frac{\partial\lambda}{\partial u_{ij}}\delta u_{ij}\right]. \tag{33.3}$$

The local variations of the velocity field are given in the relation (18.4), whereas

$$\delta u_{ij} = \frac{1}{2}\delta\left[\frac{\partial u_i}{\partial x^j} + \frac{\partial u_j}{\partial x^i}\right] = \frac{1}{2}\frac{\partial}{\partial x^j}\left[\delta u_i + \delta u_j\,\delta_{ij}\right] = \frac{\partial}{\partial x^j}\left(\delta u_i\right);$$

where the definition (10.9) was used and the differential operators δ and $\partial/\partial\boldsymbol{x}$ were interchanged because they are independent between them. Now and given that the displacement vector is only a function of the coordinates

$$\delta u_i\left(\boldsymbol{x}\right) = \frac{\partial u_i}{\partial x^k}\delta x^k;$$

in whose case

$$\delta u_{ij} = \frac{\partial}{\partial x^j}\left(u_{ik}\,\delta x^k\right). \tag{33.4}$$

Then, in (33.3) the following is obtained

$$\rho\delta\lambda = \left[\rho\frac{\partial\lambda}{\partial x^i} - \frac{d}{dt}\left(\rho\frac{\partial\lambda}{\partial v^i}\right) - \frac{\partial}{\partial x^j}\left(\rho\frac{\partial\lambda}{\partial u_{ij}}\right)u_{\circ\circ}\right]\delta x^i$$
$$+\frac{d}{dt}\left[\rho\left(\frac{\partial\lambda}{\partial v^i}\right)\delta x^i\right] + \frac{\partial}{\partial x^j}\left[\rho\frac{\partial\lambda}{\partial u_{ij}}u_{ik}\,\delta x^k\right];$$

where the relation (18.4) was used and two integrations by parts were made. Now, in the third term the first square parenthesis of the former result we have that

$$\frac{\partial}{\partial x^j}\left(\rho\frac{\partial\lambda}{\partial u_{ij}}\right)u_{\circ\circ} = \frac{\partial}{\partial x^j}\left[\rho\frac{\partial\lambda}{\partial u_{ij}}u_{\circ\circ}\right] - \rho\frac{\partial\lambda}{\partial u_{ij}}\left(\frac{\partial u_{\circ\circ}}{\partial x^j}\right);$$

where an integration by parts was made. Moreover and due to the fact that $u_{\circ\circ}$ is an invariant in any coordinate system it is possible to assume that it is not an explicit function of the coordinates so that $\partial u_{\circ\circ}/\partial \boldsymbol{x}=0$. From (32.14) and because the deformations considered are small, it can be said that

$$u_{\circ\circ} \approx \frac{1}{\rho} \tag{33.5}$$

in whose case

$$\frac{\partial}{\partial x^j}\left(\rho\frac{\partial\lambda}{\partial u_{ij}}u_{\circ\circ}\right) = \frac{\partial}{\partial x^j}\left(\frac{\partial\lambda}{\partial u_{ij}}\right). \tag{33.6}$$

Then,

$$\rho\delta\lambda = \left[\rho\frac{\partial\lambda}{\partial x^i} - \frac{d}{dt}\left(\rho\frac{\partial\lambda}{\partial v^i}\right) - \frac{\partial}{\partial x^j}\left(\frac{\partial\lambda}{\partial u_{ij}}\right)\right]\delta x^i$$
$$+\frac{d}{dt}\left[\rho\frac{\partial\lambda}{\partial v^i}\delta x^i\right] + \frac{\partial}{\partial x^j}\left[\rho\frac{\partial\lambda}{\partial u_{ij}}u_{ik}\,\delta x^k\right]. \tag{33.7}$$

However, in paragraph 18 of the chapter for ideal fluids an integral exactly equal to the one which would be obtained integrating regarding time, the second term of the right hand side of the former equation was calculated; so that it is evident that the following relation is fulfilled

$$\int_{t_1}^{t_2}\int_R \frac{d}{dt}\left[\rho\frac{\partial\lambda}{\partial v^i}\delta x^i\right]dV\,dt = -\int_{t_1}^{t_2}\int_R\left(\rho\frac{\partial\lambda}{\partial v^i}\right) div\,\boldsymbol{v}\,\delta x^i\,dV\,dt\,. \tag{33.8}$$

In order to calculate this result an integration by parts was carried out and the relation (4.4) and (5.6) were used again. On the other hand,

$$\int_{t_1}^{t_1}\int_R \frac{\partial}{\partial x^j}\left[\rho\frac{\partial\lambda}{\partial u_{ij}}u_{ik}\,\delta x^k\right]dV\,dt = \int_{t_1}^{t_2}\oint_S \rho\frac{\partial\lambda}{\partial u_{ij}}u_{ik}\,\delta x^k\,da_j\,dt\,;$$

where Green´s theorem was used. Again the surface integral becomes null because the tensor $\partial\lambda/\partial u_{ij} = 0$ over the surface, when it is assumed that it deals with an infinite continuous medium limited by an infinite surface which is not deformed there. Then, with the remains of relations (33.7) and (33.8) it is finally obtained that

$$\int_{t_1}^{t_2}\int_R\left[\rho\frac{\partial\lambda}{\partial x^i} - \mathcal{D}\left(\rho\frac{\partial\lambda}{\partial v^i}\right) - \frac{\partial}{\partial x^j}\left(\frac{\partial\lambda}{\partial u_{ij}}\right)\right]\delta x^i\,dV\,dt = 0\,. \tag{33.9}$$

So that the former result is valid it is again necessary that the integrand becomes null; that is to say

$$\mathcal{D}\left(\rho\frac{\partial\lambda}{\partial v^i}\right)-\rho\frac{\partial\lambda}{\partial x^i}+\frac{\partial}{\partial x^j}\left(\frac{\partial\lambda}{\partial u_{ij}}\right)=0\,. \tag{33.10}$$

This is the generalized momentum balance equation in terms of the specific lagrangian for viscous fluid. If the definition (7.1) for Reynolds′ differential operator is used again in the result (33.10), the following is obtained

$$\frac{d}{dt}\left(\frac{\partial\lambda}{\partial v^i}\right)-\frac{\partial\lambda}{\partial x^i}+\frac{1}{\rho}\frac{\partial}{\partial x^j}\left(\frac{\partial\lambda}{\partial u_{ij}}\right)=0; \tag{33.11}$$

where the relations (9.2) and (18.9) were used. The relationship (33.11) represents the corresponding Euler-Lagrange′s differential equation for viscous fluids. It is evident that and for consistence, the equation (33.10) must be reduced to the corresponding momentum balance equation for ideal fluids (18.8). In fact, let us consider the following calculus

$$\begin{aligned}-\frac{\partial}{\partial x^j}\left(\frac{\partial\lambda}{\partial u_{ij}}\right)&=-\frac{\partial}{\partial x^i}\left(\frac{\partial\lambda}{\partial u_{ij}}\delta_{ij}\right)=-\frac{\partial}{\partial x^i}\left(\frac{\partial\lambda}{\partial u_{\circ\circ}}\right)\\&=-\frac{\partial}{\partial x^i}\left(\frac{\partial\lambda}{\partial\frac{1}{\rho}}\right)=\frac{\partial}{\partial x^i}\left(\rho^2\frac{\partial\lambda}{\partial\rho}\right)\,.\end{aligned} \tag{33.12}$$

With this result substituted in (33.10) the equation (18.8) is recovered.

§34. Generalized energy balance equation

This equation emerges from the invariance condition for the action integral (14.1) under temporary continuous transformations. According to the lagrangian density functionality (33.1) it is easy to see that

$$\delta^+\circ=\frac{\partial\circ}{\partial x^i}\delta^+x^i+\frac{\partial\circ}{\partial v^i}\delta^+v^i+\frac{\partial\circ}{\partial u_{ij}}\delta^+u_{ij}+\frac{\partial\circ}{\partial t}\delta^+t\,. \tag{34.1}$$

Since the temporary variations of the coordinates and the velocity were previously calculated, we only have to calculate the temporary variation of the small deformation strain tensor. Then

$$\delta^+ u_{ij} = \frac{\partial}{\partial x^j}\left(\delta^+ u_i\right).$$

Now

$$\delta^+ u_i(\boldsymbol{x}) = \frac{\partial u_i}{\partial x^k}\frac{\partial x^k}{\partial t}\delta^+ t = u_{ik} v^k \delta^+ t;$$

consequently

$$\delta^+ u_{ij} = \frac{\partial}{\partial x^j}\left(u_{ik} v^k\right)\delta^+ t. \tag{34.2}$$

With the aid of this last result and with the relations (15.3) and (15.5) substituted in (34.1) the following is obtained

$$\delta^+ \circ = \left[\frac{\partial \circ}{\partial x^i} v^i + \frac{\partial \circ}{\partial v^i}\frac{dv^i}{dt} + \left(\frac{\partial \circ}{\partial u_{ij}}\right)\frac{\partial}{\partial x^j}\left(u_{ik} v^k\right) + \frac{\partial \circ}{\partial t}\right]\delta^+ t. \tag{34.3}$$

Since the mass density does not depend on the small deformations strain tensor, it is clear that the third term of the right hand side of (34.3) can be written as follows

$$\left(\rho\frac{\partial \lambda}{\partial u_{ij}}\right)\frac{\partial}{\partial x^j}\left(u_{ik} v^k\right) = \frac{\partial}{\partial x^j}\left[\rho\frac{\partial \lambda}{\partial u_{ij}} u_{ik} v^k\right] - \frac{\partial}{\partial x^j}\left[\rho\frac{\partial \lambda}{\partial u_{ij}}\right] u_{ik} v^k .$$

Now, let $v^k = v^i \delta^i_k$ be so that on the right hand side of the former relation we have

$$\frac{\partial}{\partial x^j}\left[\frac{\partial \lambda}{\partial u_{ij}} v^i\right] - \frac{\partial}{\partial x^j}\left(\frac{\partial \lambda}{\partial u_{ij}}\right) v^i \tag{34.4}$$

where the relation (33.5) was used. On the other hand and due to the fact that it is possible to use the results obtained in paragraph 19; that is to say, the relations (19.5) to (19.10) as well as the equation (34.4), in (34.3) the following is obtained

$$\begin{aligned}\delta^+\circ = &\left[\rho\frac{\partial \lambda}{\partial x^i} - \mathcal{D}\left(\rho\frac{\partial \lambda}{\partial v^i}\right) - \frac{\partial}{\partial x^j}\left(\frac{\partial \lambda}{\partial u_{ij}}\right)\right] v^i \delta^+ t \\ &+ \left[\frac{d}{dt}\left(\frac{\partial \circ}{\partial v^i} v^i\right) + \mathcal{H}\, div\mathbf{v} + \frac{\partial}{\partial x^j}\left(\frac{\partial \lambda}{\partial u_{ij}} v^i\right) + \rho\frac{\partial \lambda}{\partial t}\right]\delta^+ t.\end{aligned} \tag{34.5}$$

Evidently the first square parenthesis is zero because it deals with the generalized momentum balance equation (33.10), whereas in the other term $\mathcal{H}$ is the corresponding hamiltonian density. Introducing the remaining of (34.5) in the relation (19.1) we have that

$$\int_{t_1}^{t_2}\int_R \left[\mathcal{D}\mathcal{H} + \frac{\partial}{\partial x^j}\left(\frac{\partial \lambda}{\partial u_{ij}} v^i\right) + \rho\frac{\partial \lambda}{\partial t}\right]\delta^+ t\, dV\, dt = 0. \tag{34.6}$$

For the same arguments related to the increments $\delta^+ t, dV$, and dt that systematically were used along this book, the former relationship is only fulfilled if

$$\mathcal{D}\mathcal{H} + \frac{\partial}{\partial x^j}\left(\frac{\partial \lambda}{\partial u_{ij}} v^i\right) + \rho\frac{\partial \lambda}{\partial t} = 0; \tag{34.7}$$

and this is the generalized energy balance equation for any viscous fluid. It is a scalar equation in terms of hamiltonian density and the specific lagrangian for this case. It is clear that if the relation (33.12) is fulfilled, the equation (34.7) is reduced to the corresponding of the ideal fluid; that is to say, it is transformed into the equation (19.13).

§35 Cauchy´s and Navier-Stokes´ equations

The motion equations for any newtonian viscous fluid can be obtained from a lagrangian density identical in form to the one of ideal fluid. The sole difference between both functions radicates in that for the perfect fluid the specific internal energy of the system depends on the mass density whereas for viscous fluid turns out to be a function of the small deformations strain tensor. As a consequence let

$$\circ = \rho\left[\frac{1}{2}v^2 - \Box(u_{ij})\right] \tag{35.1}$$

be the corresponding lagrangian density. In this case

$$\lambda = \frac{1}{2}v^2 - \Box(u_{ij}) \tag{35.2}$$

is the explicit form of specific lagrangian. With this relationship substituted in (33.11) the following results are obtained

$$\frac{d}{dt}\left(\frac{\partial \lambda}{\partial \mathbf{v}}\right) = \frac{\partial \mathbf{v}}{\partial t} + (\mathbf{v} \cdot \mathbf{grad})\mathbf{v}\,; \tag{35.3}$$

where the definition (6.1) for the hydrodynamics derivative was used.

Moreover,

$$\frac{\partial}{\partial x^j}\left(\frac{\partial \lambda}{\partial u_{ij}}\right) = -\frac{\partial}{\partial x^j}\left(\frac{\partial \Box}{\partial u_{ij}}\right) = -\frac{\partial \sigma_{ij}}{\partial x^j}\,. \tag{35.4}$$

Since neither $\circ$ nor λ are explicit functions of the coordinates because ever since paragraph 20 these functions only contain hydrodynamic terms, we have that $\partial \lambda / \partial \mathbf{x} = 0$, which means that the external body forces have been ignored. With the results (35.3) and (35.4) substituted in (33.11) it is obtained that

$$\frac{\partial \boldsymbol{v}}{\partial t} + (\boldsymbol{v} \cdot \boldsymbol{grad})\boldsymbol{v} = \frac{1}{\rho} div\, \tilde{\sigma} . \tag{35.5}$$

This is the well known Cauchy´s motion equation. For the case of viscous fluids it is known that the components of stress tensor can be expressed as follows

$$\sigma_{ij} = -(p + C)\delta_{ij} + \sigma'_{ij} , \tag{35.6}$$

where $C(T)$ is some constant which depends on the temperature differences between different parts of the fluid, whereas $\tilde{\sigma}'(\boldsymbol{x}, t)$ is the viscosity stress tensor and obviously $p(\boldsymbol{x}, t)$ is the hydrostatic pressure. The most general explicit form for the viscosity stress tensor is given by the following formula

$$\sigma'_{ij} = \eta\left[\frac{\partial v^i}{\partial x^j} + \frac{\partial v^j}{\partial x^i} - \frac{2}{3}\delta_{ij}\frac{\partial v^\circ}{\partial x^\circ}\right] + \zeta\delta_{ij}\frac{\partial v^\circ}{\partial x^\circ} ; \tag{35.7}$$

where η and ζ are the coefficients of viscosity which are in general functions of pressure and temperature. Since in many practical cases these coefficients do not change noticeably in the fluid, they may be regarded as constant. With the definition (35.7) the motion equation for any viscous fluid can be expressed as follows

$$\begin{aligned} \rho\left[\frac{\partial \boldsymbol{v}}{\partial t} + (\boldsymbol{v} \cdot \boldsymbol{grad})\boldsymbol{v}\right] &= -\boldsymbol{grad}\, p + \eta\nabla^2\boldsymbol{v} \\ &+ \left(\zeta + \frac{1}{3}\eta\right)\boldsymbol{grad}\, div\, \boldsymbol{v} ; \end{aligned} \tag{35.8}$$

as it is easy to verify if the divergence of the expression (35.7) is directly calculated. If the fluid is incompressible the density is a constant so that $d\rho/dt = 0$; and from continuity equation (9.2) we have that $div\, \boldsymbol{v}=0$.

In practice this always occurs, so instead of the equation (35.8), we use the following relationship

$$\frac{\partial \boldsymbol{v}}{\partial t} + (\boldsymbol{v} \cdot \boldsymbol{grad})\boldsymbol{v} = -\frac{1}{\rho}\,\boldsymbol{grad}\,p + \vartheta \nabla^2 \boldsymbol{v}. \tag{35.9}$$

This is the well known Navier-Stokes′ equation. Normally this equation is used to study the liquid dynamics because they are incompressible viscous fluids. For such cases it is fulfilled that

$$\sigma'_{ij} = \eta\left[\frac{\partial v^i}{\partial x^j} + \frac{\partial v^j}{\partial x^i}\right]. \tag{35.10}$$

These are the components of the Stokes′ viscosity stress tensor whose form is determined from (35.7) when $\partial v^\circ/\partial x^\circ = 0$. As it is easy to see from the relationship (35.10), the viscosity of an incompressible fluid is determined by only one coefficient known by the name of dynamic viscosity. Often as in the formula (35.9), the following expression is used

$$\vartheta = \frac{\eta}{\rho} \tag{35.11}$$

known as the kinematics viscosity.

§36. Energy conservation law

In order to obtain this expression it is necessary to use the explicit form of the specific lagrangian given in (35.2). In this case and according to the general definition (35.6), the following can be calculated

$$-\frac{\partial}{\partial x^j}\left(\frac{\partial \lambda}{\partial u_{ij}} v^i\right) = -\frac{\partial}{\partial x^j}\left[(p + C)v^j - \sigma'_{ij} v^i\right].$$

Again and due to time uniformity, the specific lagrangian can not be an explicit function of time so that $\partial\lambda/\partial t = 0$; in which case in (34.7) we have that

$$\frac{\partial \mathcal{H}}{\partial t} = -div\left[\boldsymbol{v}\left(\mathcal{H} + p\right) - \boldsymbol{v}\cdot\tilde{\sigma}' + \boldsymbol{q}\right]. \tag{36.1}$$

In this relationship

$$\boldsymbol{q} \equiv C\boldsymbol{v} \tag{36.2}$$

is the vectorial heat flux density due to thermal conduction. This vector is related to the variation of temperature through the fluid. It takes into consideration the direct molecular transfer of energy from those parts of the fluid where the temperature is high to those regions where the corresponding temperature is low. The origin of the constant C must be looked for in the scope of Statistical Mechanics due to the fact that its determination does not correspond to the objectives of this book.

In Classical Fluid Mechanics it is assumed that the relation between the vector $\boldsymbol{q}$ and the temperature of the fluid is given by Fourier's law

$$\boldsymbol{q} = -\kappa\, \boldsymbol{grad}\, T\,, \tag{36.3}$$

where κ is a constant called the thermal conductivity; quantity that is always positive due to the fact that the thermal flux of energy goes from regions of the fluid with high temperatures to those parts of the system that have low temperatures. It is for this reason that $\boldsymbol{q}$ and $\boldsymbol{grad}\, T$ point out in opposite directions; hence, the sign minus of the former relationship. The thermal conductivity is in general terms a function of temperature and pressure; but in most cases studied in Fluid Dynamics turns out to be a constant.

Then and according to the general definition (19.14) of the hamiltonean density and with the relationship (21.6) for the heat function or specific enthalpy; in (36.1) it is finally obtained that

$$\frac{\partial}{\partial t}\left[\frac{1}{2}\rho v^2 + \rho\square\right] = -div\left[\rho\boldsymbol{v}\left(\frac{1}{2}v^2 + w\right) - \boldsymbol{v}\cdot\tilde{\sigma}' - \kappa\, \boldsymbol{grad}\, T\right]. \tag{36.4}$$

On the right hand side of the former expression we have the total energy flux in any newtonian viscous fluid. In fact, as in the case of the ideal fluid, the term $\rho\boldsymbol{v}(1/2\, v^2 + w)$ is the mechanical energy flux density vector due to a simple transfer of mass because of the motion effect. The next term is an energy flux due to processes of internal frictions represented by the vector $\boldsymbol{v} \cdot \tilde{\sigma}'$. Finally, the transfer of heat due to thermal conduction is given by the term $\kappa\, \boldsymbol{grad}\, T$. Since on the left hand side we have the change in time of the total energy of the system, the equation (36.4) is the energy conservation law for any newtonian viscous fluid. With this equation we have the complete set of partial differential equations describing the dynamic state for any real newtonian fluid where viscosity and thermal conductivity are the relevant dissipative processes. The other equations of the set are Cauchy´s or of Navier-Stokes´, and the scalar equation for mass density.

§37.The general equation of heat transfer

It is convenient to transform the former equation to give it another form. This can be accomplished with the aid of continuity equation and Navier-Stokes´ equation. Thus, in the energy conservation law (36.4) when the left hand side is developed, we have that

$$\frac{\partial}{\partial t}\left[\frac{1}{2}\rho v^2 + \rho\square\right] = \frac{1}{2}v^2\frac{\partial \rho}{\partial t} + \rho \boldsymbol{v}\cdot\frac{\partial \boldsymbol{v}}{\partial t} + \rho\frac{\partial \square}{\partial t} + \square\frac{\partial \rho}{\partial t}. \quad (37.1)$$

Now and with the aid of definition (6.1), the continuity equation (9.2) can be written as follows

$$\frac{\partial \rho}{\partial t} = -div\left(\rho \boldsymbol{v}\right).$$

Moreover, from Navier-Stokes´ equation (35.9) it can be seen that

$$\frac{\partial \boldsymbol{v}}{\partial t} = -\frac{1}{\rho}\boldsymbol{grad}\, p - \left(\boldsymbol{v}\cdot\boldsymbol{grad}\right)\boldsymbol{v} + \frac{1}{\rho}\frac{\partial \sigma'_{ij}}{\partial x^j};$$

so that in (37.1) the following is obtained

$$\frac{\partial}{\partial t}\left(\frac{1}{2}\rho v^2 + \rho\square\right) = -\frac{1}{2}v^2 div\left(\rho \boldsymbol{v}\right) - \rho \boldsymbol{v}\cdot\boldsymbol{grad}\left(\frac{1}{2}v^2\right) - \boldsymbol{v}\cdot\boldsymbol{grad}\, p$$
$$+ \rho\frac{\partial \square}{\partial t} - \square div\left(\rho \boldsymbol{v}\right) + v^i\frac{\partial \sigma'_{ij}}{\partial x^j}$$

Now from the thermodynamic relation $d\square = Tds + (p/\rho^2)\,d\rho$, the following result is obtained

$$\frac{\partial \square}{\partial t} = T\frac{\partial s}{\partial t} + \frac{p}{\rho^2}\frac{\partial \rho}{\partial t} = T\frac{\partial s}{\partial t} - \frac{p}{\rho^2} div(\rho \mathbf{v}).$$

According to this and with the definition (21.6) we have that

$$\frac{\partial}{\partial t}\left(\frac{1}{2}\rho v^2 + \rho\square\right) = -\left(\frac{1}{2}v^2 + w\right) div(\rho \mathbf{v})$$
$$- \rho \mathbf{v}\cdot \mathbf{grad}\left(\frac{1}{2}v^2\right) - \mathbf{v}\cdot \mathbf{grad}\, p + \rho T\frac{\partial s}{\partial t} + v^i \frac{\partial \sigma'_{ij}}{\partial x^j}. \tag{37.2}$$

From the thermodynamic relationship $dw = Tds + dp/\rho$ we have that

$$\mathbf{grad}\, p = \rho\, \mathbf{grad}\, w - \rho T\, \mathbf{grad}\, s .$$

On the other hand

$$v^i \frac{\partial \sigma'_{ij}}{\partial x^j} = \frac{\partial}{\partial x^j}\left(v^i \sigma'_{ij}\right) - \sigma'_{ij}\frac{\partial v^i}{\partial x^j} = div(\mathbf{v}\cdot\tilde{\sigma}') - \sigma'_{ij}\frac{\partial v^i}{\partial x^j};$$

where an integration by parts was made. With all the former results substituted in (37.2) the following is obtained

$$\frac{\partial}{\partial t}\left[\frac{1}{2}\rho v^2 + \rho\square\right] + div\left[\rho \mathbf{v}\left(\frac{1}{2}v^2 + w\right) - \mathbf{v}\cdot\tilde{\sigma}' - \kappa\, \mathbf{grad}\, T\right]$$
$$= \rho T\left[\frac{\partial s}{\partial t} + \mathbf{v}\cdot \mathbf{grad}\, s\right] - \sigma'_{ij}\frac{\partial v^i}{\partial x^j} - div(\kappa\, \mathbf{grad}\, T); \tag{37.3}$$

where additionally and to reach the previous result the term $div(\kappa\, \mathbf{grad} T)$ has been added and subtracted. Since the left hand side of (37.3) is zero because it is the energy conservation law (36.4), in (37.3) it is finally obtained that

$$\rho T\left[\frac{\partial s}{\partial t}+\boldsymbol{v}\cdot\boldsymbol{grad}\, s\right]=\sigma'_{ij}\frac{\partial v^i}{\partial x^j}+div\left[\kappa\, \boldsymbol{grad}\, T\right]. \quad (37.4)$$

This is the general equation of heat transfer. It is clear that if the viscosity is negligible and there is no thermal conduction, the right hand side of (37.4) is zero; so that this relation is reduced to the specific entropy conser-servation law (21.8) for the ideal fluid. The equation (37.4) has a simple interpretation. On its left hand side we have the total change in time of the specific entropy of the system multiplied by ρT. Now ds/dt is the rate of change of the specific entropy of a unit mass of fluid as it moves about in space, so that $T\, ds/dt$ is therefore the quantity of heat gained by this unit mass in unit time, so that $\rho\, T\, ds/dt$ is the quantity of heat gained per unit volume. But this is equal to the right hand side of (37.4) where the first term is the energy dissipated into heat by the processes of internal friction (viscosity), whereas the other term is the heat conducted into the volume concerned.

Selected Topics

The concept of viscosity

In the study of the dynamical state of a fluid, many of the properties of real fluids have been ignored so far. There is a great number of phenomena in the scope of continuous media for which the approach of the ideal fluid is not valid, so this model does not provide an adequate description. In an ideal fluid the only way in which a force can be generated or, equivalently, in which the momentum it can be transferred is through the pressure gradient. But forces of this type must always be normal to the surface on which they are being exerted; in such a way that if it were somehow possible to isolate only one fluid element into an ideal fluid applying a force on it, there would be no other dynamical agent but the pressure gradient to oppose the motion of the element. If it were not so, that element would be quickly accelerated.

One way of visualizing the motion of a fluid is to imagine that it is formed by a great number of layers each one moving at different velocity.

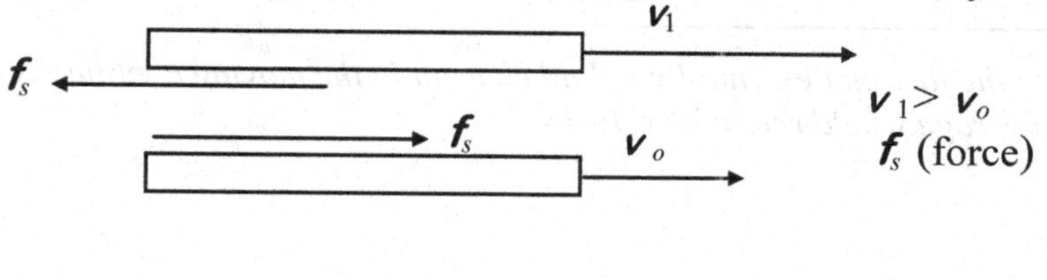

For a classical ideal fluid the present dynamical situation is the following: the layer of fluid moving with a velocity $\boldsymbol{v}_1$, grater than the velocity $\boldsymbol{v}_o$ at which the other layer moves, would have to keep moving indefinitely, even if no force were acting on it. However, it is easy to become convinced that in a real situation this layer would eventually suffer a slow down process and stop.

This means that there must be some way of exerting forces which are different from the pressure forces that clearly must act along the surface, rather than in a normal way on it. When such a situation is described it is usually stated that a real fluid is capable of exerting a certain shear force whose effect is additional to the pressure. The physical phenomenon associated with this force is known by the name of viscosity of the medium.

To understand how viscosity works in a continuous medium let us consider a collision between two elements of fluid. If in such event only pressure forces act, the momentum transfers could occur only in a direction normal to the contact surface between these elements and the momentum of each element along this surface would have to remain constant†. When the real situation is visualized in this way it is clear that the assumptions associated with the scheme of ideal fluids are rather artificial. A more realistic picture consists of imagining that the momentum could be transferred in any direction. The parts of fluid of the lower layer of the former example would, on the average, suffer an acceleration by the collision effect, while the elements in the upper layer would, on the average, be slowed down. The net result would be that the relative velocity between the two layers would be reduced and eventually its value becomes zero; mechanism that is similar to the phenomenon of friction in mechanics of particles.

Viscosity stress tensor

When the motion of a real fluid is studied, the effect of energy dissipation must be taken into account because it affects the motion itself. This process is the result of the thermodynamic irreversibility of the motion. Such irreversibility always occurs to some extent and is due to internal frictions (viscosity) and thermal conduction.

In order to obtain the equations describing the dynamical state of a viscous fluid it is necessary to include some additional terms in the equation

---

† *Due to the fact that essentially a fluid element in the moving continuous medium on the average would retain its velocity.*

of motion of an ideal fluid. Thus, one term must be added which takes into account the irreversible transfer of viscous momentum in a real fluid. This term is equal to $\tilde{\sigma}'(\boldsymbol{x}, t)$, and is known by the name of viscosity stress tensor. In this case the mechanical stress tensor in components, takes the following form

$$\sigma_{ik} = -p\delta_{ik} + \sigma'_{ik};$$

where σ'_{ik} contains the part of the momentum flux which is not due simply to the momentum transfer with the motion of mass in the mobile fluid.

The general form of the tensor $\tilde{\sigma}'(\boldsymbol{x},t)$ can be established as follows. Processes of internal friction occur in a fluid only when different fluid elements move at different velocities, so that there is a relative motion between various parts of the fluid. Hence, σ'_{ik} must depend on the space derivative of the velocity. If the velocity gradients are small, we may suppose that the momentum transfer due to viscosity depends only on the first derivatives of the velocity. Thus and within the same approximation, σ'_{ik} may be assumed as a linear function of the derivatives $\partial v^i/\partial x^k$. Additionally, it can be said that there can be no terms in σ'_{ik} independent from $\partial v^i/\partial x^k$ since the viscosity stress tensor must vanish when the velocity is a constant. On the other hand, it can be stated that σ'_{ik} must also vanish when the whole fluid is in uniform rotation, since in such a motion no internal friction occurs in the fluid. In uniform rotation with angular velocity Ω, the velocity field $\boldsymbol{v}(\boldsymbol{x}, t)$ is equal to the vector product $\Omega \times \boldsymbol{r}$. In this case, the sums

$$\frac{\partial v^i}{\partial x^k} + \frac{\partial v^k}{\partial x^i},$$

are linear combinations of the derivatives $\partial v^i/\partial x^k$ that vanish when $\boldsymbol{v}=\Omega\times\boldsymbol{r}$. In consequence, σ'_{ik} must contain just these symmetrical combinations of the velocity gradients. The only combinations of the velocity derivatives that satisfy the conditions above mentioned are

$$\frac{\partial v^i}{\partial x^k} + \frac{\partial v^k}{\partial x^i}, \quad \text{and} \quad \delta_{ik}\frac{\partial v^\circ}{\partial x^\circ};$$

and of course, a great number of terms containing second and higher order derivatives of the velocity, since there is not any reason to think that such terms are not present in σ'_{ik}. However, the theory would be much more simple if we considered that the viscous forces only depended on the first derivatives of the velocity.

This means that the most general form of a tensor of rank two satisfying the above condition is the following†

$$\sigma'_{ik} = \eta\left[\frac{\partial v^i}{\partial x^k} + \frac{\partial v^k}{\partial x^i} - \frac{2}{3}\delta_{ik}\frac{\partial v^\circ}{\partial x^\circ}\right] + \zeta\delta_{ik}\frac{\partial v^\circ}{\partial x^\circ}.$$

The expression in parenthesis has the property of vanishing upon contracting with respect to i and k. The constants η and ζ are called coefficients of viscosity. They are both positive, that is to say $\eta>0$ and $\zeta>0$. η is the shear viscosity whereas ζ is the volumetric viscosity.

The entropy

In a real fluid the entropy increases as a result of the irreversible processes of thermal conduction and internal friction. The total entropy of the fluid as a whole* is equal to the integral over the region R that contains the system, that is

$$S = \int_R \rho s dV;$$

where $\rho(\boldsymbol{x}, t)$ is the mass density, s the specific entropy, and dV the volume element in region R.

The change in entropy per unit time is given by the time derivative dS/dt; that is to say

$$\frac{d}{dt}\int_R \rho s dV = \frac{d}{dt}\int_{R_o} \rho s J dV_o = \int_R \left[\frac{d}{dt}(\rho s) + \rho s \, div\, \boldsymbol{v}\right] dV,$$

due to the fact that $dV = JdV_o$ was used here as well as Euler´s relation.

Clearly R_o is a region independent of time is such a way that the symbols of derivative and integral can be interchanged. In this case

---

† *Notice that σ'_{ik} contains a term which is proportional to δ_{ik}, that is to say, of the same form that $p\delta_{ik}$. When the mechanical stress tensor is written in this form it is necessary to specify what pressure p means.*

* *That means not the entropy of each volume element of fluid separately.*

$$\frac{dS}{dt} = \int_R \frac{\partial}{\partial t}(\rho s)dV + \oint_\Sigma \rho s \boldsymbol{v} \cdot d\boldsymbol{a};$$

where Σ is the surface that limits the region R and $d\boldsymbol{a}$ the vector differential of area. The volume integral of $div(\rho s \boldsymbol{v})$ can be transformed into the integral of the entropy flux over the surface. If it is considered an unbounded volume of fluid at rest at infinity, the bounding surface of R can be removed at infinity. Thus, the integrand in surface integral is then zero, an so is the integral itself.
On the other hand, in the general equation of heat transfer

$$\rho T\left(\frac{\partial s}{\partial t} + \boldsymbol{v} \cdot \boldsymbol{grad}\, s\right) = \sigma'_{ik} \frac{\partial v^i}{\partial x^k} + div(\kappa\, \boldsymbol{grad}\, T); \tag{1}$$

the term $\sigma'_{ik}\, \partial v^i/\partial x^k$ can be expanded using the expression for σ'_{ik} from the last paragraph in such a way that

$$\sigma'_{ik} \frac{\partial v^i}{\partial x^k} = \eta \frac{\partial v^i}{\partial x^k}\left(\frac{\partial v^i}{\partial x^k} + \frac{\partial v^k}{\partial x^i} - \frac{2}{3}\delta_{ik}\frac{\partial v^\circ}{\partial x^\circ}\right) + \zeta \frac{\partial v^i}{\partial x^k}\delta_{ik}\frac{\partial v^\circ}{\partial x^\circ}. \tag{2}$$

It is easy to verify by direct calculus that if

$$\frac{\partial v^i}{\partial x^k} \equiv \frac{1}{2}\left(\frac{\partial v^i}{\partial x^k} + \frac{\partial v^i}{\partial x^k}\right)$$

and since σ'_{ik} and δ_{ik} are symmetrical,

$$\eta \frac{\partial v^i}{\partial x^k}\left(\frac{\partial v^i}{\partial x^k} + \frac{\partial v^k}{\partial x^i} - \frac{2}{3}\delta_{ik}\frac{\partial v^\circ}{\partial x^\circ}\right) =$$

$$\frac{1}{2}\eta\left[\left(\frac{\partial v^i}{\partial x^k} + \frac{\partial v^k}{\partial x^i} - \frac{2}{3}\delta_{ik}\frac{\partial v^\circ}{\partial x^\circ}\right)\frac{\partial v^i}{\partial x^k} + \left(\frac{\partial v^i}{\partial x^k} + \frac{\partial v^k}{\partial x^i} - \frac{2}{3}\delta_{ik}\frac{\partial v^\circ}{\partial x^\circ}\right)\frac{\partial v^i}{\partial x^k}\right].$$

In the second term of the right hand side of the former expression we can change $i \to k$ as well as $k \to i$. Also, we can add and subtract the quantity 2/3 $\delta_{ik}\, \partial v^\circ / \partial x^\circ$ multiplied by the resulting parenthesis, to obtain the following

$$\frac{1}{2}\eta\left[\left(\frac{\partial v^i}{\partial x^k}+\frac{\partial v^k}{\partial x^i}-\frac{2}{3}\delta_{ik}\frac{\partial v^\circ}{\partial x^\circ}\right)^2+\frac{4}{3}\left(\frac{\partial v^\circ}{\partial x^\circ}\right)^2-\frac{4}{9}\delta_{ik}\delta_{ki}\left(\frac{\partial v^\circ}{\partial x^\circ}\right)^2\right].$$

However $\delta_{ik}\ \delta_{ki} = \delta_{ii}$ and $\delta_{ii} = 3$, so that the last two terms of the former result are canceled between them and the first expression of the right hand side of (2) is transformed into

$$\frac{1}{2}\eta\left(\frac{\partial v^i}{\partial x^k}+\frac{\partial v^k}{\partial x^i}-\frac{2}{3}\delta_{ik}\frac{\partial v^\circ}{\partial x^\circ}\right)^2;$$

whereas the second expression is

$$\zeta\frac{\partial v^i}{\partial x^k}\delta_{ik}\frac{\partial v^\circ}{\partial x^\circ}=\zeta\frac{\partial v^\circ}{\partial x^\circ}\frac{\partial v^\circ}{\partial x^\circ}\equiv\zeta(div\,\mathbf{v})^2.$$

In this case in (1) we have that

$$\rho T\left(\frac{\partial s}{\partial t}+\mathbf{v}\cdot\mathbf{grad}\,s\right)=div(\kappa\,\mathbf{grad}\,T)$$
$$+\frac{1}{2}\eta\left(\frac{\partial v^i}{\partial x^k}+\frac{\partial v^k}{\partial x^i}-\frac{2}{3}\delta_{ik}\frac{\partial v^\circ}{\partial x^\circ}\right)^2+\zeta(div\,\mathbf{v})^2.$$

According to this result and with the continuity equation,

$$\frac{\partial}{\partial t}(\rho s)=-s\,div(\rho\mathbf{v})-\rho\mathbf{v}\cdot\mathbf{grad}\,s+\frac{1}{T}div(\kappa\,\mathbf{grad}T)$$
$$+\frac{\eta}{2T}\left(\frac{\partial v^i}{\partial x^k}+\frac{\partial v^k}{\partial x^i}-\frac{2}{3}\delta_{ik}\frac{\partial v^\circ}{\partial x^\circ}\right)^2+\frac{\zeta}{T}(div\,\mathbf{v})^2.$$

Now

$$-s\,div(\rho\mathbf{v}) - \rho\mathbf{v}\cdot\mathbf{grad}\,s = -div(\rho s\mathbf{v}),$$

where an integration by parts was made. When this term is integrated the result is zero according to the arguments before given. The integral of the next term can be written as follows

$$\int_R \frac{1}{T}\,div(\kappa\,\mathbf{grad}\,T)dV = \int_R div\left(\frac{\kappa\,\mathbf{grad}\,T}{T}\right)dV + \int_R \frac{\kappa(\mathbf{grad}\,T)^2}{T^2}dV;$$

where an integration by parts was made. Assuming that the fluid temperature tends sufficiently rapid to a constant value at infinity, the integral can be transformed as follows

$$\int_R div\left(\frac{\kappa\,\mathbf{grad}\,T}{T}\right)dV = \oint_\Sigma \frac{\kappa\,\mathbf{grad}\,T}{T}\cdot d\mathbf{a}.$$

The surface integral is zero because over an infinitely remote surface we have that $\mathbf{grad}\,T = 0$. Therefore, the final result is

$$\frac{d}{dt}\int_R \rho s dV = \int_R \frac{\kappa(\mathbf{grad}\,T)^2}{T^2}dV + \int_R \frac{\eta}{2T}\left(\frac{\partial v^i}{\partial x^k} + \frac{\partial v^k}{\partial x^i} - \frac{2}{3}\delta_{ik}\frac{\partial v^\circ}{\partial x^\circ}\right)^2 dV$$
$$+ \int_R \frac{\zeta}{T}(div\,\mathbf{v})^2 dV. \tag{3}$$

The first term on the right hand side is the rate of increase of entropy due to thermal conduction, while the other two terms give the rate of increase due to internal friction. Since the value of entropy can only increase, the sum on the right hand side must always be positive. In each term, the integrand may be non zero even if the other two integrals vanish. Hence it follows that the second viscosity coefficient ζ is positive, as well as κ and η.

The thermodynamic entities in a real fluid

In the derivation of Fourier´s law for thermal conduction

$$q = -\kappa \; grad \; T,$$

it has been tacitly assumed that the heat flux depends only on the temperature gradient, and not on the pressure gradient. This assumption, which is
not evident a priori, can now be justified as follows. If q contained a term
proportional to $grad \; p$, the expression (3) of the last paragraph would
include another term having the product $grad \; p \cdot grad \; T$ in the first integrand of the right. Since the latter might be either positive or negative, the
time derivative of the entropy would not necessarily be positive and the
result obtained would be impossible. In a system which is not in thermodynamic equilibrium, such as it occurs in a fluid with velocity and temperature gradients, the usual definitions of thermodynamic quantities are no
longer meaningful as in the case of equilibrium and must be modified.

The necessary definitions are, firstly, that $\rho, \Box$ and v are defined as before; that is to say, ρ and $\rho \; \Box$ are the mass and internal energy per unit volume and the velocity field or the momentum of unit mass of fluid. The re-
mainning thermodynamic quantities are then defined as being the same
functions of ρ and $\Box$ as they are in thermal equilibrium. However, the specific entropy $s=s(\rho, \Box)$ is no longer the true thermodynamic entropy due to
the fact that the integral $\int(\rho s)dV$ will not, strictly speaking, be a quantity
that must increase with time. Nevertheless, for small velocity and temperature gradients, s is the same as the true entropy within the approximation
used here. However, if there are gradients present in the fluid, they in general lead to additional terms in the entropy $s(\rho, \Box)$. Thus, the results given
above can be altered only by linear terms in the gradients; as will be the
case, for instance, of a term proportional to the scalar $div v$. Such terms
would necessarily take both positive and negative values. But these quantities ought to be necessarily defined negatives, since the equilibrium value
$s=s(\rho, \Box)$ of the entropy is the maximum possible value. Hence, the expansion of the entropy in powers of the small gradients can contain, apart
from the zero order term, only terms of the second and higher orders.

Similar remarks should be made in the foot note of the second paragraph
of this *Selected Topics*, since the presence of even a velocity gradient as in

the case of the ideal fluid, implies the absence of thermodynamic equilibrium. The pressure p which appears in the expression for the divergence of the stress tensor in the motion equation for a viscous fluid must be taken to be the same function $p=p(\rho,\square)$ as in thermal equilibrium. However, in Fluid Dynamics the pressure p will not, strictly speaking, be the pressure in the usual sense; that is to say, it should not be taken only as the force per unit area normal to the surface element. In spite of what happens in the case of the entropy, there is here a resulting difference of the first order with respect to the small gradient in the divergence of the mechanical stress tensor. We have seen that the normal component of the force, includes, besides p, a term proportional to $div\boldsymbol{v}$; term on the other hand that is zero in an incompressible fluid, so that in this situation, the difference is found in higher order terms in the gradients.

Finally, the three coefficients η, ζ and κ which appear in the equations of motion of a viscous conducting fluid from the thermal point of view, completely determine the dynamical state of the continuous medium in the approximation considered; that is to say, when the higher-order space derivatives of velocity, temperature, etc. are neglected. The introduction of any further terms in balance equations, as for example, the inclusion of quantities associated with the mass flux density of terms proportional to the gradient of density or temperature, has no physical meaning, and would indicate at least a change in the definition of the fundamental quantities. In particular, the velocity would no longer be the momentum per unit mass of fluid. In fact, still worse, the inclusion of such terms may violate the basic conservation laws. It must be kept in mind that, whatever the definition used, the mass flux density $\boldsymbol{j}=\rho\boldsymbol{v}$ must always be the momentum of unit volume of fluid. For $\boldsymbol{j}$ is defined by the equation of continuity

$$\frac{\partial\rho}{\partial t}+div\boldsymbol{j}=0\,;$$

when this is multiplied by $\boldsymbol{r}$ and it is integrated over all the region R that contains the fluid, we have that

$$\frac{d}{dt}\left(\int_R \rho\boldsymbol{r}dV\right)=\int_R \boldsymbol{j}dV.$$

Now and since the integral

$$\int_R \rho r dV$$

determines the position of the center of mass, it is clear that the integral

$$\int_R j dV$$

is the momentum.

Fourier´s equation

The general equation of thermal conduction (37.4) for an incompressible fluid can be expressed as follows

$$\frac{\partial T}{\partial t}+\boldsymbol{v}\cdot\boldsymbol{grad}T=\chi\nabla^2 T+\frac{\vartheta}{2c_p}\left(\frac{\partial v^i}{\partial x^k}+\frac{\partial v^k}{\partial x^i}\right)^2; \tag{1}$$

where a calculus similar to that on page 149† was made and

$$\chi=\frac{\kappa}{\rho c_p} \tag{2}$$

is the thermometric conductivity.

The equation (1) has a particularly simple form if the fluid is at rest. In this case we have that

---

† *In fact,* $\sigma'_{ik}\frac{\partial v^i}{\partial x^k}=\eta\frac{\partial v^i}{\partial x^k}\left(\frac{\partial v^i}{\partial x^k}+\frac{\partial v^k}{\partial x^i}\right)$ *and according to the definition made on page* 149 *of* $\frac{\partial v^i}{\partial x^k}$, $\sigma'_{ik}\frac{\partial v^i}{\partial x^k}=\frac{1}{2}\eta\left(\frac{\partial v^i}{\partial x^k}+\frac{\partial v^k}{\partial x^i}\right)^2$

$$\frac{\partial T}{\partial t} = \chi \nabla^2 T \cdot \tag{3}$$

This equation is known in mathematical physics by the name of the equation of thermal conduction or Fourier's equation. If the temperature distribution in a non uniformily heated medium at rest is maintained constant in time by means of some external source of heat, the equation of thermal conduction becomes

$$\nabla^2 T = 0 . \tag{4}$$

That means that a steady distribution of temperature in a continuous medium at rest satisfied Laplace´s equation. In the more general case where κ cannot be regarded as a constant, the former equation must be substituted by the following relationship

$$div(\kappa\, \mathbf{grad}\, T) = 0 . \tag{5}$$

Finally, if the fluid contains any external source of heat, Fourier´s equation must contain another term. Let Q be the quantity of heat generated for that source in unit volume of the fluid per unit time, with Q in general a function of coordinates and time. In this case, the corresponding equation of thermal conduction has the following form

$$\rho c_p \frac{\partial T}{\partial t} = \kappa \nabla^2 T + Q . \tag{6}$$

This is the heat balance equation.

References

1. Fierros Palacios, Angel. "Las ecuaciones de balance para un fluido viscoso a partir de un principio variacional tipo Hamilton". Rev. Mex. Fís. 38, No. 4 (1992) 518-531.
2. Landau, L.D. and Lifshitz, E.M. "Fluid Mechanics". Addison-Wesley Publishing Co. (1959).

3. Landau, L.D. and Lifshitz, E.M. "Theory of Elasticity". Addison-Wesley Publishing Co. (1959).
4. Meyer R. "Introduction to Mathematical Fluid Dynamics". Wiley Interscience (1971).
5. Reif, F. "Fundamentals of Statistical and Thermal Physics". International Student Edition. McGraw-Hill Book Company (1965).

Chapter VII

Free Convection

§38. The start up mechanism of free convection

This phenomenon is particularly important in the terrestrial atmosphere concerning air pollution. The interest that we have is to study the generating dynamical mechanism which makes it possible that from a calm situation in some regions of the atmosphere, huge masses of air start their motion in order to take advantage of this knowledge in the fight against the atmospherical pollution. With the theoretical treatment that will be given to the problem, we pretend to reach an analytical solution that can be reinforced or refuted by the experiment.

Let us consider a huge mass of air which is under the influence of a gravitational field. It is known that when in a part of the atmosphere the mechanical equilibrium exists, the temperature distribution in those regions only depends on the height over the ground level. Also it is known that the mechanical equilibrium is not possible if the above mentioned field of temperatures is at the same time a function of the other coordinates so that obviously, the previous condition is not satisfied. Otherwise, although this requirement can be satisfied, it can happen that the mechanical equilibrium it not obtained if the temperature gradient in the upward direction is directed downwards and its magnitude is greater than a certain value.

The absence of mechanical equilibrium has as a consequence the motion of the air. Internal currents appear in it which tend to mix their parts until the temperature reaches a constant value in all the volume occupied by the air. Such an air motion in Earth´s gravitational field is known by the name of thermal or free convection.

As it is well known, in the scope of energetical processes it is not possible to contend with nature, due to the fact that the quantity of energy needed to produce and maintain by long temporary period winds of moderate intensity is of such magnitude, that any proposition in this sense must be rejected as a viable alternative to reach a substantial improvement in the air quality. However, the process of free convection generator of the atmosphere motion in certain localized regions, has to be started under certain minimal energetic conditions that may be possible reproduced in a practical way with the aid of any workmanship of low ambient impact and reasonable cost. Knowing such energy minimal conditions, some practical solution can be suggested to prevent in some places the concentration of large quantities of contaminants which obstruct the solar radiation; favouring this way the condition so that the atmospherical phenomenon called the thermal inversion may be produced. If the concentration of contaminants in the air could be stopped this would allow the solar energy to pass through as far as Earth´s surface where a large part of it would be absorbed to be afterwards radiated in the form of heat to the lower layers of the atmosphere. The air so heated, starts the free convection characteristic upward motion dispersing this way the cloud of contaminants. This is possible due to the fact that once the atmosphere starts moving, the solar radiation gives the massive expenditure of energy required to maintain the air motion and cleaning process. So it is very important to know the physical conditions and the minimal energy that have to be produced in order to generate an atmospherical stimulus capable of originating the start up mechanism of the free convection.

§39. The conditions so that the mechanical equilibrium be unstable

So that a mechanical equilibrium exists in any portion of the terrestrial atmosphere it is necessary that the following condition is fulfilled

$$\textbf{grad}\, p = \rho \mathbf{g}\,. \tag{39.1}$$

If the Z–axe of the used inertial frame of reference is vertically directed upward, the former relation can be written as follows

$$\frac{dp}{dz} = -\frac{g}{V}; \tag{39.2}$$

where g is the acceleration of gravity and $V = 1/\rho$ the specific volume. It is clear that if the temperature is not constant throughout the air, the resultant mechanical equilibrium can be stable or unstable according to certain conditions. For the present case, it is interesting to determine the conditions under which the equilibrium in some region of the atmosphere is unstable. Once knowing these conditions we can inquire if it is possible to influence on these unstable regions to favor the air motion by means of small variations of energy.

Let us consider a mass of air at height z over the ground level, having a specific volume equal to $V(p,s)$, where s is the specific entropy and p the pressure at this height respectively. Suppose that this mass of air undergoes an adiabatic upward displacement as far as a small additional altitude equal to h, where its specific volume becomes $V(p',s)$; with p' the corresponding pressure at the altitude $z+h$. For the equilibrium to be stable, it is necessary, but not in general sufficient, that the force on this displaced volume element should tend to return it to its original position. This means that the mass of air initially considered must be heavier than the air which it displaces in its new position. The specific volume of the latter is $V(p',s')$, where now s' is the specific entropy of equilibrium at height $z+h$. To make a balance of forces taking into account Newton's second law for the motion $\boldsymbol{F} = \rho V \boldsymbol{a}$, the stability condition can be written as

$$V(p',s') - V(p',s) > 0 . \tag{39.3}$$

When an expansion is made in power series of $s' - s = h ds/dz$ of the former difference we have that

$$\left(\frac{\partial V}{\partial s}\right)_p \frac{ds}{dz} > 0 . \tag{39.4}$$

The following result can be obtained from thermodynamics

$$\left(\frac{\partial V}{\partial s}\right)_p = \frac{T}{c_p}\left(\frac{\partial V}{\partial T}\right)_p ; \tag{39.5}$$

where c_p is the specific heat at constant pressure. As T and c_p are both positive quantities, the relation (39.4) can be written as follows

$$\left(\frac{\partial V}{\partial T}\right)_p \frac{ds}{dz} > 0 . \tag{39.6}$$

On the other hand, the majority of substances expand on heating so that in general $(\partial V/\partial T)_p > 0$. Thus, the condition so that the convection does not occur in the atmosphere is that

$$\frac{ds}{dz} > 0 . \tag{39.7}$$

In other words, the specific entropy must increase with height. From (39.7) the conditions that must satisfy the temperature gradient for the equilibrium so that the region of the atmosphere is unstable and we have the possibility of favoring the appearance of the free convection phenomenon can be determined. In fact, if the derivative ds/dz is expanded it is obtained that

$$\frac{ds}{dz} = \frac{c_p}{T}\frac{dT}{dz} - \left(\frac{\partial V}{\partial T}\right)_p \frac{dp}{dz} > 0 . \tag{39.8}$$

Finally and according to the result (39.2)

$$\frac{dT}{dz} > -\frac{gT}{c_p V}\left(\frac{\partial V}{\partial T}\right)_p . \tag{39.9}$$

In this case the free convection can be produced in the atmosphere if the temperature of air falls with increasing height and the magnitude of the temperature gradient is larger than certain value; that is to say, if

$$\left|\frac{dT}{dz}\right| > \frac{gT}{c_p V}\left(\frac{\partial V}{\partial T}\right)_p . \tag{39.10}$$

§40. The velocity field

Suppose that the terrestrial atmosphere behaves as a compressible viscous fluid but that thermodynamically it can be described as an ideal gas. Be $T(\boldsymbol{x}, t)=T_o+T'$ the field of temperature with T_o some constant mean value throughout the volume occupied by the air and T' the temperature to which the mechanical equilibrium is broken and the free convection process starts such that $T'<<T_o$. Let $\rho(\boldsymbol{x}, t)=\rho_o+\rho'$ be the mass density of the air with $\rho_o>>\rho'$; where ρ' is the value of mass density corresponding to the physical conditions to begin the start up mechanism and ρ_o a constant. Since T' as ρ' are small quantities it can be stated that

$$\rho' = \left(\frac{\partial \rho}{\partial T}\right)_p T' = -\rho_o \alpha T' ; \qquad (40.1)$$

where

$$\alpha \equiv -\frac{1}{\rho}\left(\frac{\partial \rho}{\partial T}\right) \qquad (40.2)$$

is the thermal coefficient of the air. On the other hand and according to the scalar equation for mass density (11.9), it can be directly demonstrated, with the aid of the proposed approximations for the mass density, that

$$\alpha = \frac{1}{T'}\left[1 - \frac{\boldsymbol{u}}{J} \cdot \boldsymbol{grad}\ J\right]. \qquad (40.3)$$

In the case of pressure we have that $p(\boldsymbol{x}, t) = p_o+p'$ with $p_o>>p'$. However, p_o is not constant but deals with the pressure corresponding to a huge mass of air in local mechanical equilibrium subjected to the influence of terrestrial gravitational field. In consequence, the pressure p_o varies with height according to the hydrostatic equation

$$p_o = \rho_o\, \boldsymbol{g} \cdot \boldsymbol{x} + constant\,. \qquad (40.4)$$

Let us consider Cauchy's motion equation for the present case, that is to say

$$\rho \frac{d\boldsymbol{v}}{dt} - div\, \tilde{\sigma} - \rho \boldsymbol{f} = 0; \tag{40.5}$$

with $\boldsymbol{f}$ the body force per unit mass which in this case is equal to $\boldsymbol{g}$. On the other hand, the most general form of the stress tensor is given by the relationship (35.6); so that it is possible to write that

$$\frac{1}{\rho}\frac{\partial \sigma_{ij}}{\partial x^j} = -\frac{1}{\rho}\frac{\partial p}{\partial x^i} + \boldsymbol{F}_v; \tag{40.6}$$

where

$$\boldsymbol{F}_v = \frac{1}{\rho} div\, \tilde{\sigma}' \tag{40.7}$$

is the viscous force with $\tilde{\sigma}'$ the viscosity stress tensor whose most general form is given by the equation (35.7). Now and according to the expansion of the hydrostatic pressure and the mass density made before respectively, it is easy to see that

$$\frac{\boldsymbol{grad}\, p}{\rho} = \frac{\boldsymbol{grad}\, p_o}{\rho_o} + \frac{\boldsymbol{grad}\, p'}{\rho_o} - \frac{\boldsymbol{grad}\, p_o}{\rho_o^2}\rho'; \tag{40.8}$$

where the calculus was made as far as first order terms of small quantities. From the hydrostatic equation (40.4) it is immediate that

$$\frac{1}{\rho_o}\boldsymbol{grad}\, p_o = \boldsymbol{g}, \tag{40.9}$$

whereas from (40.1) it is obtained

$$\frac{\rho'}{\rho_o^2}\,\boldsymbol{grad}\;p_o = -\alpha T'\boldsymbol{g}\,; \tag{40.10}$$

so that in (40.6) it can be seen that

$$\frac{1}{\rho}\frac{\partial \sigma_{ij}}{\partial x^j} = -\frac{1}{\rho_o}\,\boldsymbol{grad}\;p' - \boldsymbol{g} - \alpha\,\boldsymbol{g}\,T' + \boldsymbol{F}_v\,, \tag{40.11}$$

in such a way that in Cauchy's equation (40.5) we have that

$$\frac{d\boldsymbol{v}}{dt} = -\frac{1}{\rho_o}\,\boldsymbol{grad}\;p' - \alpha\,\boldsymbol{g}\,T' + \boldsymbol{F}_v\,. \tag{40.12}$$

Now, we require an equation that links pressure with temperature and mass density. Such relationship is the thermal equation of state. As we are assuming that the air behaves as an ideal gas from the thermodynamical point of view; we will use the corresponding equation, that is to say

$$p = \rho R' T\,, \tag{40.13}$$

where R' is the value of the gas constant for the air. According to the scalar equation for mass density (11.9) and with the expasion for the pressure and the temperature made before, it is clear that

$$p' = \frac{\rho_o R'\left(T_o + T'\right)}{J}\boldsymbol{u}\cdot\boldsymbol{grad}\;J - p_o\,. \tag{40.14}$$

Then

$$p' = \rho_o R' T'\left(1 - \alpha T_o\right), \tag{40.15}$$

where the result (40.3) was used. Moreover it is clear that

$$p_o = \rho_o R' T_o \tag{40.16}$$

is the thermal equation of state for the mechanical equilibrium situation. At making the former calculus the term $\alpha T'$ was neglected compared to αT_o. The required thermal equation of state is the relationship (40.15). In order to obtain the hydrodynamic force responsible of the creation of the start up mechanism of free convection, the gradient of this expression must be obtained, that is to say

$$\boldsymbol{f}'_H = -\frac{1}{\rho_o} \textbf{grad}\, p' = R'(\alpha T_o - 1)\, \textbf{grad}\, T' . \tag{40.17}$$

With this result in the equation (40.12) the following is obtained

$$\frac{d\boldsymbol{v}}{dt} = R'(\alpha T_o - 1)\, \textbf{grad}\, T' - \alpha\, \boldsymbol{g} T' + \boldsymbol{F}_v . \tag{40.18}$$

Let us consider next that for the start up mechanism the viscous force is irrelevant so that in (40.18) the term $\boldsymbol{F}_v$ can be neglected compared to the other ones. For the process of free convection itself is not possible to have this approach since once the air is moving, this term becomes more and more important in the proportion in which the wind velocity increases its value, due fundamentally to the fact that the viscous force increases with the velocity gradient. However, for an analysis of the start up mechanism; that is to say, for a small time scales compared to the global process of dissipation (due basically to the viscosity), the dominant process is no dissipative so that for such situation the viscous force is not important. For this reason it can be assumed that the generation process of the start up mechanism is so slow that the change of the temperature from T_o to T occurs at a stationary regime. Thus, the small temperature T'at which the mechanical equilibrium is broken and the free convection starts, only depends on the coordinates and not on the time. Integrating the remains of equation (40.18) when the term of the viscous force is ignored, the following is obtained

$$\boldsymbol{v}' = \boldsymbol{v}_o + \left[R'(\alpha T_o - 1)\textbf{grad}\, T' - \alpha \boldsymbol{g} T'\right](t' - t_o) . \tag{40.19}$$

However $\mathbf{v}_o=0$ because it is the velocity corresponding to (static) mechanical equilibrium. If now the scale of time is elected in such a way that $t_o=0$ we have that

$$\mathbf{v}' = \left[R'(\alpha T_o - 1)\,\mathbf{grad}\,T' - \alpha\,\mathbf{g}\,T'\right]t'. \qquad (40.20)$$

If the Z-direction is only considered, from the former relation the upwards velocity of the air is obtained, once the generation of the atmospheric stimulus is completed and the free convection phenomenon has begun, that is to say

$$v_z' = \left[R'(\alpha T_o - 1)\,grad_z\ T' + \alpha\,g\,T'\right]t'. \qquad (40.21)$$

§41. The general equation of heat transfer

The general equation of heat transfer for any newtonian viscous fluid was obtained in paragraph 37. If the air velocity during the free convection process is lesser than the sound velocity in the medium, the variations in the pressure occurring as a result of their motion are so small that the changes in density and the other thermodynamic quantities implied in the phenomenon may be neglected. However, the air at normal condition is not incompressible, it is still less if it is subjected to a non uniform heating due to the fact that the mass density changes with the temperature as it is easy to see from the relationship (40.1). As this variation can not be ignored, the mass density can not be considered as a constant. That conclusion turns to be very reasonable if attention is fixed on what occurs before the mechanical equlibrium is broken, that is to say; when the start up mechanism of the phenomenon under consideration is being generated. In such circumstances the different thermodynamic quantities suffers changes in such a way that

$$\frac{\partial s}{\partial t} = \left(\frac{\partial s}{\partial T}\right)_p \frac{\partial T}{\partial t} \qquad (41.1)$$

and

$$\textbf{grad}\ s = \left(\frac{\partial s}{\partial T}\right)_p \textbf{grad}\ T. \tag{41.2}$$

However,

$$c_p = T\left(\frac{\partial s}{\partial T}\right)_p \tag{41.3}$$

so that in the equation (37.4) the right hand side is transformed as follows

$$\rho T\left[\frac{\partial s}{\partial t} + \mathbf{v}\cdot\textbf{grad}\ s\right] = \rho c_p\left[\frac{\partial T}{\partial t} + \mathbf{v}\cdot\textbf{grad}\ T\right] \tag{41.4}$$

and in the general equation of heat transfer we have that

$$\left[\frac{\partial T}{\partial t} + \mathbf{v}\cdot\textbf{grad}\ T\right] = \chi\nabla^2 T + \frac{1}{\rho c_p}\sigma'_{ij}\frac{\partial v^i}{\partial x^j}, \tag{41.5}$$

where κ has been considered as a constant and

$$\chi \equiv \frac{\kappa}{\rho c_p} \tag{41.6}$$

is the thermometric conductivity.

For the conditions under which the start up mechanism is produced it is fulfilled that $\mathbf{v} = \mathbf{v}'$ so that the velocity field is the small perturbation in the velocity of air when the phenomenon begins. In consequence,

$$S'_{ij} = \eta\left[\frac{\partial v'^i}{\partial x^j} + \frac{\partial v'^j}{\partial x^i} - \frac{2}{3}\delta_{ij}\frac{\partial v'^\circ}{\partial x^\circ}\right] + \zeta\delta_{ij}\frac{\partial v'^\circ}{\partial x^\circ} \tag{41.7}$$

are the components of the corresponding viscosity stress tensor. As $\boldsymbol{v}'$ is a small quantity and the involved distances are relatively large, the gradients of the velocity field will be small and certainly S'_{ij} will be too, so that for the start up mechanism the term $S_{ij}'\partial v'^{i}/\partial x^{j}$ becomes even smaller. Consequently in the general equation of heat transfer, this quantity can be ignored in such a way that T' fulfills the following relation

$$\frac{dT'}{dt} - \chi\nabla^2 T' = 0 . \qquad (41.8)$$

This is the general equation of heat transfer for the studied phenomenon in its most general form. As it was said before, T' is only a function of the coordinates so that $\partial T'/\partial t=0$ and in the former relation the following is obtained

$$v'^{i}\frac{\partial T'}{\partial x^{i}} - \chi\frac{\partial}{\partial x^{i}}\left(\frac{\partial T'}{\partial x^{i}}\right) = 0 .$$

In this case

$$\frac{\partial}{\partial x^{i}}\left[v'^{i}T' - \chi grad_{i}\ T'\right] = 0; \qquad (41.9)$$

where an integration by parts was made and was considered that the term $T'\partial v'^{i}/\partial x^{i}$ resulting from this integration, can be neglected because it is of higher order in the approximations. Therefore, integrating the last relation we have that

$$\textbf{\textit{grad}}\ T' = \frac{T'\boldsymbol{v}'}{\chi}, \qquad (41.10)$$

and without lose of generality it is assumed that the constant of integration is zero. If in this result the value of $\boldsymbol{v}'$ given in the relation (40.20) is substituted the following is obtained

$$\textbf{grad}\, T' = \frac{-\alpha t' T'^2 \boldsymbol{g}}{\chi - R' T'(\alpha T_o - 1)t'}. \tag{41.11}$$

When giving numerical values to the different quantities appearing in the denominator of the former relation it can be demonstrated that

$$\chi << R' T' t'(\alpha T_o - 1), \tag{41.12}$$

in such a way that the thermal conductivity can be ignored compared to the other term. That condition allows contemplating the phenomenon as being frozen in time. Then, if in the remainder of this relation only the z component of the gradient is considered, we have that $\boldsymbol{g} = -\boldsymbol{k}g$, with $\boldsymbol{k}$ the unitary vector along z and g the gravity acceleration. In this case the value of the thermal gradient in the vertical direction is given by the following expression

$$\frac{dT'}{dz} = -\frac{\alpha g T'}{R'(\alpha T_o - 1)}; \tag{41.13}$$

so that its absolute value is

$$\left|\frac{dT'}{dz}\right| = \frac{\alpha g T'}{R'(\alpha T_o - 1)}. \tag{41.14}$$

The equation (41.13) can be written as follows

$$\frac{dT'}{T'} = -\frac{\alpha g dz}{R'(\alpha T_o - 1)}. \tag{41.15}$$

This last formula can be integrated to finally obtain that

$$T' = T_o exp\left[-\frac{\alpha g h}{R'(\alpha T_o - 1)}\right]; \tag{41.16}$$

where h is a characteristic altitude in the considered region of the terrestrial atmosphere with the thermodynamic conditions of the environment where the breakdown of the mechanical equilibrium occurs. The equation (41.16) is the analytical expression of the small thermal perturbation that should be generated with the minimum expenditure of energy at some unstable region of the atmosphere and by means of which slowly modifies the environmental conditions to increase the natural unstability of this region favouring the beginning of free convection phenomenon. In consequence, the thermoenergetic conditions of the start up mechanism must be very small speaking in relative terms, because we only try to stimulate the atmosphere in order to begin in motion. This means that T' must be lesser that the environmental temperature T_o and the start up mechanism must be originated between the ground level and a characteristic height of the environmental conditions of the region under consideration.

§42. Politropic atmosphere

The politropic atmosphere is characterized because it satisfies the following analytical relationship

$$T^k p^{1-k} = constant\,; \tag{42.1}$$

where

$$k \equiv \frac{c_p}{c_V}\,; \tag{42.2}$$

with c_p and c_v the specific heat per mole at a pressure and volume constant, respectively. The relationship is known by the name of Poisson´s adiabatic. It relates the pressure with the temperature when, for instance, a huge mass of air as a part of the terrestrial atmosphere experiments an adiabatic expansion.

To determine the minimal thermoenergetic conditions that must be satisfied if we want to generate an atmospheric stimulus originating the free convection phenomenon in a continuous medium described by a relation as the one given in equation (42.1), we proceed in the following way.

Given that as it was said before, the thermoenergetic conditions are very small, a small variation of the functions T and p from the formula (42.1) can be calculated; that is to say

$$k\,T'T^{k-1}\,p^{1-k} + p'(1-k)T^k\,p^{1-k-1} = 0\,;$$

where T' and p' are very small changes in the pressure and temperature of the considered mass of air, respectively. Thus,

$$p' = \frac{p\,k\,T'}{(k-1)\,T}. \tag{42.3}$$

When the temperature of the air changes, its density changes too. If the experimented variations are very small the relation between T' and ρ' is given by the equation (40.1). In this case the thermal equation of state (40.15) is fulfilled so that in (42.3) the following is obtained

$$\rho_o\,R'T'(1-\alpha\,T_o) = \frac{k\,T'}{(k-1)}\frac{p}{T}. \tag{42.4}$$

In a cuasistatic adiabatic expansion it can be assumed that both T and p can be expanded as it was done in paragraph 40. Then in (42.4) we have that

$$\rho_o R'T'T_o(1-\alpha\,T_o) + \rho_o R'T'^2(1-\alpha\,T_o) = \left(\frac{k\,T'}{k-1}\right)(p_o + p').$$

Dividing all by the quantity $kT'/k-1$ and re-arranging the terms again, the following result is reached

$$\frac{\rho_o R'T_o(k-1)(1-\alpha\,T_o)}{k} - p_o = p' - \frac{\rho_o R'T'(k-1)(1-\alpha T_o)}{k}. \tag{42.5}$$

However

$$p_o = \frac{\rho_o R' T_o (k-1)(1-\alpha T_o)}{k} \tag{42.6}$$

is the thermal equation of state for the mechanical equilibrium in the politropic atmosphere; whereas

$$p' = \frac{\rho_o R' T' (k-1)(1-\alpha T_o)}{k} \tag{42.7}$$

is the thermal equation of state for the start up mechanism of the free convection phenomenon at present conditions. If the gradient of the last result is calculated, the corresponding hydrodynamic force responsible of starting the air motion is obtained; that is to say

$$\mathbf{f}'_{pH} = \frac{R'(k-1)(\alpha T_o - 1)}{k} \,\mathbf{grad}\, T' . \tag{42.8}$$

With this expression substituted in (40.20) instead of the one that appears there, we have that

$$\mathbf{v}' = \frac{R' t' (k-1)(\alpha T_o - 1)}{k} \,\mathbf{grad}\, T' - \alpha \,\mathbf{g} T' t' ; \tag{42.9}$$

so that its z component is

$$v'_z = \frac{R' t' (k-1)(\alpha T_o - 1)}{k} grad_z \, T' + \alpha \, g T' t' . \tag{42.10}$$

This is the upwards velocity of the air in a politropic atmosphere once the generation of the atmospheric stimulus has been completed and the convective motions have been initiated. With these results and with the aid of the corresponding condition for thermometric conductivity

$$\chi << \frac{R'T't'(k-1)(\alpha T_o - 1)}{k}, \tag{42.11}$$

the equation (41.13) is modified as follows

$$\frac{dT'}{dz} = -\frac{\alpha g k T'}{R'(k-1)(\alpha T_o - 1)}. \tag{42.12}$$

For this case it is clear that

$$T' = T_o exp\left[-\frac{\alpha g h k}{R'(k-1)(\alpha T_o - 1)}\right]. \tag{42.13}$$

As it is easy to see, the obtained results for the politropic atmosphere case are very similar to those obtained before in paragraph 41.

§43. Some numerical results

With the numerical data that we have in the specialized literature, some interesting results can be obtained for two seasons of the year under extreme climatic conditions.

For the winter season the following data will be used

$$\alpha_i = 0.137^o C^{-1}; T_o = 10^o C; h = 62m. \tag{43.1}$$

With these approximate values substituted in the equation (41.16), we have that $T' = 4.6^{\circ}C$; whereas from (41.14) we obtain

$$\left|\frac{dT'}{dz}\right|_i \approx 0.6 \times 10^{-3}\ {}^oC \cdot cm^{-1}. \tag{43.2}$$

For the summer season the following values can be used

$$\alpha_v = 0.102\ {}^oC^{-1}; T_o = 15^o C; T' = 4.6^o C. \tag{43.3}$$

With these data the following results are reached: $h = 180\ m$ which is the altitude where the mechanical equilibrium is broken, and

$$\left|\frac{dT'}{dz}\right|_v \approx 0.3\times 10^{-3}\ {}^oC\cdot cm^{-1}. \tag{43.4}$$

It is interesting to compare the values obtained for the thermal gradient for the winter season and the summer season with the value of the adiabatic gradient reported in the specialized literature. It is known that the mechanical equilibrium in the atmosphere becomes unstable if the adiabatic thermal gradient is larger than a certain value. In other words, if

$$\Delta > 9.843\times 10^{-5}\ {}^oC\cdot cm^{-1} \tag{43.5}$$

the equilibrium is unstable. Then and according to the former results, for the winter season we have that

$$\left|\frac{dT'}{dz}\right|_i \approx 6\Delta \tag{43.6}$$

and for the summer

$$\left|\frac{dT'}{dz}\right|_v \approx 3\Delta . \tag{43.7}$$

The previous results point out that it is possible to satisfy the minimal thermodynamic conditions to generate an atmospheric stimulus capable of initiating the convective motions in some unstable region of the terrestrial atmosphere.

Selected Topics

The atmosphere

The atmosphere is the gaseous covering of our planet. It is formed by the air which is not a simple chemical substance but a moisture of chemical elements and compounds which do not react between them. It contains in suspension a huge variety of liquid and solid objects in the form of slender drops and particles that go from organic matter as pollen and products of combustion processes, to ions and radiative matter, going through the whole range of contaminant agents that pollute the cities and highly industrialized zones. The composition of terrestrial atmosphere remains practically constant in its low region and has the following structure

Aproximate Chemical Composition of Fresh Air		
Component	***Percentaje***	
	In Volume	***In Weight***
Nitrogen	78.03	75.58
Oxygen	20.99	23.08
Carbon dioxide	0.035	0.053
Argon	0.94	1.28
Other inert gases	0.0024	0.0017
Hydrogen	0.00005	0.000004

The *humidity* content of the air either as water vapor, liquid drops or ice crystals, changes significantly with place and time. It is responsible for many of the atmospheric phenomena that frequently oppresses us. In general, the water content in the air is determined by the *relative humidity* (*RH*) defined by the ratio of the water content of the air to the saturation water content; that is to say

$$RH = humidity / humidity\ of\ saturation\ .$$

The water content of the atmosphere, at saturation†, increases rapidly with increasing temperature. At any temperature the absolute amount of water in the air increases when the *relative humidity* increases. The atmos-

† *At any given temperature, a fixed volume of air can contain a variable quantity of water vapor until it reaches a maximum beyond which vapor is condensed in the form of little drops. It is said then the air has been* ***saturated by humidity****. The* ***absolute humidity*** *is the weight in grams of water vapor contained in a cubic meter of air, whereas the* ***relative humidity*** *is the relation in percentage between the absolute humidity and the amount of water vapor which is contained in the cubic meter of air considered if it were saturated at any temperature.*

phere has well defined boundaries in its low part that are determined by the oceans surface and the fixed earth irregularities. In its upper boundary the opposite occurs since it simply becomes subtler as height increases. Concerning its thermal properties some regions are defined in the atmosphere which from ground to space are the following.

Troposphere. This region is characterized by the fact that throughout it and in vertical direction the air temperature descends constantly at a ratio of 6.5°C for each kilometer of height. Over the equatorial regions it reaches an altitude of 16 km. In the poles its thikness is diminished as much as 8 km. In the temperate zones its thikness changes with the seasons and it reaches an average value of 13 km. The atmospheric phenomena which determines the state of time occurs in the troposphere. In it practically all the water vapor in the air as well as the condensation nuclei and the major variation of temperature are localized. These differences are the ones which determine the operation of that huge thermal machine that is the atmosphere. In general, the air motion is governed by an uneven heating of the Earth surface, modified by the effect of the *Coriolis force* originated by the planet rotation, as well as of the influence of the fix land and the ocean. The upper limit of the troposphere is the so called *time frontier* or *ceiling time*. It is known by the technical name of tropopause and it is the zone where the diminishing in temperature stops.

Stratosphere. Its principal characteristic is that in there the air temperature remains almost constant or increases its value with height. In this region important atmospheric phenomena do not occur. It is for this reason that the air motion also diminishes noticeably. Its upper frontier is found at an almost 40 km over the ground level and it is known by the name of *stratopause*.

Mesosphere. This atmospheric region extends as much as 80 km of height from the ground and it is characterized because in it an acute maximum of temperature is produced. This maximum is reached in a region called the *hot layer*, and seems to be originated by the thermal energy issued in the production an annihilation of ozone. Physical chemistry called it the *ozonesphere* or *ozone layer*. It ends in a region called the *mesopause*.

Thermosphere. Throughout this region the air temperature increases with height without interruption till it reaches 500 °C at a 500 km altitude over the ground. It ends in the *thermopause* to give place to the exosphere which is the external and unlimited region of the atmosphere. This zone is formed by freeing molecules whose concentration is progressively diminished until the region becomes *interstellar space*. The external limit of the atmosphere is fixed in a conventional way at 2000 km of altitude over the terrestrial surface

The mechanical equilibrium in the atmosphere

Let us consider again the problem of paragraph 39 under the same physical conditions. The situation can be illustrated with the aid of the following figure

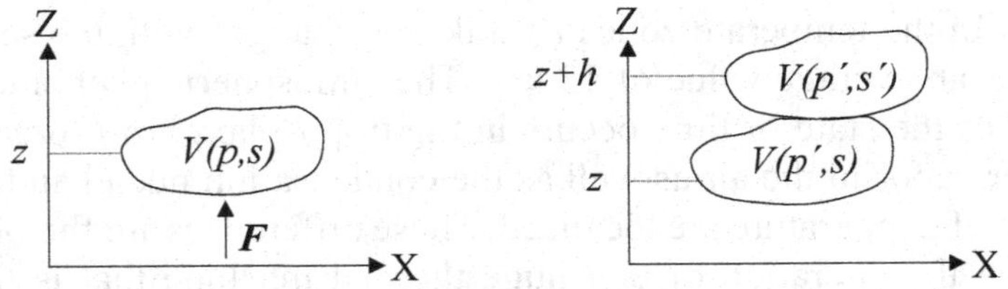

Now, considering the equation (39.3) the equilibrium condition can be expressed as follows

$$V\left(s'-s\right)>0; \tag{1}$$

due to the fact that both the mass density $\rho(\boldsymbol{x},t)$ as well as the acceleration $\boldsymbol{a}$ are the same for both air masses. If the value of the pressure remains constant and we make an expantion in power series of $(s'-s)$, we have that

$$\left(\frac{\partial V}{\partial s}\right)_p \left(s'-s\right)>0. \tag{2}$$

On the other hand and given that the specific entropy is only a function of the height over the ground level, to a small variation in the altitude corresponds a small increase in the specific entropy; that is to say

$$\Delta s=\frac{ds}{dz}\Delta z,$$

where $\Delta s \equiv s'-s$. Let $\Delta z = h$ be so that in (2) the following result is obtained

$$h\left(\frac{\partial V}{\partial s}\right)_p \frac{ds}{dz} > 0. \tag{3}$$

From thermodynamics we have that

$$\left(\frac{\partial V}{\partial s}\right)_p = \frac{T}{c_p}\left(\frac{\partial V}{\partial T}\right)_p,$$

with T the temperature and c_p the specific heat of air at a constant pressure. In this case in (3) the following is obtained

$$\frac{hT}{c_p}\left(\frac{\partial V}{\partial T}\right)_p \frac{ds}{dz} > 0. \tag{4}$$

On the other hand, h, T and c_p are positive quantities. Additionally and given that the majority of substances expand on heating, the condition so that in the atmosphere the free convection phenomenon does not occur is that the specific entropy increases its value with height; that is to say, the equation (39.7) must be fulfilled.

Now and from thermodynamics point of view,

$$s(z) = s[T(z);\, p(z)]$$

so that

$$\frac{ds}{dz} = \left(\frac{\partial s}{\partial T}\right)_p \frac{dT}{dz} + \left(\frac{\partial s}{\partial p}\right)_T \frac{dp}{dz}.$$

On the other hand, it is known that

$$\left(\frac{\partial s}{\partial T}\right)_p = \frac{c_p}{T}, \quad \text{and} \quad \left(\frac{\partial s}{\partial p}\right)_T = -\left(\frac{\partial V}{\partial T}\right)_p$$

so that

$$\frac{ds}{dz}=\frac{c_p}{T}\frac{dT}{dz}-\left(\frac{\partial V}{\partial T}\right)_p\frac{dp}{dz}$$

in such a way that the equation (39.8) is fulfilled. From the general condition for mechanical equilibrium (39.1) and in z-direction, it is obtained that $dp/d\,z=-g/V$, with g the value of gravity acceleration. Finally, with the former results, the condition (39.10) is obtained, that is to say

$$\left|\frac{dT}{dz}\right|>\frac{gT}{c_pV}\left(\frac{\partial V}{\partial T}\right)_p.$$

Air density changes with temperature and humidity

The air is a compressible viscous fluid which from thermodynamics point of view behaves as an ideal gas so that the thermal equation of state is satisfied; that is to say

$$\rho=\frac{mp}{RT}; \qquad (1)$$

where $\rho(\boldsymbol{x},t)$ is the mass density of air, $p(\boldsymbol{x},t)$ the hydrostatic pressure and T the temperature, whereas m =28.964 gr·mol^{-1} is the molecular weight of the air and R=8.314×10 erg·(mol·°C)$^{-1}$ is the universal gas constant.

It is known that at certain altitude over the ground level and hence for one particular pressure the air density is determined by the molecular weight and the temperature. Deriving the former relation and taking into account the fact that R=*constant* we have that

$$\frac{d\rho}{\rho}=\frac{dm}{m}+\frac{dp}{p}-\frac{dT}{T}; \qquad (2)$$

where systematically the thermal equation of state for the ideal gas (1) was used. If the molecular weight and the pressure are maintained constant, it is clear that (2) is reduced to the following expression

$$\frac{d\rho}{\rho} = -\frac{dT}{T}. \tag{3}$$

On the other hand, the average molecular weight of air is given by the following formula

$$< m >= m_{air} + y_{H_2O}\left(m_{H_2O} - m_{air}\right); \tag{4}$$

where y_{H_2O} is the *mol fraction* of water vapor.

To estimate the change in air density due to an increment *dT* in the temperature equal to 1 °*C* and a 1 percent increase in relative humidity *dRH*, both at 20°C, the following is done. Given that *T*(°*C*)=*T*(*K*) – 273.15, with *T*(°*C*) and *T*(*K*) the centigrade and Kelvin temperatures respectively, in (3) we have that

$$\frac{d\rho}{\rho} = -\frac{1°C}{(20 + 273.15)°C} = -0.0034; \tag{5}$$

or also, at 20 °C the former relation can be written as follows

$$\frac{d\rho / \rho}{dT} = -0.0034 \cdot °C^{-1}.$$

Moreover, at 20 °C we can see that

$$y_{water} \approx 0.023\ RH$$

so that

$$<m> \approx 29-0.023\ RH(29-18)=29-0.253\ RH$$

and then

$$\frac{d\rho}{\rho} = \frac{d < m >}{< m >} = \frac{-0.253dRH}{29 - 0.253RH} \approx -\frac{0.253dRH}{29}. \tag{6}$$

In this case,

$$\frac{d\rho/\rho}{dRH}=\frac{-0.253\left[0.01\cdot(\%RH)^{-1}\right]}{29}=-8.7\times10^{-5}\cdot(\%RH)^{-1}$$

It is easy to see that about a 40% in realtive humidity is required to produce the same effect as a 1°C increase in temperature. In fact, when the results (5) and (6) are taken into account we have that

$$dRH=\frac{3.4\times10^{-2}\times29}{2.53}=38.97\%$$

This explains why most of the vertical motion of the atmosphere is driven by changes in temperature rather than by changes in humidity.

The control of atmospheric pollution

It is desired to eliminate the cloud of contaminants that are daily formed over a city. The construction of a tunnel to send through it this contaminated air anywhere else where it does not affect the inhabitants of the metropolitan area is proposed. In order to make an objective analysis of the magnitude of the proposal, let us suppose that the contaminated air daily covers on the average a certain area A, forming a kind of cap which totally encloses the metropolitan zone. If the thickness of the cap is equal to h, with the building of such engineering construction we wish to daily remove a volume of polluted air equal to V, with $V=Ah$. Also we assume that the average velocity in the tunnel is equal to v. From what magnitude must the required construction diameter be?

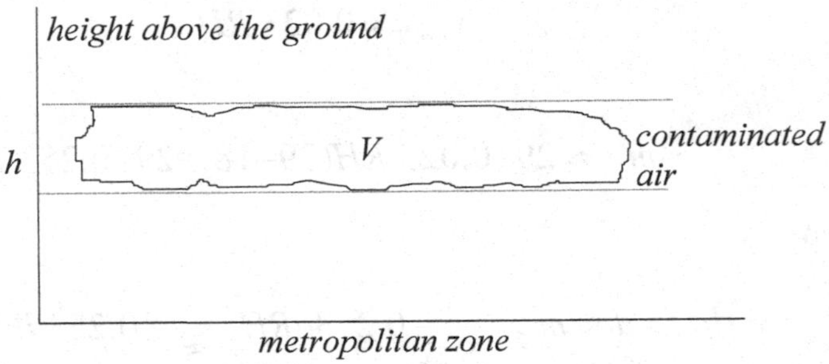

It is evident that the daily discharge of contaminated air throughout the tunnel is

$$Discharge = \frac{Volume\ of\ contaminated\ air}{time}.$$

Let D be the discharge and t the time equal to 24 hours, so that

$$D = \frac{V}{t} = \frac{Ah}{24\ hours}. \tag{1}$$

If we suppose that the air density is a constant, the continuity equation is simply

$$Area \times velocity = constant; \tag{2}$$

where the constant is the discharge and the velocity is the flux rapidity in the tunnel. Almost all the tunnels are cylindrical so that if the inlet and the outlet are equal we have that

$$A = \frac{\pi d^2}{4};$$

with d the tunnel diameter

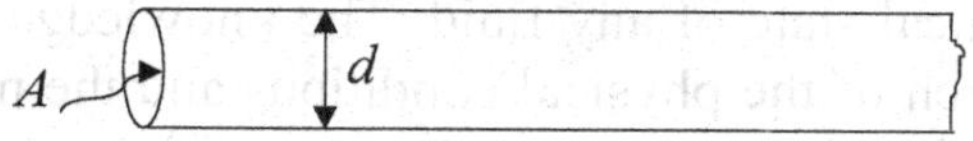

Then, from (1) and (2) it is obvious that

$$d = \sqrt{\frac{4D}{\pi v}}.$$

Let us suppose that we are dealing with Mexico City, one of the largest cities in the world, with a surface equal to 2,300 km^2, and the project consisted of perforating the Ajusco to send the contaminated air to Cuernavaca Valle. If the thickness of the cloud is of about 400 m in average and given that a day = 8.64×10^4 seconds, we have that

$$D = \frac{1.3\times10^{9}\times4.0\times10^{2}m^{3}}{8.64\times10^{4}\sec} = \frac{9.2\times10^{11}m^{3}}{8.64\times10^{4}\sec} = 1.0648\times10^{7}m^{3}\cdot\sec^{-1}$$

On the other hand, let us consider that the flux rapidity in the tunnel is of 15 m/sec so that

$$d = \sqrt{\frac{4\times1.0648\times10^{7}m^{3}\cdot\sec^{-1}}{3.14\times15\ m\cdot\sec^{-1}}} = \sqrt{\frac{4.2593\times10^{7}m^{2}}{4.7\times10}}$$
$$= \sqrt{0.9062\times10^{6}m^{2}} = 0.952\times10^{3}\ m = 952\ m.$$

Pemex is a 44-story executive tower, so that it can be said that it is about 220 m tall. In this case, the calculated diameter is 4 times larger than the altitude of that structure. Evidently, it is a construction that is far beyond any current structural engineering capabilities of any country of the world.

Wind generation

In the study of the minimal thermoenergetic conditions required to generate an atmospheric stimulus in some unstable region of the atmosphere, in order to produce via an unstability, winds of moderate intensity, it is necessary to use small perturbative terms in the relevant variables that characterize the dynamical state of any fluid. The knowledge obtained with the theoretical research of the physical conditions and the minimal energy required for producing the start up mechanism of free convection, has as an immediate consequence to know if it is possible to influence the motion of huge masses of air by means of small energy variations, in order to use that knowledge in the fight against atmospheric pollution. The exact and complete solution of the problem and its application in the case of Mexico City, has interesting and practical implications. To generate winds capable of cleaning the air it is proposed that the neighbors from the south of the city cover the roof of their lodgings with a special black mate color material. That material, besides containing ecological substances in its composition and being of easy application it is required to have the following characteristics

- **It must be a good thermal isolator.**
- **It must be a good permeating material.**

- **It must behave as a solar energy accumulator and as a thermal energy radiator.**

Summarizing, in the south of Mexico City there would be many *solar accumulators* which would store the *solar radiation* to spill it afterwards as heat to the lower layers of the southern atmosphere this way propitiating the required thermoenergetic conditions to be reached in order to generate the start up mechanism of free convection. In other words, the idea is to convert a great deal of roofs of the south of the city into an *ecological system of thermal energy accumulators-radiators*. Once the air starts moving, the solar radiation takes care of energy massive expenditure required to maintain the atmosphere motion as well as that of the dispersion of the cloud of contaminants. This proposal extracted from the analytical solution of free convection, has interesting geological antecedents. In fact, just 2,000 years ago the volcano *Xitle* located in the so called *Pico del Aguila* located at the *Ajusco Mount* made an eruption. That catastrophe not only buried some native ceremonial centers but also covered with volcanic lava a surface of nearly 100 km^2. The volcanic stone is hard, brittle and has black mate color. It absorbs the 83 percent of solar radiation and it becomes extremely hot. Besides producing little dust, it allows that some vegetation in the form of shrubs and grass besides some animal life typical of the rocky regions grow over the ground. That land will have the required properties for producing by unstability and at large scale, the thermal or free convection phenomenon, in such a way that it maintains the atmosphere of that zone clean, a reason for which it is historically known by the name of ***the most transparent region of the air***.

In the last decades the metropolitan area has spread in all directions, in particular towards the south, in such a way that this zone has been covered by houses almost in its totality despoiling the city of the natural quality it used to have and which allowed it to maintain its atmosphere clean. On the other hand, the houses have some of the qualities of the volcanic land; that is to say, they have a garden with plants and grass and its roofs are hard and produce little dust. They only require the adequate colors so that the similarity is practically complete. The proposal has other evident characteristics. It can be translated to other cities with atmospheric pollution problems like *Guadalajara* and *Monterrey*, for instance. Moreover, it is a solution that despite the fact that it fights the air pollution effects and not their causes, has the quality of giving the time required to find definite solutions for the environmental contamination problem, that as it is well known, has multiple origins.

References

1. Battan, L. J. “La naturaleza de las tormentas”. Doubleday & Company Inc. New York (1961). Colección Ciencia Joven. Editorial Universitaria de Buenos Aires (1964).
2. Biblioteca Salvat de grandes temas, “La atmósfera y la predicción del tiempo” Vol. 42(1973).
3. Bird, R.B. Steward, W.E. and Lightfoot, E.M.“Transport Phenomena”.Wiley Interscience Edition (1960).
4. Fermi, E. “Thermodynamics”. Dover Publications, Inc. New York (1936)
5. Fierros Palacios, Angel. “La generación de un estímulo atmosférico y el problema de la contaminación del aire del Valle de México”.Rev del IMP. Vol.XXVI, No. 2 (julio-diciembre, 1994) 47-56.
6. Landau, L.D. and Lifshitz, E.M. “Fluid Mechanics”. Addison -Wesley Publishing Co. (1959).
7. De Nevers, N “Air Pollution Control Engineering”, Mc Graw-Hill (1995).
8. Prandtl L. and Tietjens, O.G. “Fundamentals of Hidro-and Aeromechanics”. Dover Publications, New York (1934).

Chapter VIII

Magnetohydrodynamics

§44. Field equations in a mobile conducting continuous medium

When a conducting fluid moves in a region where there is a magnetic field, electric fields are induced in it. According to Classical Electrodynamics, the existence of those induced fields has as a consequence the production in the medium of electric currents flow. On the other hand, the external magnetic field becomes present in the system in the form of magnetic forces acting over the produced currents. That influence in turn can give as a result a considerable modification of the fluid flow. At the same time, the external magnetic field experiments changes due to the presence of the generated electric currents. In consequence, the dynamical state of a conducting continuous medium that moves in an external magnetic field is a very complex phenomenon. There is in fact a complicated interaction between the magnetic phenomena and the one corresponding to Fluid Dynamics; therefore, the resulting flux must be studied with the help of Classical Electrodynamics field equations and the balance equations of Fluid Dynamics, within the theoretical frame of Magnetohydrodynamics frequently called Hydromagnetism and often, in short as MHD.

Magnetohydrodynamics constitutes a very specialized branch of theoretical physics where other two, Hydrodynamics, and Electromagnetism are joined. It could be logical to suppose that its extension is larger than any of its components separately considered. In fact Hydrodynamics deals with all type of fluids either in gaseous state or in liquid. Therefore, Fluid Dynamics is the branch that must be fused with the Electromagnetism. On the other hand, the relevant part of Electromagnetism in MHD is that it deals with the magnetic effects of the electric currents. For the case of a conducting continuous medium that moves in an external magnetic field, Maxwell´s equations of electromagnetics used are the ones corresponding to a homogeneous conductor in a magnetic field; that is to say

$$div\, \boldsymbol{H} = 0\,, \tag{44.1}$$

and

$$\frac{\partial \boldsymbol{H}}{\partial t} = \boldsymbol{rot}\left(\boldsymbol{v} \times \boldsymbol{H}\right) + \frac{c^2}{4\pi\sigma}\nabla^2 \boldsymbol{H}\,; \tag{44.2}$$

where $\boldsymbol{H}(\boldsymbol{x}, t)$ is the external magnetic field, c the velocity of light in the empty space, and σ the electrical conductivity. In those relations it is usual to consider that the magnetic permeability of the medium differs very little from the unity so that $\mu = 1$. Also it is assumed that σ is constant in the fluid, and therefore, in particular its value does not keep any relation with the magnetic field. In what follows, we will use this same letter to designate the electrical conductivity and the components of the stress tensor in their different forms, without generating any confusion because distinction will be made between them in the way to express it.

Since the conducting continuous medium that is going to be studied is a compressible viscous fluid, it is important to determine in what way the balance equations describing the flux are modified by the external magnetic field. If as it has already been said the motion of the fluid is disturbed by the external magnetic field, the hydrodinamic regime responsible of the flux must by more complex than the one in the case of non conducting viscous fluid flux.

Somehow, the electromagnetic phenomena are shown in the continuos medium as internal stresses of magnetic origin which are added to internal mechanical stresses produced when the region containing the fluid is deformed by external agents. Its origin must be looked for in the magnetic viscosity that the external field generates in the fluid and has the effect of giving it a certain degree of rigidity capable to modify the flux process. Then, it can be assumed that due to that rigidity, the effect of external magnetic field over the considered conducting continuous medium is similar to a deformation that contributes to remove the fluid from its mechanical, as well as thermal equilibrium state propitiating the emergence of those internal magnetic stresses which are trying to return it to its original situation of equilibrium. The internal magnetic stresses can be represented by a tensor of rank two known as Maxwell's stress tensor of the magnetic field which is in components

$$m_{ij} = \frac{1}{4\pi}\left[H_i H_j - \frac{1}{2}H^2\delta_{ij}\right]. \tag{44.3}$$

Clearly it is about a symmetrical tensor, that is to say it is fulfilled that $m_{ij}=m_{ji}$ as it is easy to see from its definition. In consequence, a conducting continuous medium in an external magnetic field which suffers a mechanical deformation that removes it from its initial thermomechanical equilibrium, generates internal mechanical stresses that are added to the magnetic ones due to the magnetic viscosity created in the fluid by the field, which are opposed to the change and try to return it to its original condition of equilibrium. Those stresses can be represented by means of a tensor of rank two whose components are

$$\sigma^o_{ij} = \sigma_{ij} + m_{ij} \tag{44.4}$$

such that

$$\sigma^o_{ij} = \sigma^o_{ji}. \tag{44.5}$$

We shall name the tensor $\tilde{\sigma}^o(\boldsymbol{x}, t)$ the generalized stress tensor, whereas $\sigma(\tilde{\boldsymbol{x}}, t)$ is the corresponding mechanical stress tensor of the newtonian viscous fluid whose general form is given in paragraph 35.

If the deformations suffered by the region which contains the fluid as well as the changes experimented by the external magnetic field are small, the corresponding variations in the displacement vector are infinitesimal, so that the magnetomechanical internal stresses work. In this case the results that were obtained for the viscous fluid can be generalized and write that

$$\Delta\mathcal{R}^o = -\sigma^o_{ij}\,\Delta u_{ij}. \tag{44.6}$$

This is the work done by the generalized internal stresses. Due to the deformation process that we call magnetomechanical, the specific internal energy of the fluid changes according to the general relationship $d\mathcal{U}=Tds-d\mathcal{R}^o$; that with the aid of the former result is transformed into

$$d\square = Tds + \sigma^{o}_{ij} du_{ij}; \tag{44.7}$$

where now $\square$ is the generalized specific internal energy of the conducting continuous medium under study.

§45 The change in the generalized specific internal energy

To obtain that change it is necessary to take into account the fact that the external magnetic field does not do any work over the mobile conducting fluid. Hence, to calculate the change in the specific internal energy in that situation and to determine the work done it is necessary to close examine what occurs during the complicated electromagnetic phenomena that originate there. The latter is obtained if the following field equation that relates the variable electric and magnetic fields is used

$$\textbf{rot}\,\boldsymbol{E} = -\frac{1}{c}\frac{\partial \boldsymbol{B}}{\partial t}; \tag{45.1}$$

where $\boldsymbol{E}(\boldsymbol{x}, t)$ is the induced electric field and in general $\boldsymbol{B}=\mu\boldsymbol{H}$ is the mean magnetic field usually called the magnetic induction.

During a temporary interval Δt, the electric field $\boldsymbol{E}$ does the work $\Delta t \int \boldsymbol{j}\cdot\boldsymbol{E}\,dV$ over the conduction current density $\boldsymbol{j}(\boldsymbol{x}, t)$. That quantity with the oppo-site sign is the work $\Delta\mathcal{R}^{o}$ done over the field by the external electromotive force that maintains the currents. Now and because the conduction current is

$$\boldsymbol{j} = \frac{c}{4\pi}\,\textbf{rot}\,\boldsymbol{H} \tag{45.2}$$

it is clear that

$$\Delta\mathcal{R}^{o} = -\frac{\Delta t\, c}{4\pi}\int \boldsymbol{E}\cdot\textbf{rot}\,\boldsymbol{H}\,dV\,.$$

From vectorial anaysis we have that

$$\boldsymbol{H} \cdot \boldsymbol{rot}\, \boldsymbol{E} = div\left(\boldsymbol{E} \times \boldsymbol{H}\right) + \boldsymbol{E} \cdot \boldsymbol{rot}\, \boldsymbol{H}\,. \tag{45.3}$$

Then

$$\Delta \mathcal{R}^{o} = \frac{\Delta t\, c}{4\pi} \int div\left(\boldsymbol{E} \times \boldsymbol{H}\right) dV - \frac{\Delta t\, c}{4\pi} \int \boldsymbol{H} \cdot \boldsymbol{rot}\boldsymbol{E}\, dV\,. \tag{45.4}$$

The first integral is zero if it is transformed into an integral over a surface infinitely distant†. In the second integral, ***rot E*** can be substituted by its value given in (45.1) and write

$$\Delta \boldsymbol{B} = \Delta t \frac{\partial \boldsymbol{B}}{\partial t}$$

for a temporary change in the magnetic induction; so that

$$\Delta \mathcal{R}^{o} = \int \frac{\boldsymbol{H} \cdot \Delta \boldsymbol{B}}{4\pi} dV \tag{45.5}$$

is the work done when ***B*** experiments an infinitesimal change. Thus, the change in the specific internal energy of the conducting fluid in an external magnetic field can be written as

$$\Delta \square^{o} = \int \frac{\boldsymbol{H} \cdot \Delta \boldsymbol{B}}{4\pi} dV\,.$$

In consequence for a conducting continuous medium that moves in an external magnetic field, the specific internal energy must contain an additional term to take into account such field; that is to say

$$\square = \square + \frac{\boldsymbol{H} \cdot \boldsymbol{B}}{4\pi\rho}, \tag{45.6}$$

† *Due to the fact that the electromagnetic field becomes zero in the infinity.*

with $\square(u_{ij})$ the internal specific energy of the newtonian viscous fluid. From relations (44.7) and (45.6) it is evident that the functional relationship for the generalized internal specific energy has the following form

$$\square = \square\left(s, u_{ij}, B_j\right); \tag{45.7}$$

so that the definition (32.16) is fulfilled; whereas

$$\left(\frac{\partial \square}{\partial u_{ij}}\right)_{s,B_j} = \sigma^{o}_{ij} \tag{45.8}$$

and

$$\left(\frac{\partial \square}{\partial B_j}\right)_{s,u_{ij}} = \frac{H_j}{4\pi\rho} \tag{45.9}$$

§46. Momentum balance equation

The dynamical state of a conducting newtonian viscous fluid flowing in a region R of the three-dimensional Euclidean Space where an external magnetic field exists, can be described by the aid of a lagrangian density whose functional form is

$$\circ = \circ\left(\boldsymbol{x}, \boldsymbol{v}, \tilde{u}, \boldsymbol{B}, t\right). \tag{46.1}$$

The corresponding specific lagrangian is

$$\lambda = \lambda\left(\boldsymbol{x}, \boldsymbol{v}, \tilde{u}, \boldsymbol{B}, t\right). \tag{46.2}$$

Both functions contain all the dynamic information required to make a complete description of the motion state of any real newtonian fluid, besides they contain an additional term to take into account the effect of the external magnetic field over the flux. The action integral (14.1) will be used again but now the lagrangian density (46.1) must be considered in it.

Thus, with the aid of the invariance condition (14.2) and according to the Hamilton-Type Variational Principle the general formula (16.4) is reached again; where now the specific lagrangian (46.2) must be taken into account. The variational calculus applied to that function gives as a consequence the following

$$\rho\delta\lambda = \rho\left[\frac{\partial\lambda}{\partial x^i}\delta x^i + \frac{\partial\lambda}{\partial v^i}\delta v^i + \frac{\partial\lambda}{\partial u_{ij}}\delta u_{ij} + \frac{\partial\lambda}{\partial B_j}\delta B_j\right]. \quad (46.3)$$

Given that $\boldsymbol{B}=\boldsymbol{B}(\boldsymbol{x},t)$ we have that

$$\delta B_j\left(\boldsymbol{x},t\right) = \left(\frac{\partial B_j}{\partial t} + \frac{\partial B_j}{\partial x^i}\frac{\partial x^i}{\partial t}\right)\delta t = 0. \quad (46.4)$$

The local variation of magnetic induction is zero because time does not depend on the geometrical parameter set, so that $\delta t=0$ for every t, according to the usual definition of geometric variation.

As a consequence of that, the local variation of λ is identical to that obtained in Chapter VI for the case of viscous fluid; in such a way that all the results obtained in paragraph 33 are valid. Therefore, the generalized momentum balance equation for MHD has the same form for the case before mentioned; that is to say

$$\mathcal{D}\left[\rho\frac{\partial\lambda}{\partial v^i}\right] - \rho\frac{\partial\lambda}{\partial x^i} + \frac{\partial}{\partial x^j}\left(\frac{\partial\lambda}{\partial u_{ij}}\right) = 0. \quad (46.5)$$

Let us not overlook the fact that this relation and the one corresponding to the newtonian real fluid of paragraph 33 only has an identical form. The relationship (46.5) is a vectorial equation for a specific lagrangian whose functional dependence is given in (46.2) where the contribution of the external magnetic field is included. For the same reasons Euler-Lagrange's field differential equation of MHD in terms of the above mentioned specific lagrangian has the same form as the one for the viscous fluid, that is to say

$$\frac{d}{dt}\left[\frac{\partial\lambda}{\partial v^i}\right] - \frac{\partial\lambda}{\partial x^i} + \frac{1}{\rho}\frac{\partial}{\partial x^j}\left[\frac{\partial\lambda}{\partial u_{ij}}\right] = 0. \quad (46.6)$$

Since the difference between both cases is found in the functionality of the corresponding specific lagrangian, it is natural to assume that the influence of the external magnetic field will be shown in the generalized stress tensor and in the explicit form of the generalized specific internal energy; as it is easy to see from the results obtained in paragraphs 44 and 45.

§47. Generalized energy balance equation

The invariance of the action integral (14.1) under continuous temporary transformations has as a consequence the general result (19.1) where the lagrangian density that must appear in it now contains in its argument an additional term which does not become null in order to take into account the influence that the external magnetic field exerts over the flux. Thus and according to the functionality of the lagrangian density given in (46.1)

$$\begin{aligned}\delta^+ \circ &= \frac{\partial \circ}{\partial x^i}\delta^+ x^i + \frac{\partial \circ}{\partial v^i}\delta^+ v^i + \frac{\partial \circ}{\partial u_{ij}}\delta^+ u_{ij} \\ &+ \frac{\partial \circ}{\partial B_j}\delta^+ B_j + \frac{\partial \circ}{\partial t}\delta^+ t .\end{aligned} \tag{47.1}$$

This variational process only differs from the one made when the newtonian real fluid was studied, in the variation of the term corresponding to magnetic induction. In consequence, be

$$\frac{\partial \circ}{\partial B_j}\delta^+ B_j = \frac{\partial \circ}{\partial B_j}\frac{\partial B_j}{\partial t}\delta^+ t . \tag{47.2}$$

According to the result (45.9) the following definition is proposed

$$\frac{\partial \circ}{\partial B_j} \equiv -\frac{H_j}{4\pi}; \tag{47.3}$$

in such a way that in (47.2) the following result is reached

$$\frac{\partial \circ}{\partial B_j}\delta^+ B_j = -\frac{\boldsymbol{H}}{4\pi}\cdot\frac{\partial \boldsymbol{B}}{\partial t}\delta^+ t . \tag{47.4}$$

When the flow of a conducting real fluid in an external magnetic field is studied, it is important to recognize that it deals with a conducting continuous medium which does not carry any current. In this case, in the relation (45.2) we have that ***rot H***= 0 because ***j***=0. With this result substituted in the vectorial identity (45.3) and with the aid of (45.1), in (47.4) the following is obtained

$$-\frac{\boldsymbol{H}}{4\pi}\cdot\frac{\partial \boldsymbol{B}}{\partial t}\delta^{+}t=\frac{c}{4\pi}div\left(\boldsymbol{E}\times\boldsymbol{H}\right)\delta^{+}t\,;$$

so that

$$\frac{\partial \circ}{\partial B_j}\delta^{+}B_j=\frac{\partial S_j}{\partial x^j}\delta^{+}t, \tag{47.5}$$

where

$$\boldsymbol{S}=\frac{c}{4\pi}\boldsymbol{E}\times\boldsymbol{H} \tag{47.6}$$

is Pointing´s vector.

Finally and according to the previous results and those obtained in Chapter VI for the temporary variations of the coordinates, the time, the velocity field, and the small deformations strain tensor,

$$\begin{aligned}\delta^{+}\circ=&\left[\rho\frac{\partial\lambda}{\partial x^i}-D\left(\rho\frac{\partial\lambda}{\partial v^i}\right)-\frac{\partial}{\partial x^j}\left(\frac{\partial\lambda}{\partial u_{ij}}\right)\right]v^i\delta^{+}t\\&+\left[\frac{d}{dt}\left(\rho\frac{\partial\circ}{\partial v^i}v^i\right)+\mathcal{H}\,div\boldsymbol{v}+\frac{\partial}{\partial x^j}\left(\frac{\partial\lambda}{\partial u_{ij}}v^i+S_j\right)+\rho\frac{\partial\lambda}{\partial t}\right]\delta^{+}t.\end{aligned} \tag{47.7}$$

The first square parenthesis of the right hand side of the former result is zero because it is the momentum balance equation of the MHD. Then in the formula (19.1) we have that

$$\int_{t_1}^{t_2}\int_R\left[\mathcal{D}\mathcal{H}+\frac{\partial}{\partial x^j}\left(\frac{\partial\lambda}{\partial u_{ij}}v^i+S_j\right)+\rho\frac{\partial\lambda}{\partial t}\right]\delta^{+}t\,dV\,dt=0\,. \tag{47.8}$$

Again and due to the fact that the increments $\overset{+}{\delta} t, dV$, and dt are totally arbitrary and therefore different from zero, the former relation is only fulfilled if the integrand becomes null; that is to say if

$$\mathcal{D}\mathcal{H} + \frac{\partial}{\partial x^j}\left(\frac{\partial \lambda}{\partial u_{ij}} v^i + S_j\right) + \rho \frac{\partial \lambda}{\partial t} = 0 . \tag{47.9}$$

This is the generalized energy balance equation of MHD. It includes an additional term that is the divergence of Pointing´s vector which represents the electromagnetic energy flux density; term that clearly, does not exist in the corresponding equation for the newtonian real fluid

§48. The equation of motion

In order to obtain the equation of motion of MHD, a lagrangian density identical in form to those that have been utilized to study the cases of the ideal and newtonian real fluid is proposed. Then, be

$$\circ = \rho\left[\frac{1}{2} v^2 - \square\right] \tag{48.1}$$

the proper lagrangian density so that

$$\lambda = \frac{1}{2} v^2 - \square \tag{48.2}$$

is the corresponding specific lagrangian. In these expressions

$$\square = \square + \frac{H^2}{8\pi\rho} \tag{48.3}$$

is the generalized specific internal energy, with $H^2/8\pi\rho$ the magnetic energy per unit mass.

With these definitions in the relationship (46.6) the following is obtained

$$\rho\left[\frac{\partial \boldsymbol{v}}{\partial t} + (\boldsymbol{v} \cdot \boldsymbol{grad})\, \boldsymbol{v}\right] = \frac{\partial \sigma_{ij}^{o}}{\partial x^j} . \tag{48.4}$$

We shall call this result the generalized Cauchy′s motion equation.

According to definitions (35.6) and (35.7), as well as (44.3) and (44.4), on the right hand side of (48.4) we have that

$$\frac{\partial \sigma_{ij}^{o}}{\partial x^{j}} = -\boldsymbol{grad}\, p + \eta \nabla^{2} \boldsymbol{v} + \left(\zeta + \frac{1}{3}\eta\right) \boldsymbol{grad}\, div\, \boldsymbol{v}$$
$$+ \frac{1}{4\pi}\left[H_{i}\, div \boldsymbol{H} + (\boldsymbol{H} \cdot \boldsymbol{grad})\boldsymbol{H} - \frac{1}{2}\boldsymbol{grad}\, H^{2} \right].$$

According to (44.1) and with the following vectorial identity

$$(\boldsymbol{H} \cdot \boldsymbol{grad})\boldsymbol{H} - \frac{1}{2}\boldsymbol{grad}\, H^{2} = -\boldsymbol{H} \times \boldsymbol{rot}\, \boldsymbol{H} \tag{48.5}$$

in (48.4) finally it is obtained that

$$\frac{\partial \boldsymbol{v}}{\partial t} + (\boldsymbol{v} \cdot \boldsymbol{grad})\boldsymbol{v} = -\frac{1}{\rho}\boldsymbol{grad}\, p - \frac{1}{4\pi\rho}(\boldsymbol{H} \times \boldsymbol{rot}\, \boldsymbol{H}) + \frac{\eta}{\rho}\nabla^{2}\boldsymbol{v}$$
$$+ \frac{1}{\rho}\left(\zeta + \frac{1}{3}\eta\right)\boldsymbol{grad}\, div\, \boldsymbol{v}; \tag{48.6}$$

which is no other thing than the motion equation of MHD. If the fluid considered is incompressible, the mass density is a constant so that the first term of the right hand side of (48.6) is zero, in which case we have that

$$\frac{\partial \boldsymbol{v}}{\partial t} + (\boldsymbol{v} \cdot \boldsymbol{grad})\boldsymbol{v} = -\frac{1}{\rho}\boldsymbol{grad}\left(p + \frac{H^{2}}{8\pi}\right)$$
$$+ \frac{1}{4\pi\rho}(\boldsymbol{H} \cdot \boldsymbol{grad})\boldsymbol{H} + \vartheta\nabla^{2}\boldsymbol{v}; \tag{48.7}$$

where the definition (35.11) was used. This is the corresponding Navier-Stokes′ equation of MHD. In it, the term $H^{2}/8\pi$ is the hydrostatic magnetic pressure.

§49. Energy conservation law

The explicit form of hamiltonian density is obtained substituting in the general definition (19.14) the definitions (48.1) and (48.3); that is to say

$$\mathcal{H} = \frac{1}{2}\rho v^2 + \rho\square + \frac{H^2}{8\pi}\,. \tag{49.1}$$

On the other hand

$$\mathcal{D}\mathcal{H} = \frac{\partial}{\partial t}\left[\frac{1}{2}\rho v^2 + \rho\square + \frac{H^2}{8\pi}\right] + div\left[\rho\boldsymbol{v}\left(\frac{1}{2}v^2 + \square + \frac{H^2}{8\pi\rho}\right)\right]. \tag{49.2}$$

From the following vectorial identity

$$\frac{1}{4\pi}\left(H_i H_j v^i - \frac{1}{2}H^2\boldsymbol{v}\right) = -\frac{1}{4\pi}\boldsymbol{H}\times(\boldsymbol{v}\times\boldsymbol{H}), \tag{49.3}$$

and from the general form (35.6) of the viscosity stress tensor, the following expression can be calculated

$$-\frac{\partial}{\partial x^j}\left(\frac{\partial\lambda}{\partial u_{ij}}v^i\right) = -\frac{\partial}{\partial x^j}\left[p\boldsymbol{v} - \kappa\,\boldsymbol{grad}T - \sigma'_{ij}v^i + \frac{1}{4\pi}\boldsymbol{H}\times(\boldsymbol{v}\times\boldsymbol{H})\right].$$

On the other hand and given that for a conducting medium in a magnetic field it is fulfilled that

$$\boldsymbol{E} = \frac{c}{4\pi\sigma}\,\boldsymbol{rot\,H}, \tag{49.4}$$

the definition (47.6) can be written as follows

$$\boldsymbol{S} = -\frac{c^2}{16\pi^2\sigma}\boldsymbol{H}\times\boldsymbol{rot\,H}\,. \tag{49.5}$$

According to the most basic postulates of Theoretical Physics, the energy conservation law is a consequence of time uniformity. Thus, we say that a mechanical system is conservative if its lagrangian function is not an explicit function of time. For the problem being studied the former means that $\partial \mathcal{L}/\partial t=0$; so that in (47.9) it is finally obtained that

$$\frac{\partial}{\partial t}\left[\frac{1}{2}\rho v^2+\rho\square\right]=-div\left[\rho \boldsymbol{v}\left(\frac{1}{2}v^2+w'\right)+\frac{1}{4\pi}\boldsymbol{H}\times(\boldsymbol{v}\times\boldsymbol{H})\right.$$
$$\left.-\frac{c^2}{16\pi^2\sigma}\boldsymbol{H}\times\boldsymbol{rot}\,\boldsymbol{H}-\boldsymbol{v}\cdot\tilde{\sigma}'-\kappa\,\boldsymbol{grad}\,T\right]; \tag{49.6}$$

where $w'=\square+p/\rho$ is the generalized specific enthalpy.

The former equation is the explicit form of the energy conservation law of MHD.

§50. General equation of heat transfer

If the same procedure that was used in paragraph 37 is followed, it can be demonstrated that expanding the left hand side of the former expression the following is obtained

$$\frac{\partial}{\partial t}\left[\frac{1}{2}\rho v^2+\rho\square+\frac{H^2}{8\pi}\right]=-div\left[\rho\boldsymbol{v}\left(\frac{1}{2}v^2+w\right)\right]+\rho\boldsymbol{v}\cdot\frac{\partial\boldsymbol{v}}{\partial t}$$
$$+\rho T\frac{ds}{dt}+\boldsymbol{v}\cdot\boldsymbol{grad}\,p+\rho\boldsymbol{v}\cdot\boldsymbol{grad}\left(\frac{1}{2}v^2\right)+\frac{1}{4\pi}\boldsymbol{H}\cdot\frac{\partial\boldsymbol{H}}{\partial t}. \tag{50.1}$$

From (44.2) we have that

$$\frac{1}{4\pi}\boldsymbol{H}\cdot\frac{\partial\boldsymbol{H}}{\partial t}=\frac{1}{4\pi}\boldsymbol{H}\cdot[\nabla\times(\boldsymbol{v}\times\boldsymbol{H})]+\frac{c^2}{16\pi^2\sigma}\boldsymbol{H}\cdot(\nabla^2\boldsymbol{H}). \tag{50.2}$$

From the results obtained in paragraph 37 where all the terms of the expansion given in (50.1) referring to purely hydrodynamic quantities belonging to the viscous fluid were calculated; it can be demonstrated that

$$\rho T\frac{ds}{dt}-\sigma'_{ij}\frac{\partial v^i}{\partial x^j}-div(\kappa\,\mathbf{grad}T)+\frac{1}{4\pi}\mathbf{H}\cdot[\nabla\times(\mathbf{v}\times\mathbf{H})]$$
$$+\frac{c^2}{16\pi^2\sigma}\mathbf{H}\cdot(\nabla^2\mathbf{H})=-\frac{1}{4\pi}\nabla\cdot[\mathbf{H}\times(\mathbf{v}\times\mathbf{H})] \qquad (50.3)$$
$$+\frac{c^2}{16\pi^2\sigma}\nabla\cdot[\mathbf{H}\times(\nabla\times\mathbf{H})].$$

Let $\mathbf{B}_o\equiv\mathbf{v}\times\mathbf{H}$ be any vector so that

$$-\frac{1}{4\pi}\nabla\cdot[\mathbf{H}\times(\mathbf{v}\times\mathbf{H})]=\frac{1}{4\pi}\mathbf{H}\cdot[\nabla\times(\mathbf{v}\times\mathbf{H})]; \qquad (50.4)$$

where the vectorial identity

$$\nabla\cdot(\mathbf{A}\times\mathbf{B}_o)=\mathbf{B}_o\cdot(\nabla\times\mathbf{A})-\mathbf{A}\cdot(\nabla\times\mathbf{B}_o)$$

was used and the fact that

$$(\mathbf{v}\times\mathbf{H})\cdot(\nabla\times\mathbf{H})=0$$

because they are two vectors perpendicular between them so that their scalar product is zero.

If now in the second term of the right hand side of (50.3) the last vectorial definition is used again as well as the following result

$$\nabla\times(\nabla\times\mathbf{H})=\nabla\times(\nabla\cdot\mathbf{H})-\nabla^2\mathbf{H}=-\nabla^2\mathbf{H},$$

only because $\nabla\cdot\mathbf{H}=0$, the following is obtained

$$\nabla\cdot[\mathbf{H}\times(\nabla\times\mathbf{H})]=(\mathbf{rot}\,\mathbf{H})^2+\mathbf{H}\cdot(\nabla^2\mathbf{H}). \qquad (50.5)$$

With those last results substituted in (50.3) it is finally obtained that

$$\rho T\left[\frac{\partial s}{\partial t}+\boldsymbol{v}\cdot\boldsymbol{grad}\,s\right]=\sigma'_{ij}\frac{\partial v^i}{\partial x^j}+div\left(\kappa\,\boldsymbol{grad}\,T\right)$$
$$+\frac{c^2}{16\pi^2\sigma}\left(\boldsymbol{rot\,H}\right)^2; \tag{50.6}$$

and this is the corresponding general equation of heat transfer of MHD.

In it the term

$$\frac{j^2}{\sigma}=\frac{c^2}{16\pi^2\sigma}\left(\boldsymbol{rot\,H}\right)^2 \tag{50.7}$$

is the Joule´s heat per unit volume.

§51. Thermal equation of state

In order to complete the theoretical scheme and to be able to dispose of a set of MHD equations, it is required from a thermal equation of state to relate pressure with density and temperature of the fluid. For the physical conditions of the phenomenon it seems reasonable to suppose that the presence of the magnetic field must perform an important influence on the thermodynamic state of the conducting continuous medium; so that this presence must substantially modify such relationship. Therefore be

$$p=p(\rho,T,\boldsymbol{H}) \tag{51.1}$$

the required functional form. If the total differential of that relation is calculated, the following result is obtained

$$dp=c_o^2\,d\rho+\frac{\varepsilon}{k}dT+\frac{\boldsymbol{H}}{4\pi}\cdot d\boldsymbol{H}; \tag{51.2}$$

where

$$c_o^2=\left(\frac{\partial p}{\partial\rho}\right)_{T,\boldsymbol{H}} \tag{51.3}$$

is the square of the sound velocity in the medium. Moreover

$$\frac{\varepsilon}{k} = \left(\frac{\partial p}{\partial T}\right)_{\rho,\boldsymbol{H}}; \tag{51.4}$$

with

$$\varepsilon = \frac{1}{V}\left(\frac{\partial V}{\partial T}\right) \tag{51.5}$$

the volume coefficient of expansion and

$$k = -\frac{1}{V}\left(\frac{\partial V}{\partial p}\right) \tag{51.6}$$

the isothermal compressibility that as it is well known is always positive.

On the other hand

$$\frac{\boldsymbol{H}}{4\pi} \equiv \left(\frac{\partial p}{\partial \boldsymbol{H}}\right)_{\rho,T} \tag{51.7}$$

is the change in the hydrostatic pressure due to the presence of the external magnetic field. In this case the last term of the right hand side of (51.2) can be written as follows

$$\frac{\boldsymbol{H}}{4\pi} \cdot d\boldsymbol{H} = d\left(\frac{H^2}{8\pi}\right). \tag{51.8}$$

According to all the former results it is proposed that the MHD generalized thermal equation of state has the following form

$$p = c_o^2 \rho + \frac{\varepsilon}{k} T + \frac{H^2}{8\pi}; \tag{51.9}$$

that with the aid of the expression (11.9) can be written as follows

$$p = \frac{c_o^2 \rho_o}{J} \boldsymbol{u} \cdot \boldsymbol{grad}\, J + \frac{\varepsilon}{k} T + \frac{H^2}{8\pi}. \tag{51.10}$$

In the former relationship the last term, as it has already been said is the hydrostatic magnetic pressure. It is the contribution of the external magnetic field to pressure; whereas the other two terms represent the contribution of geometrical changes and of temperature.

Selected Topics

Magnetohydrodynamics and plasma physics

Magnetohydrodynamics and plasma physics† are scientific disciplines that deal with the behavior of conducting gases and liquids that move in electromagnetic fields. In particular, the electric conduction occurs when there are free or cuasi free electrons which move under the action of applied fields. In a solid substance, electrons are actually bound, but can move considerable distance on atomic scale within the crystal lattice before colliding.

Dynamical effects such as conduction currents and Hall effect are observed when fields are applied to a solid conductor, but mass motion does not in general occur. Those effects act over the crystal lattice in the form of mechanical stresses. In the case of a fluid, the field acts on both electrons and ionized atoms to produce dynamical effects, including bulk motion of the medium itself. The dynamical effects can be taken up as internal stresses of electromagnetic character that are added to the mechanical stresses produced by the mass motion, so that added effects produce noticeable modifications in the present electromagnetic field. It deals therefore with a complex dynamical interaction between fields and matter.

The distinction between Magnetohydrodynamics and Plasma Physics is not a sharp one. A separation can not be clearly established between them. One way of seeing the distinction is to look at the way in which the Ohm´s law is established for a conducting substance.

In the more simple model that can be constructed, the electrons are imagined to be accelerated by the present fields, but without collisions altering

† *The word plasma emerges in science in 1922 as a suggestion of I. Langmuir, who was carrying out certain experiments consisting of electric discharges through gases and he required a name for the gas partially ionized that was produced. As the gas contained high velocity electrons, molecules and impurities, it reminded him of the way the sanguineous plasma carried red and white globulets and germs, for this reason he proposed this term. At present the word is accepted to name the fourth state of matter.*

their direction in such a way that their motion in the direction of the field is opposed by an effective frictional force; so that this law just represents a balance between the applied force and the frictional drag. When the frequency of the applied fields is comparable to the collision frequency, the electrons have time to accelerate and decelerate between collisions. Then, inertial effects enter and the conductivity becomes complex. At these same frequencies the description of collisions in terms of a frictional force tends to lose its validity due to the fact that the whole process becomes more complicated. At frequencies well above the collision frequency, another thing happens. Due to the acceleration produced by the induced electric field, the electrons and ions are accelerated in opposite directions and tend to separate, propitiating this way that strong electrostatic restoring forces are set up between these charges of opposite sign. The resulting process gives place to high-frequency oscillations known by the name of plasma oscillation; that must be distinguished from those of lower frequency which involve motion of the fluid, but no charge separation. These low-frequency oscillations are called magnetohydrodynamic waves.

In conducting fluids or dense ionized gases the collition frequency is sufficiently high even for very good conductors, in such a way that there is a wide frequency range where Ohm´s law in its simplest form is valid. Under the action of applied fields, the electrons and ions move in such a way that there is no separation of charge. Electric fields arise from motion of the fluid which causes a current flow. Also, they arise as a result of time-varying magnetic fields or charge distributions external to the fluid. The mechanical motion of the system can therefore be described in terms of a single conducting fluid with the usual set of hydrodynamic variables such as mass density, velocity field, and others.

Moreover, at low frequencies it is customary to neglect the displacement current in Ampere´s law. This approximation is known by the name of Magnetohydrodynamics.

In less dense gases the collision frequency is smaller. However, there may still be a low-frequency domain where the magnetohydrodynamic equations are applicable to quasi-stationary processes, as it occurs in Astrophysics. At higher frequencies, the neglect of charge separation and the displacement current effects are not allowable. In these cases, the separate inertial effects of charge must be included in the description of the motion. This is the domain of Plasma Physics. A plasma is by definition, an ionized gas in which the length which divides the small-scale individual particle behavior from the large-scale collective behavior is small compared to certain characteristic magnitudes that define the plasma criteria. These are the Debye´s lenght or Debye screening radius λ_D, the number of particles in the Debye sphere N_D, and a third condition that deals with the collisions. If ω is the frequency of typical plasma oscillations and τ is the mean time between collisions with neutral atoms, it is required that $\omega\tau$ 1 for the gas to behave like a plasma rather than as a neutral gas.

Generalized Bernoulli´s equation

Let us consider the motion equation for a perfect fluid that flows in an external magnetic field. If the viscosity is irrelevant, from the MHD Navier-Stokes′ equation (48.7) the following result is obtained

$$\frac{\partial \mathbf{v}}{\partial t} + (\mathbf{v}\cdot\mathbf{grad})\mathbf{v} = -\frac{1}{\rho}\,\mathbf{grad}\left(p + \frac{H^2}{8\pi}\right) + \frac{1}{4\pi\rho}(\mathbf{H}\cdot\mathbf{grad})\mathbf{H}\,; \qquad (1)$$

where the vectorial identity (48.5) was used again. This is the generalized Euler′s equation. If the flux is steady, the velocity field only depends on the coordinates and not on time, so that $\partial\mathbf{v}/\partial t=0$. As on the other hand

$$(\mathbf{v}\cdot\mathbf{grad})\mathbf{v} = \frac{1}{2}\,\mathbf{grad}\,v^2 - \mathbf{v}\times\mathbf{rot}\,\mathbf{v}\,,$$

we have that

$$\frac{1}{2}\,\mathbf{grad}\,v^2 - \mathbf{v}\times\mathbf{rot}\,\mathbf{v} + \frac{1}{\rho}\,\mathbf{grad}\left(p + \frac{H^2}{8\pi}\right) = \frac{1}{4\pi\rho}(\mathbf{H}\cdot\mathbf{grad})\mathbf{H}\,. \qquad (2)$$

From the thermodynamic relationship

$$dw = Tds + Vdp\,, \qquad (3)$$

where $s(\rho, \square t)$ is the specific entropy, T the temperature, p the pressure, $V=1/\rho$ the specific volume, and w is the specific enthalpy (21.6), the following result is obtained

$$\frac{1}{\rho}\,\textbf{\textit{grad}}\,p = \textbf{\textit{grad}}\,w;$$

due to the fact that the entropy by unit mass is not an explicit function of the coordinates. Since for the present case there is an external magnetic field $\boldsymbol{H}(\boldsymbol{x},t)$, the total pressure takes the following form

$$p' = p + p_M;$$

where $p_M = H^2/8\pi$ is the hydrostatic magnetic pressure. Then, the thermodynamic relationship (3) becomes

$$dw' = Tds + Vdp'; \tag{4}$$

with

$$w' = w + \frac{H^2}{8\pi\rho}$$

the generalized specific enthalpy. Then, from (4) it is obtained that

$$\textbf{\textit{grad}}\,w' = \frac{1}{\rho}\,\textbf{\textit{grad}}\,p'$$

and (2) takes the following form

$$\textbf{\textit{grad}}\left(\frac{1}{2}v^2 + w'\right) = \boldsymbol{v}\times\textbf{\textit{rot}}\,\boldsymbol{v} + \frac{1}{4\pi\rho}(\boldsymbol{H}\cdot\textbf{\textit{grad}})\boldsymbol{H}. \tag{5}$$

Let us consider what occurs along a streamline. The scalar product of equation (5) with a unitary vector tangent to the streamline in each point, has the following result. The projection of the gradient in any direction is the derivative in that direction, so that the projection of the term $\textbf{\textit{grad}}(1/2v^2+w')$ in the $\boldsymbol{m}$ direction is equal to

$$\frac{\partial}{\partial m}\left(\frac{1}{2}v^2 + w'\right).$$

On the other hand, the vector $\boldsymbol{v} \times \boldsymbol{rot}\ \boldsymbol{v}$ is perpendicular to $\boldsymbol{v}$ and its projection in the $\boldsymbol{m}$ direction is zero because $\boldsymbol{v}$ and $\boldsymbol{m}$ are parallel between them; so that the angle formed by $\boldsymbol{m}$ and $\boldsymbol{v} \times \boldsymbol{rot}\ \boldsymbol{v}$ is of 90° and *cosine* 90° = 0. Now, $\boldsymbol{H}$ and $\boldsymbol{grad}\ \boldsymbol{H}$ point out in the same direction, that is to say, they are vectors parallel between them. But $\boldsymbol{H}$ and $\boldsymbol{grad}\ \boldsymbol{H}$ are perpendicular to $\boldsymbol{v}$ and $\boldsymbol{v}$ is parallel to $\boldsymbol{m}$ so that the projection of $(\boldsymbol{H} \cdot \boldsymbol{grad})\boldsymbol{H}$ in the $\boldsymbol{m}$ direction is zero for the same previous reason. Therefore, from equation (5) it is obtained that

$$\frac{\partial}{\partial m}\left(\frac{1}{2}v^2 + w'\right) = 0. \tag{6}$$

From here if follows that along a streamline

$$\frac{1}{2}v^2 + w' = constant. \tag{7}$$

In general, the constant takes different values for different streamlines. The relationship (7) is the generalized Bernoulli's equation.

If the flux occurs in a gravitational field, on the right hand side of (5) the acceleration $\boldsymbol{g}$ of gravity must be added. According to what was obtained in the *Selected Topics* of Chapter IV, the projection of $\boldsymbol{g}$ over $\boldsymbol{m}$ is equal to $-gdz/dm$, so that in (6) we have that

$$\frac{\partial}{\partial m}\left(\frac{1}{2}v^2 + w' + gz\right) = 0,$$

and the corresponding generalized Bernoulli's equation takes the following form

$$\frac{1}{2}v^2 + w' + gz = constant. \tag{8}$$

References

1. Fierros Palacios, Angel "Las ecuaciones de balance para un fluido viscoso a partir de un principio variacional Tipo Hamilton". Rev. Mex. de Fís., 38, No. 4 (1982) 518-531.
2. Fierros Palacios, Angel. "Las ecuaciones diferenciales de campo de la magnetohidrodinámica". Rev. Mex. de Fís., 44, No. 2, (1998) 120-127.
3. Jackson, J.D. "Classical Electrodynamics". John Wiley & Sons, Inc. New York, London (1962).
4. Landau, L.D. and Lifshitz, E.M. "Electrodynamics of Continuous Media". Addison-Wesley Publishing Co. (1960).
5. Zemansky M.W. y Dittman, R.H. "Calor y Termodinámica". McGraw-Hill 6a. edición (1990).

Chapter IX

Potential Flow in a Magnetic Field

§52. The equation of motion

The propagation of an oscillatory disturbance of small amplitude generated in a conducting homogeneous and continuous medium that moves in a uniform magnetic field, can be studied within the theoretical frame developed in this book. In order to properly setup the problem, it is necessary to propose some approximations. Thus, it is assumed that the viscosity, the thermal conductivity, and the electric resistance of the continuous medium considered are so small that their effect, due to process of energy dissipation on the propagation of the perturbation, may be neglected in a first approach. If these general conditions are satisfied, the small perturbations will be propagated as undamped waves. When all dissipative terms are ignored, the basic equations of MHD can be written as follows

$$\frac{\partial \boldsymbol{H}}{\partial t} = \boldsymbol{rot}\left(\boldsymbol{v} \times \boldsymbol{H}\right), \tag{52.1}$$

together with the general condition to any magnetic field given by the relationship (44.1). The motion equation for this case takes the following form

$$\frac{\partial \boldsymbol{v}}{\partial t} + \left(\boldsymbol{v} \cdot \boldsymbol{grad}\right)\boldsymbol{v} = -\frac{1}{\rho}\boldsymbol{grad}\, p + \frac{1}{4\pi\rho}\left(\boldsymbol{rot}\,\boldsymbol{H}\right) \times \boldsymbol{H}. \tag{52.2}$$

The thermal equation of state is obtained from the general form (51.10) for the case in which the thermal conductivity is irrelevant, that is to say

$$p = \frac{c_o^2 \rho_o}{J} \boldsymbol{u} \cdot \boldsymbol{grad}\, J' + \frac{H^2}{8\pi}. \tag{52.3}$$

For the proposed conditions the MHD generalized equation of heat transfer (50.6) is reduced to the condition for adiabatic flow (21.8). Additionally, we have the continuity equation (9.2) as a suplementary condition externally imposed on the system.

Let us consider now that the Jacobian can be written as it was done in paragraph 29; whereas for the magnetic field we have that

$$\boldsymbol{H}(\boldsymbol{x},t) = \boldsymbol{H}_o + \boldsymbol{h}, \tag{52.4}$$

where $\boldsymbol{H}_o$ is a uniform magnetic field and $\boldsymbol{h}(\boldsymbol{x}, t)$ a small perturbation such that $\boldsymbol{h} \ll \boldsymbol{H}_o$.

In what follows the decompositions (22.2) and (22.3) for the hydrostatic pressure and the mass density will be used. A periodical motion of small amplitude in a conducting continuous medium that moves in an external magnetic field satisfaying the general conditions that have been imposed, occurs at constant specific entropy, in such a way that the flux is isentropic. Therefore the changes in pressure and mass density are related to the external magnetic field as follows

$$p' = c_o^2 \rho' + \frac{\boldsymbol{H}_o \cdot \boldsymbol{h}}{4\pi}. \tag{52.5}$$

It is convenient to express the former relation in terms of the Jacobian perturbation. According to the result (29.1)

$$p' = -\frac{c_o^2 \rho_o}{J_o} J' + \frac{\boldsymbol{H}_o \cdot \boldsymbol{h}}{4\pi}. \tag{52.6}$$

The gradient of the last expression is equal to

$$\begin{aligned} -\frac{1}{\rho_o} \boldsymbol{grad}\, p' &= \boldsymbol{grad}\left(\frac{c_o^2 J'}{J_o}\right) - \frac{1}{4\pi\rho_o}(\boldsymbol{H}_o \cdot \boldsymbol{grad})\boldsymbol{h} \\ &= \boldsymbol{grad}\left[\frac{c_o^2 J'}{J_o} - \frac{\boldsymbol{H}_o \cdot \boldsymbol{h}}{4\pi\rho_o}\right] + \frac{1}{4\pi\rho_o} \boldsymbol{H}_o \times \boldsymbol{rot}\, \boldsymbol{h}; \end{aligned} \tag{52.7}$$

where the vectorial identity (48.5) was used. Thus, in the motion equation (52.2) the following is obtained

$$\frac{\partial \boldsymbol{v}'}{\partial t}+\left(\boldsymbol{v}_o\cdot \boldsymbol{grad}\right)\boldsymbol{v}'=\boldsymbol{grad}\left[\frac{c_o^2 J'}{J_o}-\frac{\boldsymbol{H}_o\cdot \boldsymbol{h}}{4\pi\rho_o}\right]. \tag{52.8}$$

As in the simple case of potential flow, $\boldsymbol{v}'$ which is the small perturbation in the velocity field, can be expressed in terms of the velocity potential $\phi(\boldsymbol{x}, t)$ according to definition (22.1). Again, as it was done in Chapter V, the term $(\boldsymbol{v}_o\cdot \boldsymbol{grad})\boldsymbol{v}'$ will be neglected because it is of higher order in the approximations and we use again $\boldsymbol{v}$ insted of $\boldsymbol{v}'$ without confusion. Then in the former equation the following result is obtained

$$\boldsymbol{grad}\left[\frac{\partial\phi}{\partial t}+\frac{\boldsymbol{H}_o\cdot \boldsymbol{h}}{4\pi\rho_o}-\frac{c_o^2 J'}{J_o}\right]=0. \tag{52.9}$$

This equation can be integrated to obtain that

$$\frac{\partial\phi}{\partial t}=\frac{c_o^2 J'}{J_o}-\frac{\boldsymbol{H}_o\cdot \boldsymbol{h}}{4\pi\rho_o}; \tag{52.10}$$

where again the constant of integration has been considered as zero without lose of generality. If the last equation is derived again with respect to time, the following is obtained

$$\frac{\partial^2\phi}{\partial t^2}=\frac{c_o^2}{J_o}\frac{\partial J'}{\partial t}-\frac{\boldsymbol{H}_o}{4\pi\rho_o}\cdot\frac{\partial \boldsymbol{h}}{\partial t}. \tag{52.11}$$

Now and according to (52.1)

$$\frac{\boldsymbol{H}_o}{4\pi\rho_o}\cdot\frac{\partial \boldsymbol{h}}{\partial t}=\frac{\boldsymbol{H}_o}{4\pi\rho_o}\cdot\left[\boldsymbol{rot}\left(\boldsymbol{v}\times \boldsymbol{H}_o\right)\right].$$

However,

$$
\begin{aligned}
\boldsymbol{rot}\left(\boldsymbol{v}\times\boldsymbol{H}_o\right) &= \left(\boldsymbol{H}_o\cdot\boldsymbol{grad}\right)\boldsymbol{v}-\boldsymbol{H}_o div\,\boldsymbol{v}-\left(\boldsymbol{v}\cdot\boldsymbol{grad}\right)\boldsymbol{H}_o \\
&+\boldsymbol{v}\,div\,\boldsymbol{H}_o = \left(\boldsymbol{H}_o\cdot\boldsymbol{grad}\right)\boldsymbol{v}-\boldsymbol{H}_o div\,\boldsymbol{v}.
\end{aligned}
\tag{52.12}
$$

The second term of the right hand side of (52.12) is proportional to $\partial\rho'/\partial t$, whereas the first one has the same form as the term $(\boldsymbol{v}_o\cdot\boldsymbol{grad})\boldsymbol{v}'$. However $(\boldsymbol{H}_o\cdot\boldsymbol{grad})\boldsymbol{v}\approx\boldsymbol{grad}(\boldsymbol{H}_o\ \boldsymbol{v})$, where an integration by parts was made and the definition (44.1) was used. The remaining term is of higher order in the approximations whereas $-\boldsymbol{H}_o\,div\,\boldsymbol{v}=(\boldsymbol{H}_o/\rho_o)\,\partial\rho'/\partial t$. In this case, the first term is neglected so that only the term of the divergence survives. Therefore in a first approach

$$\frac{\boldsymbol{H}_o}{4\pi\rho_o}\cdot\frac{\partial\boldsymbol{h}}{\partial t}=-\frac{H_o^2}{4\pi\rho_o}div\,\boldsymbol{v}=-\frac{H_o^2}{4\pi\rho_o}\nabla^2\phi. \tag{52.13}$$

Thus, in (52.11) we have that

$$\frac{\partial^2\phi}{\partial t^2}-\frac{c_o^2}{J_o}\frac{\partial J'}{\partial t}=\frac{H_o^2}{4\pi\rho_o}\nabla^2\phi. \tag{52.14}$$

On the other hand from the equations (29.5) and (22.1) it is easy to see that

$$\frac{\partial J'}{\partial t}=J_o\nabla^2\phi. \tag{52.15}$$

In this case in (52.14) the motion equation sought is finally obtained:

$$\nabla^2\phi-\frac{1}{a_\circ^2}\frac{\partial^2\phi}{\partial t^2}=0. \tag{52.16}$$

This is the wave equation for the velocity potential that governs the propagation of a periodical perturbation of small amplitude in a conducting perfect fluid that moves in an external magnetic field. In it

$$a_\circ=\sqrt{c_o^2+v_a^2} \tag{52.17}$$

is the phase velocity of an MHD longitudinal wave, whereas

$$\boldsymbol{v}_a = -\frac{\boldsymbol{H}_o}{\sqrt{4\pi\rho_o}} \qquad (52.18)$$

is Alfven's vectorial velocity in terms of the uniform magnetic field $\boldsymbol{H}_o$.

§53. Field differential equations

Due to the fact that the influence of external magnetic field is only present in the specific internal energy of the fluid, the problem to obtain the field differential equations is reduced to repeating the variational calculus made for potential flow in Chapter V. On the other hand and given that in the former chapter it was demonstrated that the local variation of $\boldsymbol{B}$ is zero, the lagrangian density, as well as the specific lagrangian are the same functions that we use for the description of potential flow within the theoretical frame of Hamilton-Type Variational Principle; together with another term which takes into account the external magnetic field. Hence, for the potential flow in an external magnetic field, the field equations are the relations (23.6); that is to say

$$\frac{\partial}{\partial x^i}\left[\frac{\partial \lambda}{\partial(\nabla_i \phi)}\right] + \frac{\partial}{\partial t}\left[\frac{\partial \lambda}{\partial\left(\frac{\partial \phi}{\partial t}\right)}\right] = 0 .$$

The difference between both problems lies in the functionality of the respective lagrangians. In one case the external magnetic field must be included but not in the other[†].

§54. The lagrangian functions

To obtain the explicit form of these functions we must start from the definitions (48.1) to (48.3) of the previous chapter. At the passing of disturbance

---

[†] *The lagrangian density and the specific lagrangian must contain the magnetic induction in their argument. However and given that $\delta\boldsymbol{B}=0$, these functions and the calculus of variations lead again to the field equations* (23.6)

the lagrangian density (48.1) changes according to the relation (25.1). Then it is clear that

$$\circ_o = \rho_o\left(\frac{1}{2}v^2 - \square_o - \frac{H_o^2}{8\pi\rho_o}\right) \tag{54.1}$$

is the equilibrium lagrangian density, whereas

$$\circ' = -\rho_o \square' \tag{54.2}$$

is its variation at a constant specific entropy. On the other hand and according to (22.2) and (52.3) it is easy to become convinced that

$$p_o = c_o^2\rho_o + \frac{H_o^2}{8\pi} \tag{54.3}$$

whereas from (29.1) and (52.10) we have that

$$\rho' = -\frac{\rho_o}{c_o^2}\left(\frac{\partial\phi}{\partial t}\right) - \frac{\boldsymbol{H}_o\cdot\boldsymbol{h}}{4\pi c_o^2}\,. \tag{54.4}$$

With these results substituted in the equation (25.9) and before doing some algebra, the following is obtained

$$\square = \frac{1}{c_o^2}\left[\left(\frac{\partial\phi}{\partial t}\right)^2 + \left(\frac{\partial\phi}{\partial t}\right)\left(\frac{\boldsymbol{H}_o\cdot\boldsymbol{h}}{2\pi\rho_o}\right)\right]. \tag{54.5}$$

Then and according to the definition (25.6) the following expression for potential energy density is obtained

$$u = \frac{\rho_o}{2c_o^2}\left[\left(\frac{\partial\phi}{\partial t}\right)^2 + \left(\frac{\partial\phi}{\partial t}\right)\left(\frac{\boldsymbol{H}_o\cdot\boldsymbol{h}}{2\pi\rho_o}\right)\right]. \tag{54.6}$$

According to the definition (22.1) the corresponding kinetic energy density turns out to be the relation (25.12); that is

$$t = \frac{1}{2}\rho_o \left|\boldsymbol{grad}\,\phi\right|^2 .$$

Consequently and on a first approach the form of lagrangian density for the potential flow in an external magnetic field is the following

$$\circ = \frac{1}{2}\rho_o \left[\left|\boldsymbol{grad}\,\phi\right|^2 - \frac{1}{c_o^2}\left(\frac{\partial \phi}{\partial t}\right)^2 - \frac{1}{c_o^2}\left(\frac{\partial \phi}{\partial t}\right)\left(\frac{\boldsymbol{H}_o \cdot \boldsymbol{h}}{2\pi \rho_o}\right) \right] ; \qquad (54.7)$$

whereas

$$\lambda = \frac{1}{2}\left[\left|\boldsymbol{grad}\,\phi\right|^2 - \frac{1}{c_o^2}\left(\frac{\partial \phi}{\partial t}\right)^2 - \frac{1}{c_o^2}\left(\frac{\partial \phi}{\partial t}\right)\left(\frac{\boldsymbol{H}_o \cdot \boldsymbol{h}}{2\pi \rho_o}\right) \right] \qquad (54.8)$$

is the corresponding specific lagrangian. Thus, in the field differential equations (23.6) we have that

$$\frac{\partial}{\partial x^i}\left[\frac{\partial \lambda}{\partial(\nabla_i \phi)}\right] = \nabla^2 \phi$$

and

$$\frac{\partial}{\partial t}\left[\frac{\partial \lambda}{\partial\left(\dfrac{\partial \phi}{\partial t}\right)}\right] = -\frac{1}{c_o^2}\left\{\left[\frac{\partial^2 \phi}{\partial t^2}\right] - \frac{H_o^2}{4\pi\rho_o}\nabla^2 \phi\right\}.$$

Adding those results, the wave equation (52.16) is recovered. On the other hand and according to (54.4) and (52.5) it is easy to see that

$$-\frac{1}{c_o^2}\left(\frac{\partial \phi}{\partial t}\right)^2 = -\frac{p'^2}{\rho_o^2 c_o^2} \qquad (54.9)$$

and obviously,

$$-\frac{1}{c_o^2}\left(\frac{\partial \phi}{\partial t}\right)\left(\frac{\boldsymbol{H}_o \cdot \boldsymbol{h}}{2\pi\rho_o}\right)=\frac{p'}{\rho_o c_o^2}\left(\frac{\boldsymbol{H}_o \cdot \boldsymbol{h}}{2\pi\rho_o}\right); \tag{54.10}$$

so that the following is fulfilled

$$\frac{\partial \phi}{\partial t}=-\frac{p'}{\rho_o};$$

which is no other thing than the result (22.8). On the other hand, it is evident that

$$v^2=\left|\boldsymbol{grad}\,\phi\right|^2.$$

With all the former results substituted in the relation (54.8) the following equivalent expression for the specific lagrangian is obtained

$$\lambda_{p'}=\frac{1}{2}\left[v^2-\frac{p'^2}{\rho_o^2 c_o^2}+\frac{p'}{\rho_o c_o^2}\left(\frac{\boldsymbol{H}_o \cdot \boldsymbol{h}}{2\pi\rho_o}\right)\right]. \tag{54.11}$$

This is a specific lagrangian in terms of the small variation of pressure, in the velocity perturbation, in the uniform magnetic field, and in the small perturbation ***h***. According to (22.1) and (22.8), in the equations (23.6) it can be seen that in terms of perturbations in the velocity field and the pressure, the following totally equivalent field differential equation can be obtained

$$\frac{\partial}{\partial x^i}\left[\frac{\partial \lambda_{p'}}{\partial v^i}\right]-\frac{\partial}{\partial t}\left[\frac{\partial \lambda_{p'}}{\partial\left(\frac{p'}{\rho_o}\right)}\right]=0. \tag{54.12}$$

It can be demonstrated that if in this last equation the specific lagrangian (54.11) is substituted, the linearized continuity equation (22.10) is obtained. In fact,

$$\frac{\partial}{\partial x^i}\left[\frac{\partial \lambda_{p'}}{\partial v^i}\right] = div\,\mathbf{v}$$

whereas

$$\frac{\partial}{\partial t}\left[\frac{\partial \lambda_{p'}}{\partial\left(\dfrac{p'}{\rho_o}\right)}\right] = -\frac{1}{\rho_o}\frac{\partial}{\partial t}\left[\frac{p'}{c_o^2} - \frac{\boldsymbol{H}_o \cdot \boldsymbol{h}}{4\pi c_o^2}\right] = -\frac{1}{\rho_o}\frac{\partial \rho'}{\partial t}$$

as it is easy to see from (52.5)

§55. The energy conservation law

In order to obtain the energy conservation law we can start from the results previously obtained in Chapter V, given that for the same reasons mentioned in paragraph 53, the calculus of temporary variations leads directly to the energy balance equation (27.4) for the potential flow, that is to say[†]

$$\frac{\partial \mathcal{H}_\rho}{\partial t} - div\left(\rho_o^2 \frac{\partial \lambda_\rho}{\partial \rho'}\mathbf{v}\right) = 0\,;$$

where the lagrangian density $\circ_\rho$ and the specific lagrangian λ_ρ are given in their functional form by the relations (26.1) and (26.2) respectively. Now and doing some algebra it can be demonstrated that

$$-\frac{1}{c_o^2}\left(\frac{\partial \phi}{\partial t}\right)^2 - \frac{1}{c_o^2}\left(\frac{\partial \phi}{\partial t}\right)\left(\frac{\boldsymbol{H}_o \cdot \boldsymbol{h}}{2\pi\rho_o}\right) = -\frac{c_o^2 \rho'^2}{\rho_o^2} + \left(\frac{\boldsymbol{H}_o \cdot \boldsymbol{h}}{4\pi\rho_o c_o}\right)^2 . \quad (55.1)$$

Then in terms of the velocity field variations and those of the mass density, the explicit form of the specific lagrangian is

---

† *For this case, in* (45.1) *it is obtained that* **rot** $\boldsymbol{E}_o$=0; *with* $\boldsymbol{E}_o \gg \boldsymbol{e}$ *a uniform electric field. Notice that* $\boldsymbol{E}=\boldsymbol{E}_o+\boldsymbol{e}$. *Consequently,* $\partial \boldsymbol{B}/\partial t$=0 *and clearly,* $c/4\pi$ *div* $(\boldsymbol{E}_o \times \boldsymbol{H}_o)$ = *div* $\boldsymbol{S}_o$=0.

$$\lambda_{\rho} = \frac{1}{2}\left[v^2 - \frac{c_o^2 \rho'^2}{\rho_o^2} + \left(\frac{\boldsymbol{H}_o \cdot \boldsymbol{h}}{4\pi \rho_o c_o} \right)^2 \right]; \tag{55.2}$$

whereas the corresponding lagrangian density is

$$\circ_{\rho} = \frac{1}{2}\rho_o \left[v^2 - \frac{c_o^2 \rho'^2}{\rho_o^2} + \left(\frac{\boldsymbol{H}_o \cdot \boldsymbol{h}}{4\pi \rho_o c_o} \right)^2 \right]. \tag{55.3}$$

In consequence the respective explicit form of hamiltonian density is

$$\mathcal{H}_{\rho} = \frac{1}{2}\rho_o v^2 + \frac{c_o^2 \rho'^2}{2\rho_o} - \frac{1}{2\rho_o}\left(\frac{\boldsymbol{H}_o \cdot \boldsymbol{h}}{4\pi c_o} \right)^2 . \tag{55.4}$$

On the other hand,

$$-\frac{\partial}{\partial x^i}\left[\rho_o^2 \frac{\partial \lambda_{\rho}}{\partial \rho'} v^i \right] = \frac{\partial}{\partial x^i}\left(c_o^2 \rho' v^i \right) = \frac{\partial}{\partial x^i}\left[\left(p' - \frac{\boldsymbol{H}_o \cdot \boldsymbol{h}}{4\pi} \right) v^i \right]; \tag{55.5}$$

as it is easy to see from (52.5). With all the former results substituted in the equation (27.4) it is finally obtained that

$$\frac{\partial}{\partial t}\left[\frac{1}{2}\rho_o v^2 + \frac{c_o^2 \rho'^2}{2\rho_o} - \frac{1}{2\rho_o}\left(\frac{\boldsymbol{H}_o \cdot \boldsymbol{h}}{4\pi c_o} \right)^2 \right] = -div\left[\left(p' - \frac{\boldsymbol{H}_o \cdot \boldsymbol{h}}{4\pi} \right) \boldsymbol{v} \right]. \tag{55.6}$$

This is the analytical expression for the energy conservation law for the waves originating in a conducting continuous medium that moves in an external magnetic field. As in the case of the simple potential flow, this law is valid for the flux at any instant.

Selected Topics

The drift of the lines of force

Given that the motion of a perfect fluid occurs at constant specific entropy, the thermodynamic relation for the change in the specific enthalpy $dw = Vdp$

can be utilized in that situation to write Euler's equation (20.5) in terms of the velocity field only. As $\mathbf{grad}\, w = \mathbf{grad}\, p/\rho$ we have that

$$\frac{\partial \mathbf{v}}{\partial t} + \mathbf{grad}\left(\frac{1}{2}v^2\right) - \mathbf{v} \times \mathbf{rot}\ \mathbf{v} = -\mathbf{grad}\, w\,; \tag{1}$$

where the vectorial identity

$$(\mathbf{v} \cdot \mathbf{grad})\mathbf{v} = \mathbf{grad}\left(\frac{1}{2}v^2\right) - \mathbf{v} \times \mathbf{rot}\ \mathbf{v}$$

was used. Because $\mathbf{rot}\,\mathbf{grad} = 0$, from equation (1) the following result is obtained

$$\frac{\partial \boldsymbol{\Omega}}{\partial t} = \mathbf{rot}\ (\mathbf{v} \times \boldsymbol{\Omega}), \tag{2}$$

with

$$\boldsymbol{\Omega} \equiv \frac{1}{2}\mathbf{rot}\ \mathbf{v} \tag{3}$$

the vorticity. The relationship (2) is the vortex equation in the theory of non viscous fluids flux. Its physical interpretation implies the fact that the vortex lines are dragged by the fluid motion as if they were frozen in it.

Let us consider now the equation (44.2) for the case in which the electric conductivity σ is so big that the second term of the right hand side of that expression can be neglected, and then the temporary behavior of the magnetic field is given by

$$\frac{\partial \mathbf{H}}{\partial t} = \mathbf{rot}(\mathbf{v} \times \mathbf{H}). \tag{4}$$

When comparing the relations (2) and (4) we see that both equations are identical in form, so that it can be assumed that also the magnetic lines of force are frozen in the conducting fluid and they are dragged by it as it flows.

When assuming that the electric conductivity of the medium is infinitely large, the velocity $\boldsymbol{v}$ of the lines of force can be calculated. In general, this velocity is perpendicular to $\boldsymbol{H}$. From the equation (45.1) which relates the variable magnetic and electric fields, the following is obtained

$$\frac{\partial \boldsymbol{H}}{\partial t} = -c\,\textbf{rot}\,\boldsymbol{E}\,, \tag{5}$$

with $\boldsymbol{E}(\boldsymbol{x},t)$ the induced electric field and c the velocity of light in the empty space. With this result substituted in (4) we have that

$$\textbf{rot}\left(c\boldsymbol{E} + \boldsymbol{v} \times \boldsymbol{H}\right) = 0\,.$$

In consequence, under the action of the fields $\boldsymbol{E}$ and $\boldsymbol{H}$ the fluid moves in such a way that the following relation is fulfilled†

$$\boldsymbol{E} + \frac{1}{c}\boldsymbol{v} \times \boldsymbol{H} = 0\,. \tag{6}$$

If this result is multiplied vectorialy by $\boldsymbol{H}$, the following is obtained

$$c\left(\boldsymbol{H} \times \boldsymbol{E}\right) + \boldsymbol{H} \times \left(\boldsymbol{v} \times \boldsymbol{H}\right) = 0\,. \tag{7}$$

According to the vectorial identity

$$\boldsymbol{a} \times \boldsymbol{b} \times \boldsymbol{c} = \boldsymbol{b}\left(\boldsymbol{a} \cdot \boldsymbol{c}\right) - \boldsymbol{c}\left(\boldsymbol{a} \cdot \boldsymbol{b}\right),$$

it can be demonstrated that

$$\boldsymbol{H} \times \left(\boldsymbol{v} \times \boldsymbol{H}\right) = \boldsymbol{v}\left(\boldsymbol{H} \cdot \boldsymbol{H}\right) - \boldsymbol{H}\left(\boldsymbol{H} \cdot \boldsymbol{v}\right) = H^2 \boldsymbol{v}$$

only, due to the fact that if $\boldsymbol{H}$ and $\boldsymbol{v}$ are perpendicular between them, their scalar product is zero. Consequently from (7) the following result is obtained

---

† *The expression enclosed in the parenthesis is just* $\boldsymbol{j}/\sigma$, *with* $\boldsymbol{j}$ *the current density. This term becomes zero when* $\sigma \to \infty$.

$$\mathbf{v} = c\frac{(\mathbf{E} \times \mathbf{H})}{H^2}. \tag{8}$$

This relationship is known as the drift $\mathbf{E} \times \mathbf{H}$ of the lines of fluid and those of the magnetic force.

Magnetic diffusion

According to the former discussion it can be assumed that the dynamical behavior of a conducting fluid in an external magnetic field is governed by the magnitude of the electric conductivity. There is however another different limit when considering that the conducting continuous medium is at rest. For such situation, the equation (44.2) takes the following form

$$\frac{\partial \mathbf{H}}{\partial t} = \gamma \nabla^2 \mathbf{H}, \tag{1}$$

where

$$\gamma \equiv \frac{c^2}{4\pi\sigma}. \tag{2}$$

The relationship (1) is a diffusion equation. It can be solved using the method of separation of variables. Nevertheless, a dimensional analysis of that equation can be made so as to determine that the magnetic field dissipates in the form of Joule´s heat within a temporary interval equal to

$$\tau = \frac{4\pi\sigma\circ^2}{c^2}; \tag{3}$$

where $\circ$ is a characteristic length of the order of magnitude of the dimensions of the region where the induced currents circulate. The solution of the temporary part of equation (1)

$$T = T_o e^{-t/\tau} \tag{4}$$

indicates that the magnetic field diminishes with time because it has loses in the continuous medium. For laboratory conductors the decay time is small but for cosmic conductors it can be very long because of its huge dimensions.

The method of separation of variables

Let us consider a recipient that contains a plasma which decays by diffusion when it collides against the walls. Electrons and ions are recombined there giving as a result that the plasma density be essentially zero near the walls. It can be demonstrated that for this physical situation the plasma behavior is governed by the following relationship

$$\frac{\partial n}{\partial t} + div\,\boldsymbol{f} = 0\,; \tag{1}$$

where $\boldsymbol{f}$ is the total flux and n the density of particles. According to Fick´s law

$$\boldsymbol{f} = -D\,\boldsymbol{grad}\,n\,, \tag{2}$$

with D the diffusion coefficient. In this case in (1) the following is obtained

$$\frac{\partial n}{\partial t} = D\nabla^2 n\,. \tag{3}$$

To solve that diffusion equation we use the method of separation of variables that consists of the following. Be

$$n(\boldsymbol{x},t) = T(t)S(\boldsymbol{x}), \tag{4}$$

so that in the equation (3) the following result is obtained

$$\frac{1}{T}\frac{dT}{dt} = \frac{D}{S}\nabla^2 S\,. \tag{5}$$

Since the left hand side of (5) is a function of time only and the right hand side a function of space only, they must both be equal to the same constant. Let $-1/\iota$ be that constant so that the function $T(t)$ takes the following form

$$\frac{dT}{T} = -\frac{dt}{\tau}, \tag{6}$$

whose solution is

$$T = T_o e^{-t/\tau}. \tag{7}$$

The spatial part $S(\boldsymbol{x})$ obeys to the following equation

$$\frac{d^2S}{dx^2} = -\frac{S}{D\tau}; \tag{8}$$

which in general has as a solution a sine and cosine combination of the following form

$$S = A\cos\frac{x}{\sqrt{D\tau}} + B\,sen\frac{x}{\sqrt{D\tau}}. \tag{9}$$

According to the next figure

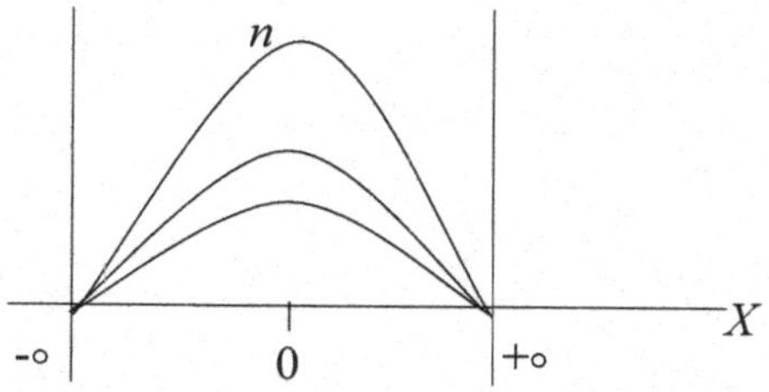

the density is expected to be nearly zero in the container walls reaching one or more peaks in between. The simplest solution consists of considering a simple maximum. By symmetry the odd term can be rejected in the equation (9) so that we only have the following

$$S = A\cos\frac{x}{\sqrt{D\tau}}. \tag{10}$$

The boundary condition S=0 at $x = \pm\circ$ then it requires that $\pi/2 = \circ/\sqrt{D\tau}$ so that

$$\tau = \left(\frac{2\circ}{\pi}\right)^2 \frac{1}{D}. \tag{11}$$

Combining the relations (4), (7), (10) and (11) the solution sought is obtained

$$n = n_o e^{-t/\tau} cos \frac{\pi x}{2\circ}. \tag{12}$$

References

1. Chen F. Francis. "Introduction to Plasma Physics" Plenum Press, New York and London (1974).
2. Fierros Palacios, Angel. "El flujo potencial en presencia de un campo magnético". Rev. Mex. de Fís. 47(1) 87-92(2001).
3. Jackson, J.P. "Classical Electrodynamics". John Wiley & Sons, Inc. New York, London (1962).
4. Landau, L.D. and Lifshitz, E.M. "Electrodynamics of Continuous Media", Addison-Wesley Publishing Co., Inc.(1960).

Chapter X

Magnetohydrodynamic Waves

§56. Alfven´s waves

Let us consider a physical system formed by a conducting continuous medium and an external magnetic field. If the fluid moves with a certain velocity in a direction perpendicular to the field, it will cut the magnetic field. The forces that act over the matter in motion are approximately due to the motion of matter and to the magnetic lines of force. In general terms, the field is modified as if the lines of force were dragged by the motion of the matter. This fact can originate that the transversal displacements of the conducting medium in the longitudinal magnetic field become propagated in the form of waves. When the conductivity of the medium tends to infinity, the magnetic lines of force behave as fluid lines. In fact, the relation (52.1) is identical for the case of $\boldsymbol{H}$ to the vortex equation in the theory of the non viscous fluids flux and its interpretation in that theory implies the dragging of the vortex lines by the fluid.

Then, and according to the vectorial identity (52.12) and in the limit of very high conductivity, the equation (52.1) is transformed into

$$\frac{\partial \boldsymbol{H}}{\partial t} = (\boldsymbol{H} \cdot \boldsymbol{grad})\boldsymbol{v} - (\boldsymbol{v} \cdot \boldsymbol{grad})\boldsymbol{H} - \boldsymbol{H}\, div\, \boldsymbol{v}\,; \tag{56.1}$$

where the condition (44.1) was used. From the continuity equation (9.2) we have that

$$div\, \boldsymbol{v} = -\frac{1}{\rho}\frac{\partial \rho}{\partial t} - \frac{\boldsymbol{v} \cdot \boldsymbol{grad}\, \rho}{\rho},$$

in such a way that in (56.1) it is obtained that

$$\frac{d\boldsymbol{H}}{dt}-\frac{\boldsymbol{H}}{\rho}\frac{d\rho}{dt}=(\boldsymbol{H}\cdot \boldsymbol{grad})\boldsymbol{v}. \tag{56.2}$$

If the last result is multiplied by $1/\rho$ we have that

$$\frac{d}{dt}\left(\frac{\boldsymbol{H}}{\rho}\right)=\left(\frac{\boldsymbol{H}}{\rho}\cdot \boldsymbol{grad}\right)\boldsymbol{v}. \tag{56.3}$$

Let us now consider some fluid line in the continuous medium; that is to say a line which moves with the fluid. Let $\Delta\circ$ be an element of length of this line and we shall know how $\Delta\circ$ varies with time. Let us suppose that $\boldsymbol{v}$ is the fluid velocity at one end of this element of length so that in the other end its change will be $\boldsymbol{v}+(\Delta\circ\cdot\boldsymbol{grad})\boldsymbol{v}$. During a time interval dt, the length of $\Delta\circ$ therefore changes to $dt(\Delta\circ\cdot\boldsymbol{grad})\boldsymbol{v}$; in other words

$$\frac{d}{dt}(\Delta\circ)=(\Delta\circ\cdot \boldsymbol{grad})\boldsymbol{v}. \tag{56.4}$$

If the equations (56.3) and (56.4) are compared, we see that the time rate of change of the vectors $\Delta\circ$ and $\boldsymbol{H}/\rho$ is given by identical formulae. Hence, it follows that, if these vectors are oriented in the same direction, they will remain parallel and their lengths will remain in the same ratio. Therefore, it can be said that in the limit when the conductivity of the conducting medium tends to infinity, the magnetic lines are frozen in the fluid and move with it.

On the other hand, when the magnetic forces are calculated it is usual to consider that $\mu=1$, in which case we have that

$$\boldsymbol{f}=\boldsymbol{j}\times\frac{\boldsymbol{H}}{c}. \tag{56.5}$$

When there are conducting currents, the field equation (45.2) is satisfied by the magnetic field, so that in the former relation we have that

$$\boldsymbol{f}=\frac{1}{4\pi}[(\boldsymbol{H}\cdot\boldsymbol{grad})\boldsymbol{H}]-\boldsymbol{grad}\left(\frac{H^2}{8\pi}\right)=\frac{1}{4\pi}\left[H_i\frac{\partial H_j}{\partial x^i}\right]-\boldsymbol{grad}\left(\frac{H^2}{8\pi}\right)$$

$$=div\left[\frac{H_iH_j}{4\pi}\right]-\boldsymbol{grad}\left(\frac{H^2}{8\pi}\right); \tag{56.6}$$

where the vectorial identity (48.5) as well as the condition (44.1) were used and an integration by parts was made. As it is easy to see in (56.6), the magnetic force is equivalent to the gradient of a pressure plus the divergence of a tension along the line of force. In other words, that is the same as a longitudinal tension plus a pressure of the same value, which is the normal way of giving Maxwell's tensions.

According to the former, the mechanical effects of a magnetic field are equivalent to a hydrostatic pressure $H^2/8\pi$ plus a magnetic tension $H^2/4\pi$ along the lines of force. In the case of an incompressible fluid, the hydrostatic mechanical and magnetic pressures may compensate each other so that only the effect of the magnetic tension persists. This tension can give place to the existence of transversal waves that propagate along the lines of force with a velocity $\pm \boldsymbol{v}_a$. The possibility of the existence of such waves was discussed by Hannes Alfven for the first time in 1942.

§57. Alfven's wave equation

In order to give a solution of the problem of obtaining the motion equation that governs the transmission of transversal waves along the lines of magnetic force, let us consider the flow of a perfect fluid in an external magnetic field. This is a particular case of potential flow previously studied in chapters V and IX; but now the mechanical effects of the magnetic field concerning the production and transmission of the above mentioned waves will be investigated. Consequently, all the approximations and the calculus carried out in those chapters are valid, so they will not be repeated here. Thus, in terms of the small perturbation $\boldsymbol{h}(\boldsymbol{x}, t)$ in the external magnetic field the fundamental equations of Classical Electrodynamics will take the following form

$$div\, \boldsymbol{h} = 0, \tag{57.1}$$

and

$$\frac{\partial \boldsymbol{h}}{\partial t} = \boldsymbol{rot}(\boldsymbol{v} \times \boldsymbol{H}_o). \tag{57.2}$$

On the other hand, the corresponding equation of motion is

$$\frac{\partial \boldsymbol{v}}{\partial t} = -\frac{1}{\rho_o} \boldsymbol{grad}\,(p_o + p') + \frac{1}{4\pi\rho_o} (\boldsymbol{rot\,h}) \times \boldsymbol{H}_o\,. \tag{57.3}$$

As in this case $c_o = 0$ because it is not considered that the propagation of any sonic disturbance should exist, in the equation of state (52.3) it is proposed that

$$p_o = v_a^2 \rho_o + \frac{H_o^2}{8\pi} \tag{57.4}$$

and

$$p' = v_a^2\, \rho'\,; \tag{57.5}$$

because the flux is barotropic. In those relations $\boldsymbol{v}_a$ is the Alfven's velocity when $\boldsymbol{H} \approx \boldsymbol{H}_o$, whose general form is given by (52.18). Now and given that in a first approach

$$\frac{1}{4\pi\rho_o} (\boldsymbol{rot\,h}) \times \boldsymbol{H}_o = \left(\frac{\boldsymbol{H}_o}{4\pi\rho_o} \cdot \boldsymbol{grad} \right) \boldsymbol{h}\,, \tag{57.6}$$

in the motion equation (57.3) it is only obtained that

$$\frac{\partial \boldsymbol{v}}{\partial t} = -\boldsymbol{grad} \left[\frac{v_a^2\, \rho'}{\rho_o} - \frac{\boldsymbol{H}_o \cdot \boldsymbol{h}}{4\pi\rho_o} \right] \tag{57.7}$$

where the relations (57.5) and (57.6) were used and the fact that p_o is a constant. On the other hand

$$\boldsymbol{rot}(\boldsymbol{v} \times \boldsymbol{H}_o) = (\boldsymbol{H}_o \cdot \boldsymbol{grad})\boldsymbol{v} - \boldsymbol{H}_o\, div\, \boldsymbol{v}\,; \tag{57.8}$$

in which case

$$\frac{\partial \boldsymbol{h}}{\partial t} = (\boldsymbol{H}_o \cdot \boldsymbol{grad})\boldsymbol{v} - \boldsymbol{H}_o\, div\, \boldsymbol{v}\,. \tag{57.9}$$

Deriving again (57.9) regarding time and taking into account the result (57.7), as far as terms of the first order in the approximations, the following result is finally obtained

$$\frac{\partial^2 \boldsymbol{h}}{\partial t^2} = H_{oj}\frac{\partial}{\partial x^j}\left[-\frac{\partial}{\partial x^i}\left\{\frac{v_a^2 \rho'}{\rho_o} - \frac{\boldsymbol{H}_o \cdot \boldsymbol{h}}{4\pi\rho_o}\right\}\right]$$
$$= -\frac{v_a^2 H_{oj}}{\rho_o}\frac{\partial^2 \rho'}{\partial x^j \partial x^i} + \frac{H_o^2}{4\pi\rho_o}\nabla^2 \boldsymbol{h} \approx v_a^2 \nabla^2 \boldsymbol{h}; \tag{57.10}$$

only, due to the fact that the other term is of higher order in the approximations. Besides, the result (52.18) was used. In this case, it is easy to see that from (57.10) the following result is finally obtained

$$\nabla^2 \boldsymbol{h} - \frac{1}{v_a^2}\frac{\partial^2 \boldsymbol{h}}{\partial t^2} = 0. \tag{57.11}$$

This is Alfven´s wave equation for MHD waves in terms of the small perturbation $\boldsymbol{h}(\boldsymbol{x},t)$ on the external magnetic field.

§58. Kinetic and potential energy densities

In order to obtain the field differential equations for Alfven´s waves from Hamilton-Type Variational Principle we proceeded this way.

Firstly, it is necessary to know which are the relevant entities which the lagrangian density as well as the specific lagrangian depend from, to later use the theoretical frame developed along the present book. Then, if in the motion equation (57.7) the divergence is calculated on the left hand side, we have that in a first approach

$$\nabla \cdot \left(\frac{\partial \boldsymbol{v}}{\partial t}\right) = 0; \tag{58.1}$$

so that on its right hand side it is obtained that

$$\nabla^2\left[\frac{\boldsymbol{H}_o \cdot \boldsymbol{h}}{4\pi} - p'\right] = 0\,, \tag{58.2}$$

where the relation (57.5) was used. The former equation means that what is enclosed in the parenthesis is a constant solution of Laplace′s equation in the external part of the disturbed region where $\boldsymbol{h} = 0$, and therefore such constant solution should be a solution at any point of the region occupied by the fluid. In consequence from (58.2) we have that

$$p' = -\rho_o \frac{\partial \varphi}{\partial t} + \frac{\boldsymbol{H}_o \cdot \boldsymbol{h}}{4\pi}\,; \tag{58.3}$$

with $\varphi(\boldsymbol{x}, t)$ some scalar function of space and time. Thus, from (57.5) it is clear that

$$\rho' = \frac{1}{v_a^2}\left(-\rho_o \frac{\partial \varphi}{\partial t} + \frac{\boldsymbol{H}_o \cdot \boldsymbol{h}}{4\pi}\right). \tag{58.4}$$

In this case the change in the specific internal energy can be obtained from the relation (25.9) but now it will only be

$$\square \approx \frac{p'\rho'}{\rho_o^2} = \frac{1}{v_a^2}\left(\frac{\boldsymbol{H}_o \cdot \boldsymbol{h}}{4\pi\rho_o}\right)^2; \tag{58.5}$$

because it is the most relevant term that contains the perturbation of the magnetic field. Hence, it is evident that the potential energy density should be expressed as follows

$$u = \frac{\rho_o}{2v_a^2}\left(\frac{\boldsymbol{H}_o \cdot \boldsymbol{h}}{4\pi\rho_o}\right)^2. \tag{58.6}$$

Now, the value of the kinetic energy density is required. In order to obtain this quantity we shall proceed as follows. The equation of motion (57.7) can be multiplied scalarlly by $\boldsymbol{v}$ to obtain the following

$$\frac{\partial}{\partial t}\left(\frac{\rho_o v^2}{2}\right) = 0\,, \tag{58.7}$$

because on the right hand side we have that

$$\boldsymbol{v} \cdot \boldsymbol{grad}\left(\frac{\partial \varphi}{\partial t}\right) \approx 0;$$

where the relation (58.4) to define $\partial\varphi/\partial t$ was used. The last equation is zero due to the fact that the calculus was made as far as terms of the first order in the approximations and clearly this is a quantity of higher order. In this case the equation (58.7) can be integrated to obtain that

$$\frac{1}{2}\rho_o v^2 = \frac{h^2}{8\pi}, \tag{58.8}$$

where $h^2/8\pi$ is the value of the magnetic energy per unit volume in terms of the small perturbative term of the magnetic field. Hence, it is evident that the kinetic energy density in terms of $\boldsymbol{h}(\boldsymbol{x}, t)$ takes the simple following form

$$t = \frac{h^2}{8\pi}\cdot \tag{58.9}$$

As the lagrangian density must be of the form T–V, according to (58.9) and (58.6)

$$\circ = \frac{1}{2}\rho_o\left[\frac{h^2}{4\pi\rho_o} - \frac{1}{v_a^2}\left(\frac{\boldsymbol{H}_o \cdot \boldsymbol{h}}{4\pi\rho_o}\right)^2\right]; \tag{58.10}$$

whereas the specific lagrangian is

$$\lambda = \frac{1}{2}\left[\frac{h^2}{4\pi\rho_o} - \frac{1}{v_a^2}\left(\frac{\boldsymbol{H}_o \cdot \boldsymbol{h}}{4\pi\rho_o}\right)^2\right]. \tag{58.11}$$

On the other hand from (58.8) it is evident that

$$\boldsymbol{v}(\boldsymbol{x},t)=\frac{\boldsymbol{h}(\boldsymbol{x},t)}{\sqrt{4\pi\rho_o}}\equiv\boldsymbol{v}_h(\boldsymbol{x},t). \qquad (58.12)$$

Let us suppose that $\boldsymbol{v}_h(\boldsymbol{x},t)$ can be written in the same form as the small perturbative term in the velocity field that was used for the case of potential flow; that is to say

$$\boldsymbol{v}_h(\boldsymbol{x},t)=\boldsymbol{grad}\ \varphi(\boldsymbol{x},t). \qquad (58.13)$$

The function $\varphi(\boldsymbol{x},t)$ is the velocity potential corresponding to the perturbed magnetic field and obviously is the term introduced in (58.3). Then

$$\frac{\partial \boldsymbol{v}_h}{\partial t}=\boldsymbol{grad}\left(\frac{\partial\varphi}{\partial t}\right). \qquad (58.14)$$

Now from (57.5) and (58.3) we have that

$$\frac{\partial\varphi}{\partial t}=\frac{\boldsymbol{H}_o\cdot\boldsymbol{h}}{4\pi\rho_o}-\frac{v_a^2\rho'}{\rho_o}, \qquad (58.15)$$

so deriving again this result with respect to time we have that

$$\frac{\partial^2\varphi}{\partial t^2}=\frac{\boldsymbol{H}_o}{4\pi\rho_o}\cdot\frac{\partial\boldsymbol{h}}{\partial t}-\frac{v_a^2}{\rho_o}\frac{\partial\rho'}{\partial t}. \qquad (58.16)$$

From the linearized continuity equation (22.10) we can see that

$$\frac{\partial\rho'}{\partial t}=-\rho_o\ div\,\boldsymbol{v}_h=-\frac{\rho_o}{\sqrt{4\pi\rho_o}}\nabla\cdot\boldsymbol{h}=0. \qquad (58.17)$$

Now, according to the first term of the right hand side of (57.9) and with the results (52.18) and (58.16) it is obtained that

$$\frac{\partial^2 \varphi}{\partial t^2} = v_a^2 \frac{\partial}{\partial x^j}\left(\frac{\partial \varphi}{\partial x^i}\right) = v_a^2 \nabla^2 \varphi \qquad (58.18)$$

because the calculus was made as far as first order in the approximations. Let us note that in (57.9) the term of the divergence of $\boldsymbol{v}_h$ is zero because in complete agreement with (58.12), $div\ \boldsymbol{v}_h = div\ \boldsymbol{h}/\sqrt{4\pi\rho_o} = 0$. Rearranging the terms in (58.18) it is easy to see that the following result is obtained

$$\nabla^2 \varphi - \frac{1}{v_a^2}\frac{\partial^2 \varphi}{\partial t^2} = 0. \qquad (58.19)$$

This is Alfven´s wave equation for MHD waves in terms of the proposed velocity potential. It is an equation totally equivalent to relationship (57.11) as it will be demonstrated next. In fact, if the relation (58.13) is derived twice with respect to time and taking into account the definition (58.12) it is clear that we obtain

$$\frac{1}{\sqrt{4\pi\rho_o}}\frac{\partial^2 \boldsymbol{h}}{\partial t^2} = \frac{\partial}{\partial x^j}\left(\frac{\partial^2 \varphi}{\partial t^2}\right); \qquad (58.20)$$

whereas from (58.18) it is obvious that

$$\frac{\partial^2 \varphi}{\partial t^2} = \frac{v_a^2}{\sqrt{4\pi\rho_o}} \nabla \boldsymbol{h}. \qquad (58.21)$$

If we substitute (58.21) in (58.20) Alfven´s wave equation (57.11) is recovered.

§59. Field differential equations

If the results (58.13) and (58.3) are taken into account kinetic and potential energy densities can be expressed in terms of the velocity potential. In fact,

$$t = \frac{1}{2}\rho_o |\boldsymbol{grad}\ \varphi|^2$$

whereas if the calculus is made as far as terms of the first order it is clear that

$$\left(\frac{\boldsymbol{H}_o\cdot\boldsymbol{h}}{4\pi\rho_o}\right)^2=\left(\frac{\partial\varphi}{\partial t}\right)^2+\frac{v_a^4\,\rho'^2}{\rho_o^2}+2\left(\frac{\partial\varphi}{\partial t}\right)\left(\frac{v_a^2\,\rho'}{\rho_o}\right)\approx\left(\frac{\partial\varphi}{\partial t}\right)^2;$$

because it is the dominant term. Then

$$u=\frac{\rho_o}{2\,v_a^2}\left(\frac{\partial\varphi}{\partial t}\right)^2.$$

In consequence

$$\text{o}=\frac{1}{2}\rho_o\left[\left|\boldsymbol{grad}\ \varphi\right|^2-\frac{1}{v_a^2}\left(\frac{\partial\varphi}{\partial t}\right)^2\right]$$

is the corresponding lagrangian density and

$$\lambda=\frac{1}{2}\left[\left|\boldsymbol{grad}\ \varphi\right|^2-\frac{1}{v_a^2}\left(\frac{\partial\varphi}{\partial t}\right)^2\right]$$

is the specific lagrangian. Also and given that they have the same form that the corresponding lagrangian for the case of potential flow, the field differential equation for Alfven´s waves is again the relationship (23.6); that is to say†,

$$\frac{\partial}{\partial x^i}\left[\frac{\partial\lambda}{\partial(\nabla_i\,\varphi)}\right]+\frac{\partial}{\partial t}\left[\frac{\partial\lambda}{\partial\left(\frac{\partial\varphi}{\partial t}\right)}\right]=0.$$

† *See foot note of paragraph* 53.

The difference resides in the velocity potential because in one case it is related to the velocity field variation, whereas in the other one it is the change in the magnetic field which is related to this function. If in (58.12) the definition (58.13) is used and the result is integrated it can be seen that in a first approach

$$\varphi(\boldsymbol{x},t)=\frac{\boldsymbol{x}\cdot\boldsymbol{h}}{\sqrt{4\pi\rho_o}}; \qquad (59.1)$$

so that the following is fulfilled

$$\textbf{grad}\ \varphi=\frac{\boldsymbol{h}}{\sqrt{4\pi\rho_o}} \qquad (59.2)$$

only if $\boldsymbol{x}$ and $\textbf{grad}\ \boldsymbol{h}$ are perpendicular between them. On the other hand, from (59.1) it is clear that

$$\nabla^2\varphi=\frac{\boldsymbol{x}}{\sqrt{4\pi\rho_o}}\cdot\nabla^2\boldsymbol{h}; \qquad (59.3)$$

only because $\nabla^2\boldsymbol{x}=0$. Moreover,

$$\frac{\partial^2\varphi}{\partial t^2}=\frac{\boldsymbol{x}}{\sqrt{4\pi\rho_o}}\cdot\frac{\partial^2\boldsymbol{h}}{\partial t^2} \qquad (59.4)$$

simply because the other terms of the expansion are of higher order in the approximations. Then and given that

$$\frac{\boldsymbol{x}}{\sqrt{4\pi\rho_o}}\cdot\left[\nabla^2\boldsymbol{h}-\frac{1}{v_a^2}\frac{\partial^2\boldsymbol{h}}{\partial t^2}\right]=0$$

Alfven´s wave equation (57.11) is recovered again due to the fact that the term $\boldsymbol{x}/\sqrt{4\pi\rho_o}\neq 0$.

§60. Energy conservation law

From the results obtained in the former paragraph it is easy to see that the kinetic energy density can be expressed in terms of velocity v_h; that is to say

$$t = \frac{1}{2}\rho_o v_h^2, \tag{60.1}$$

whereas the potential energy density can be written as follows

$$u = \frac{1}{2}\rho_o\left[\frac{v_a^2\rho'^2}{\rho_o^2}\right]. \tag{60.2}$$

In this case the lagrangian density can be expressed in terms of (60.1) and (60.2) so that

$$\circ_\rho = \frac{1}{2}\rho_o\left[v_h^2 - \frac{v_a^2\rho'^2}{\rho_o^2}\right] \tag{60.3}$$

and of course,

$$\lambda_\rho = \frac{1}{2}\left[v_h^2 - \frac{v_a^2\rho'^2}{\rho_o^2}\right] \tag{60.4}$$

is the proper specific lagrangian. Hence it is obvious that the functional form of these entities is

$$\circ_\rho = \circ_\rho\left(v_h^i, \rho'\right) \tag{60.5}$$

and

$$\lambda_\rho = \lambda_\rho\left(v_h^i, \rho'\right). \tag{60.6}$$

The calculus of variations and the general formula (19.1) can only give as a result the energy balance equation for potential flow (27.4), but in it the velocity $\boldsymbol{v}_h$ in the second term of the left hand side must appear; that is to say, for this case, we would have that

$$\frac{\partial \mathcal{H}_\rho}{\partial t} - div\left[\rho_o^2 \frac{\partial \lambda_\rho}{\partial \rho'} \boldsymbol{v}_h\right] = 0 .$$

In this case and according to (60.3) and (60.4), at a first approach we have that

$$\mathcal{H}_\rho = \frac{h^2}{8\pi} + \frac{v_a^2 \rho'^2}{2\rho_o} ;$$

where the relation (58.8) was used. Moreover

$$\rho_o^2 \frac{\partial \lambda_\rho}{\partial \rho'} \boldsymbol{v}_h = -\frac{p' \boldsymbol{h}}{\sqrt{4\pi\rho_o}} ;$$

where all these quantities have been expressed in terms of the small perturbation $\boldsymbol{h}$. Thus, the MHD energy conservation law for Alfven's waves has the following form

$$\frac{\partial}{\partial t}\left(\frac{h^2}{8\pi}\right) + div\left[\frac{p' \boldsymbol{h}}{\sqrt{4\pi\rho_o}}\right] = 0, \tag{60.7}$$

only because the term

$$\frac{\rho' v_a^2}{\rho_o} \frac{\partial \rho'}{\partial t} = 0, \tag{60.8}$$

according to the relationship (58.17). The equation (60.7) is valid for the flux at any instant.

Selected Topics

Transversal displacements

A very simple physical interpretation can be given to the fact that in a longitudinal magnetic field, the transversal displacements of the conducting fluid are propagated in the form of waves. It was demonstrated in paragraph 56 that when the electric conductivity of the conducting medium tends to infinity, the magnetic lines of force behave as fluid lines. Their transversal displacements give as a result that these lines curve propitiating that at some points they lengthen and at others they compress. The mechanical effects of a magnetic field are translated into Maxwell's tensions that seem forces as those generated if the magnetic lines of force suffered contractions and also repeled among them. Consequently, a curvature of the lines could produce cuasi-elastic forces that would constrain them to strengthen themselves originating this way further oscillations

The pressure and the velocity of sound

Let us consider the propagation of small disturbance in a homogeneous conducting medium which moves in a uniform constant magnetic field. If the viscosity, thermal conductivity, and electric resistance $1/\sigma$ of the medium are very small, their effects on the propagating of perturbations may be neglected in a first approximation. Those effects are basically due to the dissipation of energy. Then, the perturbations would be propagated as undamped waves. Moreover, since we assume that the external magnetic field and the velocity of propagation are perpendicular between themselves, the motion equation (52.2) in one dimension takes the following form

$$\frac{\partial v}{\partial t}+v\frac{\partial v}{\partial x}+\frac{1}{8\pi\rho}\frac{\partial H^2}{\partial x}=-\frac{1}{\rho}\frac{\partial p}{\partial x}; \tag{1}$$

where $v = v_x$, $v_y = v_z = 0$ and in the same way $H_x = H_z = 0$ and $H = H_y$. The former equation can be written as follows

$$\frac{\partial v}{\partial t}+v\frac{\partial v}{\partial x}=-\frac{1}{\rho}\frac{\partial}{\partial x}\left(p+\frac{H^2}{8\pi}\right). \tag{2}$$

This relationship differed from the one-dimensional motion equation for a plane sound wave which is propagated in X-direction

$$\frac{\partial v}{\partial t}+v\frac{\partial v}{\partial x}=-\frac{1}{\rho}\frac{\partial p}{\partial x}, \tag{3}$$

only in the form that the thermal equation of state has. In the case of ordinary Fluid Dynamics the true pressure $p(\rho)$ is only a function of mass density because for barotropic flow the specific entropy is a constant. For the present case, the total pressure also depends on the magnitude of the external magnetic field. Consequently, both equations will have the same form if

$$p'(\rho, H) = p(\rho) + \frac{H^2}{8\pi}; \qquad (4)$$

where $H^2/8\pi$ is the hydrostatic magnetic pressure. The so called Riemann´s solution gives the formulae that give the exact solution for the motion equation for traveling waves; solution that can be applied for this case.

The velocity of sound in the medium can be calculated following the line of reasonning of paragraph 56 and taking into account the relation given between H/ρ and fluid lines; as well as the time invariability of the length of these lines. Let us consider that for the present situation the magnetic field is such that it can be written as $H = \rho b$ with b a constant.

For the given conditions the equation (44.1) is satisfied identically; whereas when calculating the rotational of the second member of (52.1) using the tensorial notation and only considering the motion in one dimension; the equation above mentioned takes the following form

$$\frac{\partial H}{\partial t} + \frac{\partial}{\partial x}(vH) = 0. \qquad (5)$$

In this case it can be seen that b satisfies the relation $db/dt = 0$ so that if the fluid is homogeneous at some initial instant, b must remain constant at all moment. The velocity of sound in the medium is represented by the following expression

$$\begin{aligned} c'_o &= \sqrt{(\partial p'/\partial \rho)_s} = \sqrt{(c_o^2 + \rho b^2/4\pi)} \\ &= \sqrt{(c_o^2 + H^2/4\pi\rho)} \end{aligned} \qquad (6)$$

where c_o is the velocity of sound as it is usually defined.

References

1. Cowling, T.G. "Magnetohidrodinámica". Editorial Alhambra, S.A., Madrid, B.Aires, México (1968).
2. Fierros Palacios, Angel. "Las ecuaciones diferenciales de campo para las ondas MHD y la ecuación de onda de Alfvén". Rev. Mex. de Fís. 47(1) 93-97 (2001).
3. Landau, L.D. and Lifshitz, E.M. "Electrodynamics of Continuous Media". Addison-Wesley Publishing Co, Inc. (1960).
4. Landau, L.D. and Lifshitz, E.M. "Fluid Mechanics". Addison-Wesley Publishing Co. (1959).

Chapter XI

The Sunspots

§61. The problem of the sunspots

The Sun is a huge concentration of fluid at very high temperature whose parts are kept together because of gravitational attraction being in dynamic equilibrium with the gaseous pressure of solar plasma. They are also under the influence of a self-generated magnetic field whose mission is to maintain the form and structure of the heavenly body. The lines of force of the magnetic field can be imagined as if they formed a kind of superstructure which confers the solar matter an appreciable rigidity. As the Sun revolves in itself and in a non uniform manner, the lines of force of the solar magnetic field are dragged by the rotational motion as if they were frozen in the matter.

Careful observation of the solar surface shows a granular texture formed by short-lived objects known as Granules, somewhat brighter than their neighbors due to the fact that their temperature overpasses in 100 or 200 K the temperature of the darker bottom. The diameter of granule elements reaches about 725 km and their mean life is approximately about 8 minutes.

Also easily observed are the so called Sunspots which are colder and darker regions than the solar photosphere. Discovered first by Chinese astronomers, then by Galileo, and observed by himself and his contemporaries. These phenomena have a somewhat nebulous origin along with very peculiar properties and general behavior, which lead to assume that under the surface of the Sun´s photosphere certain local processes take place destabilizing and creating the phenomenon. In this chapter the thermo-energetic conditions which create a thermal unstability in some regions of the photosphere and their relationship with the appearance, properties, and life time of the Sunspots will be studied.

A large amount of observational data has been accumulated on the subject. In particular, it is well known that the presence of Sunspots on the solar disk is related to a substantial increase in Sun's activity. The most important expression of that activity is the emission of large quantities of charged particles which conform the so called Solar Wind. The strong interaction of the Solar Wind with the Earth's magnetic field and the Ionosphere, produces on the one hand the gigantic and most beautiful Aurora Boriallis; and on the other hand, serious perturbations of the Earth's telecomunication systems.

The vast majority of the Sunspots are observed at certain regions of the Sun. They appear predominantly in two stripes of equal latitude to the north and south of the solar equator in which could be called the tropical regions. They have the tendency to appear in big groups or in couples; in each of which their two members always have opposite magnetic polarity.

A typical Sunspot has the following observed structure and dimensions. It possess a dark nucleus called Umbra of about 18,000 km in diameter, and a somewhat lighter halo called Penumbra, 20,000 km wide. The lower luminosity of the spot as compared to the photosphere is due to a decrease in its temperature. George Ellery Hale (1868-1938) discovered in the Sunspots magnetic fields whose magnitudes are of about 4,000 gauss.

Later it was found that both spots or couples of the so called bipolar group of spots always present one of them, a north pole and the other one a south pole the same as it occurs with the poles of a horse-shoe magnet.

A remarkable fact that should be noted is that the polarity of the Sunspot couples in the northern hemisphere is always opposite to the polarity of the Sunspot couples in the southern hemisphere. This disparity changes periodically with a semiperiod of about 11 years.

The problem is similar to that of the mechanical unstability which occurs in some regions of the terrestrial atmosphere; problem that has been studied and solved in Chapter VII. It is known that in the case of the Sunspots, the fluid is under the influence of an intense magnetic field.

This fact makes it difficult for the theoretical treatment of the problem since it deals with a complicated interaction between the flow of compressible and conducting viscous fluid, and the electromagnetic phenomena derived from the influence of the magnetic field. As it is well known, this interaction should be dealt with the aid of the Magnetohydrodynamic equations.

A thermal unstability which triggers convection currents can be produced in any real fluid with the above mentioned characteristics, as long as the fluid is exposed to non uniform heating from below. Nevertheless, this tendency to unstability can be substantially depressed when dealing with a real conducting fluid under the influence of an intense magnetic field, which gives the fluid some kind of magnetic viscosity and a certain degree of stiffness. This is an important fact because if the magnetic forces generated in a Sunspot can substantially reduce convective movements, such reduction must take place at the expense of the thermal energy in those regions, resulting in a cold and dark spot in Sun´s tropical latitudes. In any case, the persistence of the magnetic field in the Sunspots proves the existence there of a magnetic mechanism for the regulation of thermal convection. This process has the mission to delay as much as possible the onset of the start up mechanism for convective movements.

§62. Dynamic equilibrium between regulatory and startup mechanisms

The condition for magnetomechanical equilibrium to exist in some regions of solar photosphere can be obtained from the MHD thermal equation of state whose general form is given by the equation (51.9). That equation can be written as follows

$$\left(p-\frac{H^2}{8\pi}\right)=p_i\,; \tag{62.1}$$

where and by definition

$$p_i \equiv c_o^2\rho+\frac{\varepsilon}{k}T \tag{62.2}$$

is the hydrostatic pressure inside of the region occupied by each one of the members of couples of Sunspots. As the solar fluid is subjected to the Sun´s gravitational attraction, p_i is not a constant; as it is easy to see from (62.2). It is expected that its magnitude varies with the altitude according to hydrostatic equation (40.4). In this case we have that

$$\left(p-\frac{H^2}{8\pi}\right)=\rho_o\boldsymbol{g}\cdot\boldsymbol{x}+constant\,. \tag{62.3}$$

When the gradient of the former expression is calculated, the result obtained can be written in the following general form

$$\boldsymbol{grad}\left(p - \frac{H^2}{8\pi}\right) = \rho \boldsymbol{g}\,; \tag{62.4}$$

where obviously $H^2/8\pi$ is the hydrostatic magnetic pressure, and $\boldsymbol{H}(\boldsymbol{x}, t)$ the magnetic field in that region. In this case, $\boldsymbol{g}$ is the gravity acceleration in Sun´s surface whereas $\rho(\boldsymbol{x}, t)$ is the mass density of the solar fluid which is not a constant. In fact, according to the relation (22.3) and given that ρ' is a very small quantity, in (62.4) we have that $\rho_0\boldsymbol{g}=\rho\boldsymbol{g}-\rho'\boldsymbol{g}\approx\rho\boldsymbol{g}$, because it is the dominant term. Therefore, it is reasonable to propose the relation (62.4) as a condition for the magnetomechanical equilibrium to exist in some regions of the solar photosphere. Again if the Z-axis of the reference inertial frame used points vertically upwards, the former relation can be written as follows

$$\frac{dp}{dz} = -\frac{g}{V} + grad_z\left(\frac{H^2}{8\pi}\right); \tag{62.5}$$

where as always $V=1/\rho$ is the specific volume. It is known that if temperature is not a constant throughout a fluid the resulting mechanical equilibrium could be stable or unstable depending on certain conditions. In the case of the Sunspots it is necessary to determine the conditions under which an unstability could exist in those regions and to find out its consequences.

The theoretical treatment for an adiabatic displacement of a mass of solar fluid in the photosphere is identical to the one given in Chapter VII for an air mass adiabatically displaced in the terrestrial atmosphere. The general condition of stability which is obtained in both cases is the same and it is represented by the relationship (39.7); that is, the specific entropy of the system increases with altitude, so that in the present situation the condition (39.8) is also fulfilled. Then and according to (62.5) the above mentioned condition directly leads to the following relation

$$grad_z H^2 > \frac{8\pi c_P \, grad_z T}{T\left(\frac{\partial V}{\partial T}\right)_p} + \frac{8\pi g}{V}. \qquad (62.6)$$

In this case, the condition to be satisfied so that a thermal unstability can be produced in some regions of the solar photosphere strongly regulated by the magnetic field that generates the Sunspots, can be expressed as follows

$$\left|grad_z H^2\right| > \frac{8\pi c_P \left|grad_z T\right|}{T\left(\frac{\partial V}{\partial T}\right)_p} + \frac{8\pi g}{V}. \qquad (62.7)$$

This is the condition that must be fulfilled so that the thermal convection in the spots is magnetically regulated. It simply means that the persistence of the magnetic field in those regions, assures the permanency in time of the Sunspots in the solar disk. In fact, as the magnetic field fades away, the mechanism of magnetic regulation weakens, allowing the thermal process to take over the situation; as a result, the mechanism for the onset of convection begins to generate and becomes more dominant so that, when $\boldsymbol{H} = 0$, convective movements are produced which mix the fluid of the spots with the surrounding fluid until the temperature becomes uniform. At that point the Sunspots disappear and the solar activity comes to and end in those regions. Therefore, free convection is produced when the magnitude of the temperature gradient in the *Z*-direction is larger than a certain value; this is, when

$$\left|\frac{dT}{dz}\right| > \frac{gT}{Vc_P}\left(\frac{\partial V}{\partial T}\right)_p, \qquad (62.8)$$

as it is easily seen from (62.7) when $\boldsymbol{H} = 0$. That condition is identical in form to the one which was obtained in Chapter VII for the case of free convection in the terrestrial atmosphere.

When (62.7) and the thermodynamic relationship (39.5) are used and the result is integrated, the following result is obtained

$$H^2 > \frac{8\pi\, c_P}{\left(\dfrac{\partial V}{\partial T}\right)_p} + \frac{8\pi\, gz}{V}, \tag{62.9}$$

where the relation (39.5) was used again and z is some characteristic height in the photosphere where the Sunspots are produced. Let $\boldsymbol{H}_+$ and $\boldsymbol{H}_-$ be two magnetic fields with opposite polarity such that

$$\boldsymbol{H}_+ \cdot \boldsymbol{H}_- = H^2 . \tag{62.10}$$

Furthermore, let

$$Q_+ = \left[\frac{8\pi\, c_P}{\left(\dfrac{\partial V}{\partial T}\right)_p} + \frac{8\pi gz}{V}\right]^{1/2} \tag{62.11}$$

be and

$$Q_- = -\left[\frac{8\pi c_P}{\left(\dfrac{\partial V}{\partial T}\right)_p} + \frac{8\pi gz}{V}\right]^{1/2} . \tag{62.12}$$

The persistence of each couple of Sunspots on Sun´s surface is assured when the magnitude of the magnetic field which generates and maintains them is grater than a certain amount, that is if

$$|\boldsymbol{H}_+| > 2\left[\frac{2\pi\, c_P}{\left(\dfrac{\partial V}{\partial T}\right)_p} + \frac{2\pi gz}{V}\right]^{1/2} \tag{62.13}$$

and

$$|\boldsymbol{H}_-| > 2\left[\frac{2\pi c_P}{\left(\frac{\partial V}{\partial T}\right)_p} + \frac{2\pi g z}{V}\right]^{1/2}. \tag{62.14}$$

Clearly, both magnetic fields have the same magnitude. It can further be reasonably assumed that beyond sharing the previous quality, they also have the same direction but opposite polarities. In other words, if for instance $\boldsymbol{H}_+$ is a magnetic field with the (N-S) polarity; $\boldsymbol{H}_-$ will be a magnetic field with opposite polarity, that is (S-N). Then, both magnetic fields have the same magnitude and direction but opposite polarities.

This means that couples of regions darker and colder than their surroundings, with opposite magnetic polarities and closely related between themselves must appear in the solar photosphere: one, where the lines of force of the magnetic field point outwards from Sun's surface; and another, where they point inwards. In other words, the Sunspots must appear at the same heliographic latitude and in the form of magnetic bipolar couples; in full agreement with the results of astronomical observations.

§63. The velocity of the fluid in the sunspots

Let us consider a huge mass of viscous compressible and conducting fluid, which moves in a magnetic field and is under the influence of a gravitational field. Let $T(\boldsymbol{x}, t)=T_o+T'$ be the corresponding field of temperatures in the photosphere, T_o the temperature of the Sunspots and T' the temperature at which the dynamical equilibrium between the mechanisms of magnetic regulation and the onset of thermal convection is broken, so that $T'<<T_o$. Let $\rho(\boldsymbol{x}, t)=\rho_o+\rho'$ be the mass density of the solar fluid in the photosphere, with $\rho_o>>\rho'$ the mass density of the fluid in the Sunspots and ρ' the change in density while the mechanism for the onset of convection develops.

$p(\boldsymbol{x}, t) = p_o+p'$ with $p_o>>p'$ is once more obtained for the hydrostatic pressure. As it is seen in Chapter VII, p_o is not a constant either since it deals with the pressure corresponding to the dynamic equilibrium in the inner regions of the Sunspots whose temperatures and densities are constant and equal to T_o and ρ_o, respectively. Since the fluid is under the gravitational attraction of the Sun, it will be assumed that p_o varies with height according to the hydrostatic equation (40.4).

In the motion equation of MHD, or generalized Cauchy's equation that was obtained in Chapter VIII and whose functional form is given by the relationship (48.6), some results of Chapter VII can be used. Thus and according to the equations (40.8) to (40.11), in (48.6) the following result is obtained

$$\frac{d\boldsymbol{v}}{dt} = -\frac{1}{\rho_o}\,\boldsymbol{grad}\; p' - \alpha\,\boldsymbol{g}T' - \frac{1}{4\pi\rho_o}\boldsymbol{H}_o \times \boldsymbol{rot}\;\boldsymbol{h} + \boldsymbol{F}_v\,; \tag{63.1}$$

where the relations (40.1) to (40.4) have also been used assuming that the external magnetic field could be splitted in a uniform magnetic field plus the characteristic magnetic field of the Sunspots; exactly as it was proposed in the relationship (52.4). Moreover the vectorial identity (48.5) was used again. On the other hand, $\boldsymbol{F}_v$ is the viscous force whose explicit form is given in the relationship (40.7).

Now, an equation is required which relates pressure with density, temperature, and magnetic field. Such a relation is the generalized thermal equation of state that was developed in Chapter VIII and whose general form is given in (51.9). With all the approximations previously made substituted in the above mentioned relationship, it can be demonstrated that

$$p' = -\rho_o\gamma\, T' + \frac{h^2}{8\pi} - p_o + c_o^{\,2}\rho_o + \frac{\varepsilon}{k}T_o + \frac{H_o^{\,2}}{8\pi}\,; \tag{63.2}$$

where the relation (40.3) was used and the following quantity was defined

$$\gamma \equiv \alpha\, c_o^2 - \frac{\varepsilon}{k\rho_o} \tag{63.3}$$

as a constant with units of square velocity divided by grade. On the other hand and since for the equilibrium situation between the regulatory and start up mechanisms of free convection it is fulfilled that

$$p_o = c_o^2 \rho_o + \frac{\varepsilon T_o}{k} + \frac{H_o^2}{8\pi}, \tag{63.4}$$

and in the relation (63.2) we reach the following result

$$p' = -\rho_o \gamma \, T' + \frac{h^2}{8\pi}. \tag{63.5}$$

This is the explicit form of the required thermal equation of state. From this result it can be demonstrated that

$$-\frac{1}{\rho_o} \textbf{\textit{grad}} \ p' = \gamma \ \textbf{\textit{grad}} \ T' - \left(\frac{\textbf{\textit{h}}}{4\pi \rho_o} \cdot \textbf{\textit{grad}} \right) \textbf{\textit{h}}. \tag{63.6}$$

Then in the relation (62.1) the following is obtained

$$\frac{d\textbf{\textit{v}}}{dt} = \gamma \ \textbf{\textit{grad}} T' - \alpha g T' - \frac{1}{4\pi \rho_o} \left[(\textbf{\textit{h}} \cdot \textbf{\textit{grad}}) \textbf{\textit{h}} + \textbf{\textit{H}}_o \times \textbf{\textit{rot}} \textbf{\textit{h}} \right] + \textbf{\textit{F}}_v. \tag{63.7}$$

However and according to the vectorial identity (48.5) and within the approximations that have been worked with in the former relation we obtain that

$$\frac{d\textbf{\textit{v}}}{dt} = \gamma \ \textbf{\textit{grad}} \ T' - \alpha g T' - \textbf{\textit{grad}} \ p_M^o - \frac{\textbf{\textit{H}}_o}{4\pi \rho_o} \times \textbf{\textit{rot}} \ \textbf{\textit{h}} + \textbf{\textit{F}}_v; \tag{63.8}$$

due to the fact that it is possible to neglect the term $\textbf{\textit{h}} \times \textbf{\textit{rot}}\, \textbf{\textit{h}}$ compared to $\textbf{\textit{H}}_o \times \textbf{\textit{rot}}\, \textbf{\textit{h}}$ because this last one is the dominant quantity. Moreover

$$p_M^o \equiv \frac{h^2}{8\pi \rho_o}, \tag{63.9}$$

it is the hydrostatic magnetic pressure characteristic of the Sunspots divided by ρ_o which is its typical mass density. On the other hand, the magnetic field of the Sunspots should satisfy the field equation (45.2), where $\boldsymbol{j}(\boldsymbol{x}, t)$ is the conduction current density. Nevertheless and according to what we assumed in paragraph 47, the solar plasma is a conducting and continuous medium which conducts no current at all. In this case, in the above mentio-mentioned field equation we have that $\boldsymbol{rot\ h} = 0$ because $\boldsymbol{j} = 0$. In consequence the relation (63.8) is reduced to the following

$$\frac{d\boldsymbol{v}}{dt} = \gamma\ \boldsymbol{grad}\ T' - \alpha\, \boldsymbol{g} T' - \boldsymbol{grad}\ p_M^o + \boldsymbol{F}_v . \qquad (63.10)$$

From the previous expression it is clear that the magnetohydrodynamic force which is responsible for the dynamic equilibrium between the mechanisms of magnetic regulation and start up of thermal convection is represented by the following expression

$$\boldsymbol{F}_{MHD} = \gamma\ \boldsymbol{grad}\ T' - \boldsymbol{grad}\ p_M^o . \qquad (63.11)$$

Consider now that for the condition of dynamical equilibrium, the viscous force is not relevant so that in (63.10) the term $\boldsymbol{F}_v$ can be ignored. This approach is possible due to the following: while conditions (62.13) and(62.14)are fulfilled, the dominant mechanism is the magnetic regulation of thermal convection and therefore, the convective movements are very much under control and clearly, sufficiently diminished; so that it is reasonable to assume that the viscous force can be ignored in comparison to the MHD force (63.11) and the term which contains Sun´s acceleration of gravity. As it was said before in paragraph 61, the presence of an intense magnetic field in the Sunspots confers the fluid a certain rigidity which makes the movement difficult in those regions, so that it is possible to assume that the velocity of the fluid in the Sunspots is small compared to the velocity of the warmer fluid found outside of then. If the velocity field inside the spots is small, relatively speaking, and the distances dealt with are very large, the velocity gradients should be even smaller. Since $\boldsymbol{F}_v$ depends on the gradients of the velocity field, then it is expected to be a small quantity in comparison with the other terms. When the magnetic field starts to decay and weakens, influence of the start up mechanism of thermal convection is initiated. It can be said that while the dynamic equilibrium between both mechanisms exists, the conditions to produce convective movements have been very slowly arising; so that it is possible to assert that the change in the temperature of the Sunspots from T_o to T occur in the steady state. Then, the temperature T' at which the dynamical equilibrium is broken and convection starts, depends only on the coordinates and not on the time. In consequence, throughout the whole life of the Sunspots the viscous force is not important and it can be ignored for the whole analysis of the problem.

Once the equilibrium is broken, convective movements are rapidly established mixing the fluid of the Sunspots with that of its surroundings until the average temperature T of the photosphere is reached. It is possible that during this process, the viscosity of the medium and in consequence the viscous force, become very important; but by this time the Sunspots have already disappeared because the magnetic field which generates them is zero. Clearly, the thermal gradient becomes null and, hence, the solar activity in those regions comes to an end.

As the remaining terms of the relation (63.10) are not explicitly depending on time, that equation can be integrated to obtain that

$$\boldsymbol{v}' = \boldsymbol{v}_o + \left[\gamma\, \boldsymbol{grad}\, T' - \alpha \boldsymbol{g} T' - \boldsymbol{grad}\, p_M^o\right](t' - t_o). \qquad (63.12)$$

However, $\boldsymbol{v}_o = 0$ because it is the average vectorial sum of the velocity inside the Sunspots, corresponding to the dynamic equilibrium, when the magnetic field that generates them reaches its maximum intensity and the convective movements are magnetically regulated and very depressed; that is, in the mature stage of the spots. Under this situation, it can be said that $\boldsymbol{v}_o$ is the vector addition of all the possible fluid movements inside those regions and that such vector sum is zero. Moreover, even when the dynamic equilibrium between the mentioned mechanisms may exist and the movement of the fluid in the central regions of the spots may be very depressed, there will be great activity in their edges. The fluid there over flows diverging from the center to the edge with a movement which is practically parallel to the surface of the Sun due to the gravity pull. This is due to the combination of two effects; one, the dominant effect of the solar gravity, and the other, the fact that at the edges of the spots the control of thermal convection by the mechanism of magnetic regulation is not complete enough as in the central region, so that it exists a residual influence of the thermal gradient over the gradient of the magnetic hydrostatic pressure. Thus and due to the effect of thermal stirring, the fluid flows through the edges of the spot at a certain velocity $\boldsymbol{v}'$ but it almost immediately collapses towards the surface of the Sum, giving the spots their characteristic appearance. On the other hand, the thermal fluctuations which are likely to occur at the edges, should give origin to large splashing of fluid which raises to a great altitude over the photosphere, probably constituting the so called Faculae which are the usual companions of the spots. The overflow of the solar fluid in those regions is very similar to what occurs when a viscous liquid, such as the milk, boils and overflows over the edges of its container. In such process splashing is also very common. In any case when $\boldsymbol{v}_o = 0$ and the scale of time is chosen so that $t_o = 0$, in the former expression we have that

$$\boldsymbol{v}' = \left[\gamma \,\boldsymbol{grad}\, T' - \alpha \boldsymbol{g} T' - \boldsymbol{grad}\, p^o_M\right] t'. \tag{63.13}$$

If only the Z-direction is considered, the following result is obtained

$$v'_z = \left[\gamma \frac{dT'}{dz} + \alpha g T' - \frac{d p^o_M}{dz}\right] t'. \tag{63.14}$$

When the start up mechanism of the thermal convection begins to dominate the phenomenon, the magnetic part begins to decrease so that the term $\gamma\,\boldsymbol{grad}\,T'$ takes over the situation. In the limit when $\boldsymbol{h}=0$ the vectorial velocity at which the convective movements begins is obtained in (63.13), that is

$$\boldsymbol{v}'_c = \left(\gamma \,\boldsymbol{grad}\, T' - \alpha \boldsymbol{g} T'\right) t' \tag{63.15}$$

and, its vertical component clearly is

$$v'_{zc} = \left(\gamma \frac{dT'}{dz} + \alpha g T'\right) t'; \tag{63.16}$$

where the subscript c indicates that the process of thermal convection is being considered.

§64. General equation of heat transfer

So far, the velocity field and the magnetohydrodynamic force responsible for the dynamical equilibrium, have been calculated. Also available is the scalar equation for the mass density which is valid for any fluid, and of course, the appropriate thermal equation of state for the whole phenomenon. The general equation of heat transfer remains missing in order to have an analytical solution to the problem of origin, persistence, and disappearance of the Sunspots. This last equation can be obtained from the MHD energy conservation law whose general form is given by the relationship (49.6). If the left hand side of that expression is developed in the same way as it was done in chapter VIII, the following result is obtained

$$\rho_o T_o \left[\frac{\partial s}{\partial t} + \boldsymbol{v} \cdot \boldsymbol{grad}\, s \right] = S'_{ij} \frac{\partial v'^i}{\partial x^j} + div\left(\kappa\, \boldsymbol{grad}\, T'\right) \qquad (64.1)$$

only, due to the fact that Joule´s heat per unit volume is zero because $\boldsymbol{rot\, h} = 0$ as seen before. To calculate the last expression the approximations for the hydrostatic pressure, mass density, the velocity field, and the external magnetic field that were done in Chapters V, VII, and IX, were considered. In the previous relation, s is again the specific entropy whereas

$$S'_{ij} = \eta \left[\frac{\partial v'^i}{\partial x^j} + \frac{\partial v'^j}{\partial x^i} - \frac{2}{3} \delta_{ij} \frac{\partial v'^\circ}{\partial x^\circ} \right] + \zeta \delta_{ij} \frac{\partial v'^\circ}{\partial x^\circ}, \qquad (64.2)$$

are the components of the viscosity stress tensor in terms of the perturbation in the velocity field.

According to the previous arguments given in the former paragraph, the velocity of the fluid inside the spots is found very diminished basically because of two effects: one, the fact that the temperature in those regions is smaller than that of the photosphere, and another because the solar plasma as a result of the magnetic viscosity created by the magnetic field is more rigid there. Both effects depress the thermal stirring so that inside the spots the solar fluid moves more slowly that outside of then. Hence, it is possible to assume that in those regions the fluid velocity is smaller than the speed of sound in the medium. Therefore, it could be thought that the variations in hydrostatic pressure occurring as a consequence of the motion are so small, that the changes in density and in other thermodynamic quantities involved in the description of the phenomenon, should be neglected. Nevertheless, it is clear that the solar fluid is subjected to a non uniform heating so that its density undergoes changes which can not be ignored, so that it is not possible to say that it is a constant. In consequence, it can be shown that

$$\frac{dT'}{dt} - \chi \nabla^2 T' = 0; \tag{64.3}$$

where χ is again the thermometric conductivity introduced in Chapter VII. In order to reach the last relation the results previously obtained in paragraph 41 were used. This is the final form of the general equation of heat transfer for the whole phenomenon of the generation, persistence, and disappearance of the Sunspots.

§65. Magnetic field of the sunspots

As long as the life time of the Sunspots is concerned, it has been assumed that the temperature T' only depends on the coordinates so that $\partial T'/\partial t = 0$. As a consequence of that, from the relation (64.3) it is possible to deduce the following result

$$\frac{\partial}{\partial x^i}\left[T' v'^i - \chi\, grad_i\, T'\right] = 0, \tag{65.1}$$

where an integration by parts has been made and the term $T'\partial v'^i/\partial x^i$ has been neglected because it is a term of higher order in the approximations. The previous equation can be integrated so that

$$\chi\, \mathbf{grad}\, T' = T' \mathbf{v}'; \tag{65.2}$$

where the integration constant has been assumed to be zero without lose of generality. Now and according to relation (63.13)

$$\mathbf{grad}\, T'\left[\chi - \gamma\, T' t'\right] = -\left[\alpha \mathbf{g} T' + \mathbf{grad}\, p_M^o\right] T' t'. \tag{65.3}$$

Let us assume that for the condition of dynamic equilibrium the thermometric conductivity can be neglected, so that only the other term in the left hand side of the former expression is preserved. This is due to the fact that it could be very interesting to observe the phenomenon of the Sunspots as if it were frozen in time, condition that is practically fulfilled in their stage of maturity. In general, it is not convenient to make such approach as it will be seen later on. Nevertheless, if for the time being that term is ignored, from (65.3) the following result is obtained

$$\mathbf{grad}\ p^o_M = \gamma\ \mathbf{grad}\ T' - \alpha \mathbf{g} T' . \tag{65.4}$$

This is the analytical expression for the process of dynamic equilibrium in the Sunspots. If only the z-component of the previous equation is considered, we get again that $\mathbf{g} = -\mathbf{k}g$, with $\mathbf{k}$ the unitary vector along the Z axis of the inertial frame of reference used and of course, g is the average constant value of the gravity at the surface of the Sun. In that case, it is proposed that the convective movements in the spots can be damped by magnetic regulation, thus assuring their permanency in the solar disk, if the following condition of dynamic equilibrium is satisfied

$$\left|grad_z\, h^2\right| > 8\pi\rho_o \left[\left|\gamma\ grad_z\, T'\right| + \alpha g T'\right]. \tag{65.5}$$

From the z-component of relation (65.4) we have that

$$\frac{d}{dz}\left[\gamma\ T' - p^o_M\right] = -\alpha g T' ; \tag{65.6}$$

in which case

$$h^2 = 8\pi\rho_o\, T'(\gamma + \alpha g z), \tag{65.7}$$

where z is any characteristic height in the photosphere in the region where the Sunspots appear. From the previous relationship, the following expressions can be obtained

$$\mathbf{h}_+ = \mathbf{k}\lambda_+ \tag{65.8}$$

and

$$\boldsymbol{h}_- = -\boldsymbol{k}\lambda_- ; \qquad (65.9)$$

where

$$\lambda_+ = \left[8\pi\rho_o T'(\gamma + \alpha g z)\right]^{1/2} \qquad (65.10)$$

and

$$\lambda_- = -\left[8\pi\rho_o T'(\gamma + \alpha g z)\right]^{1/2} . \qquad (65.11)$$

In (65.8) and (65.9) it has been proposed that $\boldsymbol{h}_+$ and $\boldsymbol{h}_-$ are two magnetic fields which have opposite polarity and such that $\boldsymbol{h}_+ \cdot \boldsymbol{h}_- = h^2$. Thus the persistence of the Sunspots on the surface of the Sun is assured if the magnitude of the magnetic field for each member of the couple is assumed to have the following value

$$h_+ = 2\left[2\pi\rho_o T'(\gamma + \alpha g z)\right]^{1/2} \qquad (65.12)$$

and

$$h_- = 2\left[2\pi\rho_o T'(\gamma + \alpha g z)\right]^{1/2} . \qquad (65.13)$$

Clearly, both magnetic fields have the same magnitude. On the other hand, from (65.9) and (65.11) it is easy to see that $\boldsymbol{h}_- = \boldsymbol{k}\lambda$ so that those fields are also oriented in the same direction. On the other hand, if $\boldsymbol{h}_+$ is a magnetic field with (*N-S*) polarity, $\boldsymbol{h}_-$ will have the opposite polarity, that is to say (*S-N*). That means that those magnetic fields have the same magnitude and direction but opposite polarities; so that if observed from Earth it would look as if these were the poles of a horse-shoe shaped magnet.

It can be seen from equation (65.4) that as the intensity of the magnetic field generating the Sunspots weakens, its influence declines in favor of the thermal gradient, basically because the other term is a constant for a particular couple of spots. Under such circumstances the conditions for the start up mechanism of thermal convection are given, so that the thermal processes become more important taking over the situation little by little. Again and in the limit when $\boldsymbol{h} = 0$, we have that

$$\frac{dT'}{dz} = -\frac{\alpha g T'}{\gamma}. \tag{65.14}$$

As usual, the conditions to generate the convective movements that propitiate the disappearance of the Sunspots turn out to be similar to those obtained in paragraph 62; that is to say, that the magnitude of the thermal gradient is greater than a given value. In other words that

$$\left|\frac{dT'}{dz}\right| > \frac{\alpha g T'}{\gamma}. \tag{65.15}$$

The equation (65.14) can be integrated to obtain the following result

$$T' = T_o exp\left[-\frac{\alpha g z}{\gamma}\right]. \tag{65.16}$$

This relation gives the temperature at which the thermal convection starts, with T_o the typical temperature of the Sunspots. According to some numerical data from specialized literature, it is possible to make an approximate calculation of that temperature.

Let $T_o = 3{,}700\ K$; $\alpha = 0.5\times10^{-4}\cdot K^{-1}$; $g = 2.74\times10^{4}\ cm\cdot sec^{-2}$; $z = 8\times10^{7}\ cm$ and $\gamma = 2.4\times10^{8}\ cm^{2}\cdot K^{-1}\ sec^{-2}$. In this case from (65.16) it is obtained that

$$T' = 2{,}344\ K. \tag{65.17}$$

So that

$$T = T_o + T' = 6{,}044\ K, \tag{65.18}$$

which is the approximate mean temperature of the solar photosphere. Additionally and with the help of the previous data, it is easy to see that convective movements are generated when

$$\left|\frac{dT'}{dz}\right| > 1.32\times10^{-5}\,K\cdot cm^{-1}. \qquad (65.19)$$

Finally, considering that if $\rho = 10^{-5}\,gr\cdot cm^{-3}$ in the solar photosphere; the mass density inside the Sunspots would have to be slightly smaller than that value. Assuming that for the region of the Sunspots and according to condition (22.3), $\rho_o = 0.715\times10^{-6}\,gr\cdot cm^{-3}$. Thus

$$h = 3{,}836\ gauss, \qquad (65.20)$$

as it is easy to see from any of the equations (65.12) or (65.13). Certainly, the obtained value is within the order of magnitude of the magnetic fields of the Sunspots measured with the aid of the Zeeman Effect.

§66. Persistency of the sunspots

From the z-component of relation (65.3) the following result can be obtained

$$t' = \frac{\chi q}{T'\left[\gamma\, q - \dfrac{dp^o_M}{dz} + \alpha g T'\right]}; \qquad (66.1)$$

where

$$q \equiv \frac{dT'}{dz} \qquad (66.2)$$

is the thermal gradient in the z-direction.

Persistence of the Sunspots seems to depend mainly on the z-components of both the thermal gradient and the gradient of the square of the magnetic field which generates them, respectively. The dominant mechanism at the beginning of the phenomenon is that of magnetic regulation of the convective movements; whereas at the end of it, disappearance of the Sunspots is determined by the predominance of the thermal gradient over the other quantities. It is possible that the initiation as well as the end of the phenomenon occurs very rapidly due to the magnitude of the implicated variables, so that it seems reasonable to assume that average life time of the Sunspots can be determined from the dynamical equilibrium between the magnetic regulation and startup mechanism of thermal convection. So that and for the case in which $F_{MHD} \approx 0$, in (66.1) we only have

$$\tau \approx \frac{\chi q}{\alpha g T'^2} \quad , \tag{66.3}$$

which is not other than the average life time of the phenomenon. The former relationship must at least provide an order of magnitude for the time duration of the Sunspots. Astronomical observations of the phenomenon made by other researchers, show that their mean life time could be of days, weeks or even several months. In general it is asserted that the average life time is somewhat greater than one solar revolution; that is approximately 28 days. Let us consider for instance that for the most superficial layers of the Sun´s photosphere $\chi \approx 10^{18}\ cm^2 \cdot sec^{-1}$. Then, with the numerical data from the previous paragraph and given that according to specialized literature $q=2\times10^{-5}\ K{\cdot}cm^{-1}$, the following is obtained

$$\tau \approx 31\ days\ . \tag{66.4}$$

The permanency in time of a particular couple of spots in the solar disk depends on the difference between its temperature T_o and that of the photosphere T; this is, it depends on T'. Thus, the higher the spots temperature the smaller T' and, consequently, the larger the spot average life time.

§67. Origin, permanency, disappearance, and properties of the sunspots

From the results obtained, it can be assumed that the Sunspots are phenomena very much related to processes occurring on the surface of the Sun, at a very shallow depth in the solar photosphere. In other words they are basically surface phenomena. Their generation, evolution, and disappearance can be explained in terms of the competition between two dynamical mechanisms: one of magnetic nature which controls and delays thermal convection, and the other one of thermal nature, which propitiates convective movements. It is possible that what occurs is as follows. The magnetic field characteristic of solar spots is created at some moment in certain regions of the photosphere. While its intensity increases, a growing magnetic viscosity is generated in the fluid which makes it more rigid. Such rigidity damps the thermal agitation in the fluid and makes if difficult for the fluid to move in those regions. As a consequence, the temperature in those regions decreases creating the required conditions for the appearance of a thermal gradient whose value is adjusted to the increasing intensity of the magnetic field being created. Thus, as the magnetic field intensity grows, the rigidity in the fluid increases and the magnitude of the thermal gradient grows bigger, firmly opposing the establishment of the square of magnetic field gradient. The fight between these two mechanisms leads to a gradual loss of thermal energy in those regions, resulting in areas which are darker and colder that their surroundings, which constitute the Sunspots. On the other hand, the solar activity in those regions begins to grow and reaches its maximum value when the creation of the thermal and of the square of magnetic field gradients is completed. Next, an evolution period of the spots takes place, which is characterized by the existence of a state of dynamic equilibrium between both mechanisms, and by the fact that Sun's activity in those regions is at its maximum. In that intermediate stage, the dominant force is the huge solar gravity which acts over the fluid.

Throughout this period, the hot fluid flows more or less parallel to the surface of the Sun due to its gravitational attraction, diverging from the center of the spots and overflowing through its boundaries. At the same time different varieties of the so called Faculae are produced that is, as sparks, torches, and flames, depending on theirs size. It is possible that the origin of such things can be found in the splashing caused by thermal fluctuations in the fluid which occurs at the edges of the spots.

In the final stage, the magnitude of the magnetic field begins to decrease. Meanwhile the importance of the thermal processes increase until they become the dominant part when $\boldsymbol{h} = 0$. At this moment the start up mechanism of thermal convection is completed and violent convective movements begin, which mix the fluid inside the spots free of its magnetic stiffness, with the fluid from the photosphere; thus, equalizing the temperature in those regions with that of its surroundings. With that, the spots disappear at those latitudes and the solar activity in them is depleted.

Another interesting result explains the polarity of the Sunspots, as well as their tendency to appear as magnetic bipolar couples. Apparently, they can be considered as huge electromagnets produced by gigantic solenoids formed under the surface of the photosphere by the ionized fluid which rotates at high speed and creates the monstrous currents which generate and maintain the characteristic magnetic field in those regions. The following hypothesis is suggested by the polarity in the spots, their appearance, permanency in the solar disk, and subsequent disappearance. The highly ionized solar fluid must be basically formed by an non homogeneous mixture of positive and negative ions; as well as by electrically neutral atoms. It can be assumed that throughout one solar cycle, the contents of positive ions is larger than that of negative ions; so that the electric current produced by such an excess of charge would determine the polarity of the spots in each hemisphere. On the other hand, and due to the fact that the surface where the spots appear is relatively small as those regions move through the solar disk, an at the end of a certain time which must coincide with their average life time, it can be expected that locally, the number of positive and negative ions which rotate in the solenoid become equal.

Both kinds of ion must move in the same way. As a positive particle in motion is equivalent to a negative particle which moves in the opposite direction, the net current in the solenoid is zero and, of course, the magnetic field is cancelled. This fact is enough to give a heuristic explanation about the origin, duration, and disappearance of the spots, as well as their polarity and the end of the solar activity in those regions.

To explain the change in polarity at both hemispheres in each cycle, it is enough to assume that at Sun′s latitudes where the spots are usually observed, gigantic turbulent fluid flows take place, in a way similar to what occurs in the Earth at the cyclonic zones. The flows and the turbulent eddies in the northern hemisphere move in opposite way as they do in the southern hemisphere due to the Coriolis Force. If throughout one cycle the highly ionized fluid has a net charge of a given sign, the polarity of the electromagnets would be inverse in one hemisphere with respect to the other one.

For instance, couples of spots in the northern hemisphere would have (*N*-*S*) polarity, while in the southern hemisphere their polarity would be (*S*-*N*). At the end of 11.5 years a new cycle begins and that disparity is inverted, so that northern couples would have (*S*-*N*) polarity and the southern couples (*N*-*S*) polarity. This fact suggests that in one cycle the huge electric currents which feed the gigantic solenoids which create the magnetic field of the spots, are generated by an excess of charge of the same sign in both solar hemispheres; and in the following cycle, by an excess of charge of the opposite sign. The question is: what would be the physical mechanism which originates such change in polarity with so much regularity and in such a short time ? Consider the following argument. Due to the high temperatures and pressures in the deep regions of the Sun, the conditions are given for the production of the solar plasma, basically formed by a mixture of positive and negative ions. Assume that the highly ionized fluid under the influence of a tremendous thermal stirring is slowly pushed against the solar gravity and the disorderly thermal movement, towards regions of the photosphere until the turbulent tropical regions of the Sun are reached. Here they would feed the solenoids responsible for the production of the intense magnetic fields of the couples of Sunspots.

The powerfuel schock waves coming from the convective zone and from the oven itself could be responsible of the drag of ions from this zone up to the surface of the Sun. This drag is responsible of removing both the huge quantity of heat produced and products from the combustion processes feeding at the same time the thermonuclear oven with new fuel. However, shock waves would drag positive and negative ions in a different way and that difference can depend on the mass of each type. In fact, even in the ease of ionized hydrogen (H^+), the difference between its mass and that of the free electrons (e^-) produced during the process of ionization is around 2000. For other elements ionized by the loss of one or more electrons, that difference could be even larger, such is the case of H_e^+ whose mass could be four times bigger than that of H^+. Then, the migration of positive ions is slower than that of negative ions and such difference could make the former to reach the photosphere with a delay of 11.5 years with respect to the latter in each solar cycle.

Thus, in one cycle there could be surges of ionized fluid mainly of negative charge and in the following mainly of positive charge. This last argument is only one hypothesis derived from the results previously obtained and is subjected to validation. It is put forward here as a heuristic attempt to explain the above mentioned phenomenon because it is difficult to imagine some other process occurring inside the Sun that could have so regular effects.

Nevertheless, maybe one could have an indirect proof of the periodicity and origin of this phenomenon by determining whether in the Solar Wind of the present solar cycle there is excess of a certain type of charge reaching the Earth and coming from both hemispheres of the Sun and if the sign of that charge coincides with the present polarity of the spots. The next step would be to measure whether in the next cycle there is also the said excess of charge and if it is of the opposite sign that the excess of charge of the previous cycle. If the answer is affirmative the polarity of the spots in both hemispheres has to be as expected and clearly, it must be opposite to that of the previous cycle. If the previous hypothesis is proven, the problem of finding a theoretical solution to the phenomenon of ionic migration in a star like the Sun, remains unsolved. In the opposite case, another possible explanation would be searched for.

Selected Topics

The spontaneous magnetic field produced by a turbulent motion

Turbulent motion of conducting fluid has the remarkable property that may lead to spontaneous magnetic fields which can be quite strong. It is assumed that there are always small perturbations in a conducting fluid, resulting from causes extraneous to the fluid motion itself. For example, the magnetomechanical effect in rotating parts of a fluid, or even thermal fluctuations. These effects are accompanied by very weak electric and magnetic fields. The question is whether these perturbations are, on the average, amplified or damped by the turbulent motion in the course of time. The following argumentation shows that either may occur, depending on the properties of the fluid itself.

The way of variation in time of magnetic field perturbations, once they have arisen, is determined by two physical mechanisms. One is the dissipation of magnetic energy which is converted into Joule´s heat of the induced currents which tends to diminish the field; the other one refers to the following: the magnetic field tends to increase, on the other hand, by the purely magnetic effect of the stretching of the lines of force. It is known that when a fluid of sufficiently high conductivity is in motion, the lines of magnetic force move as fluid lines, and the magnetic field varies proportionally to the stretching at each point on each line of force† . In turbulent motion any two neighbouring lines of force move apart, on the average, in the course of time. As a result of this effect, the lines of force are stretched and the magnetic field is strengthened. Under certain conditions, these two opposite tendencies may balance, and this will provide a criterion distinguishing the cases where the magnetic field perturbations increase from those where they are damped. In order to give a solution to the problem it

† *See chapter X.*

is assumed that while the magnetic field resulting from the turbulent motion remains weak its reciprocal effect on the motion can be neglected. In other words, we may consider ordinary fluid turbulence as providing some kind of background on which the magnetic perturbations develop.

Also, we assume a steady turbulent velocity distribution. The concept of steady will be used in the usual sense in turbulence theory; that is to say, as average values over the time which is consumed by the motion. In this case the average is over times which are of the order of the periods of the corresponding turbulent fluctuations but are small compared to the total time during which the system is observed.

The mathematical treatment is made from Navier-Stokes′ equation (35.9)

$$\frac{\partial \boldsymbol{v}}{\partial t}+(\boldsymbol{v}\cdot \boldsymbol{grad})\boldsymbol{v}=-\boldsymbol{grad}\left(\frac{p}{\rho}\right)+\vartheta \nabla^2 \boldsymbol{v}\,,$$

for incompressible fluids. If in the former equation the following vectorial identity $(\boldsymbol{v}\cdot\boldsymbol{grad})\boldsymbol{v}=1/2\,\boldsymbol{grad}\,\boldsymbol{v}-\boldsymbol{v}\times\boldsymbol{rot}\,\boldsymbol{v}$ is used and the rotational of the result is calculated, the following is obtained

$$\frac{\partial \Omega}{\partial t}=\boldsymbol{rot}\,(\boldsymbol{v}\times\Omega)+\vartheta\,\nabla^2\Omega\,; \tag{1}$$

where again

$$\Omega\equiv\frac{1}{2}\boldsymbol{rot}\,\boldsymbol{v}$$

is the vorticity.

Let us compare the equation (1) with the relationship (44.2)

$$\frac{\partial \boldsymbol{H}}{\partial t}=\boldsymbol{rot}\,(\boldsymbol{v}\times\boldsymbol{H})+\frac{c^2}{4\pi\sigma}\nabla^2\boldsymbol{H}\,, \tag{2}$$

which for a given velocity distribution determines the time variation of the magnetic field.

It can be seen that Ω and $\boldsymbol{H}$ satisfy equations of the same form; which become identical if

$$\vartheta = \frac{c^2}{4\pi\sigma} . \tag{3}$$

In the case of the former relationship it is fulfilled there is a solution of equation (2) for which

$$\boldsymbol{H} = constant \times \boldsymbol{\Omega} ; \tag{4}$$

so it can be stated that in the system a steady magnetic field can exist. This field, on the average, neither increases nor decreases, whatever the value of the constant coefficient in (4). We may say that there is a neutral equilibrium in which the dynamic mechanisms, mentioned above as determining the magnetic field, are exactly balanced.

In consequence, if the conductivity of the fluid exceeds $c^2/4\pi\vartheta$, the dissipative loss of electromagnetic energy will be insufficient to compensate the increase of the magnetic field by the stretching of the lines of force. Thus, the inequality

$$\frac{4\pi\vartheta\sigma}{c^2} > 1 \tag{5}$$

is the expression proposed as the condition for the spontaneous appearance of magnetic fields by the growth of small magnetic perturbations. We can say that the inequality (5) is the condition for turbulent motion to be unstable with respect to infinitesimal magnetic perturbations. However, it is a very stringent criterion. It may only be fulfilled in Sun´s photosphere and corona, and in the ionized interstellar gas, due basically to the fact that in those regions σ and ϑ increase with the mean free paths of the corresponding carriers of charge and mass. Finally, the condition (5) as a criterion of the behavior of the field is valid so long as it is reasonable to ignore the reciprocal effect of the induced magnetic field on the flow. What happens is that the field must be increased until the time when some steady state is established in which the reciprocal effect of the flux over the field can not be ignored.

Diffusion of magnetic field into a plasma

A problem which often arises in Astrophysics is the diffusion of a magnetic field into a plasma. If there is a boundary between a region with a plasma but no field and a region with field but no plasma, the regions will stay separated if the plasma has no resistivity, for the same reason that flux cannot penetrate a superconductor. Any emf that the moving lines of force generate will create an infinite electric current, and this is not physically possible. As the plasma moves around, pushes therefore the magnetic lines of force and can bend and twist them. This may be the reason for the filamentary structure of the gas in the Crab Nebula. If the resistivity is finite, the plasma can move through the field and vice versa. This diffusion process takes a certain amount of time, and if the movements are slow enough, the lines of force need not be distorted by the plasma motions. The diffusion time is easily calculated from the equation (44.2) for the case in which $\boldsymbol{v}=0$; that is to say

$$\frac{\partial \boldsymbol{H}}{\partial t}=\frac{c^2}{4\pi\sigma}\nabla^2\boldsymbol{H}. \tag{1}$$

This is a diffusion equation which can be solved by the well known method of separation of variables. However, if it is considered that x is a characteristic length of the spatial variation of the magnetic field $\boldsymbol{H}$, an approximate calculus can be made. In this case, it can be said that the laplacian operator can be grossly written as $1/x^2$ so that in (1) we have the following

$$\frac{\partial \boldsymbol{H}}{\partial t}=\frac{c^2}{4\pi\sigma x^2}\boldsymbol{H}. \tag{2}$$

The solution of this equation is

$$\boldsymbol{H}=\boldsymbol{H}_o e^{\pm t/\tau}; \tag{3}$$

where

$$\tau=\frac{4\pi\sigma x^2}{c^2} \tag{4}$$

is the characteristic time for magnetic field penetration into a plasma. It can be interpreted as the time annihilation of the magnetic field. As the field lines move through the plasma, the induced currents cause ohmic heating of the plasma and this energy comes from the energy of the field. The time τ for the average Earth magnetic field is about 10^4 years and for the typical magnetic field of the Sun is about 10^{10} years. For the case of the Sunspots and due to the high conductivity of the solar plasma, the magnetic field of a spot in a solar atmosphere at rest needs 10^3 years in order to disappear by diffusion; which means that the typical magnetic field of a Sunspot is not dissipated in the form of Joule´s heat by a diffusion process.

The thermal equation of the sunspots

In Astrophysics it is stated that from a thermodynamic point of view the stellar plasma behaves as an ideal gas, so that it must satisfy the equation $p = R\rho T$, with R the universal gas constant. In the case of the Sun, from the relations (51.3) and (51.4) we have that for the ideal gas

$$\left(\frac{\partial p}{\partial \rho}\right)_{T,H} = RT\,, \quad \text{and} \quad \left(\frac{\partial p}{\partial T}\right)_{\rho,H} = R\rho\,;$$

so that in (51.9) the following is obtained

$$p_i(\rho,T) = R'\rho T\,; \tag{1}$$

where

$$p_i(\boldsymbol{x},t) = p - \frac{H^2}{8\pi}$$

is the hydrostatic pressure inside the region occupied by each of the Sunspots of bipolar group of spots and $R' = 2R$. That pressure satisfies the hydrostatic equation (40.4). In consequence, the MHD thermal equation of state (51.9) can be written as the relation (1) which has the same form as the thermal equation of state for the ideal gas. Let us note that the presence of the external magnetic field modifies the functionality of the hydrostatic pressure.

References

1. Cowling, T.G. "Magnetohidrodinámica". Editorial Alhambra, S.A., Madrid, B.Aires, México (1968).
2. Fierros Palacios, Angel. "The sunspots". Enviado para su publicación a Rev. Mex. de Astronomía y Astrofísica (2002).
3. Gamow, G. "Una estrella llamada *Sol*". Espasa Calpe, S.A. Madrid (1967).
4. Landau, L.D. and Lifshitz, E.M. "Electrodynamics of Continuous Media". Addison-Wesley Publishing Co. (1960).
5. Landau, L.D. and Lifshitz, E.M. "Fluid Mechanics". Addison-Wesley Publishing Co. (1959).
6. Unsöld, A. "El Nuevo Cosmos". Editorial Siglo Veintiuno, S.A. (1979).
7. Wilson, P.R. "Solar and Stellar Activity Cycles". Cambridge Astrophysics Series: 24. Cambridge University Press (1994).

Chapter XII

The Hamilton Equations of Motion

§68. Legendre´s transformation

The lagrangian formulation of Fluid Dynamics developed with all the theoretical precision in this monograph is not the only scheme that can be proposed to face the dynamical problem of the flux of any fluid. In the scope of Theoretical Mechanics we also have Hamilton´s formulation, which is another alternative method equally powerful as that of Lagrange´s to work with the physical principles already established. In this scheme, a generalization of Legendre´s transformation is proposed by Hamilton as the adequate analytical procedure for the theoretical treatment of mechanical problems. Hamilton´s transformation reduces the differential equations which are obtained in the lagrangian formulation, to other mathematical relationships which have a particularly simple structure, called by Jacobi, the canonical form. The equations of motion which are associated to this new scope, replace the N second order differential equations of the lagrangian formulation by $2N$ first order differential equations which have very simple and symmetrical form. Here N represents the number of degrees of freedom which the mechanical system has that will be reformulated. Let us consider next that for an N particles set as the one described in paragraph 8, the classical lagrangian has the following functional form

$$L = L(q_i, \dot{q}_i, t). \tag{68.1}$$

According to Hamilton´s procedure, the first thing to be done is to introduce the generalized momenta p_i as the new variables defined in the following way

$$p_i = \frac{\partial L}{\partial \dot{q}_i}. \tag{68.2}$$

Next, a certain function H, which for conservative systems is identified as the total energy is proposed, so that

$$H = p_i \dot{q}_i - L . \qquad (68.3)$$

Finally and after solving the equations (68.2) for the variables $\dot{q}_i$ and substituting in (68.3) the obtained results, the new function known as the classical hamiltonian is expressed in terms of time t, the generalized coordinates q_i and the generalized momenta p_i, that is to say

$$H = H(q_i, p_i, t). \qquad (68.4)$$

The total differential of the former functional relationship has the following form

$$dH = \frac{\partial H}{\partial q_i} dq_i + \frac{\partial H}{\partial p_i} dp_i + \frac{\partial H}{\partial t} dt . \qquad (68.5)$$

On the other hand from (68.3) we have that

$$dH = \dot{q}_i dp_i + p_i d\dot{q}_i - \frac{\partial L}{\partial q_i} dq_i - \frac{\partial L}{\partial \dot{q}_i} d\dot{q}_i - \frac{\partial L}{\partial t} dt . \qquad (68.6)$$

However and in virtue of the definition (68.2) the terms $d\dot{q}_i$ cancel each other; whereas from Lagrange´s equations

$$\frac{d}{dt}\left(\frac{\partial L}{\partial \dot{q}_i}\right) - \frac{\partial L}{\partial q_i} = 0,$$

it is clear that

$$\frac{\partial L}{\partial q_i} = \dot{p}_i ;$$

so that the result (68.6) is reduced to the following relationship

$$dH = \dot{q}_i \, dp_i - \dot{p}_i \, dq_i - \frac{\partial L}{\partial t} dt \,. \tag{68.7}$$

The equations (68.7) and (68.5) are comparable if the following relationships are fulfilled

$$\begin{aligned} \dot{q}_i &= \frac{\partial H}{\partial p_i} \\ -\dot{p}_i &= \frac{\partial H}{\partial q_i} \end{aligned} \tag{68.8}$$

and

$$-\frac{\partial L}{\partial t} = \frac{\partial H}{\partial t} \,. \tag{68.9}$$

The set of results given in (68.8) is known as Hamilton's canonical equations; whereas the entities q_i, p_i are known by the name of canonical conjugate coordinates. Hamilton's canonical equations form a homogeneous system of $2N$ first order differential equations for the $2N$ variables q_i, p_i as a function of time t. The N second order Lagrange's differential equations describe the dynamical state of the mechanical system in an N-dimensional space called Configuration Space, where the generalized coordinates form the coordinated N axis. The instantaneous configuaration of the system is represented by a single particular point in that Cartesian space. As long as time goes by, the dynamic state of the system changes so that the representative point moves in Configuration Space tracing out a curve called the path of motion of the system. The motion of the system as used above, refers then to the motion of the representative point along this path in Configuration Space. Time can be considered as a parameter of the curve so that to each point on the path one or more values of the time can be associated. It must be emphasized that Configuration Space has no necessary connection with the physical three-dimensional space; just as the generalized coordinates are not necessarily position coordinates. The path of motion in Configuration Space will not have any resemblance to the path in three-dimensional space of any actual particle; due to the fact that each point on the curve represents the entire system configuration at some given instant of time. On the other hand, the $2N$ canonical equations describe the same previous situation but in terms of the motion of a representative point of the dynamical state of the whole system, in a $2N$-dimensional space constructed with the N variables q_i and the N variables p_i. That space is known by the name of Phase Space. Finally, it is possible to extract a fundamental theorem from the set of Hamilton´s canonical equations. That is the energy theorem. In fact, it can be demonstrated that if the classical lagrangian (68.1), and according to (68.9), also the hamiltonian, are not explicit functions of time, the total time derivative of the hamiltonian takes the following form

$$\frac{dH}{dt} = \frac{\partial H}{\partial q_i}\frac{\partial H}{\partial p_i} - \frac{\partial H}{\partial p_i}\frac{\partial H}{\partial q_i}\,; \tag{68.10}$$

where the relationships (68.8) were used. In consequence,

$$\frac{dH}{dt} = 0 \tag{68.11}$$

so that

$$H = constant\,. \tag{68.12}$$

If $L = T - V$, with T a cuadratic function in the generalized velocities and V a function which does not depend on the velocity, it is fulfilled that

$$H = T + V = E \tag{68.13}$$

with E the total energy of the system and T is the kinetic energy whereas V is the potential energy. Therefore, the relationship (68.12) is an expression of energy conservation law.

§69. Hamilton´s canonical equations

To obtain the set of canonical equations of motion of Fluid Dynamics we proceed this way. Let $\mathcal{H} = \rho h$ be the hamiltonian density whose general form is given by (19.14). Here, ρ is again the mass density whereas h is the specific hamiltonian such that

$$h = \frac{\partial \lambda}{\partial v^i} v^i - \lambda \,, \qquad (69.1)$$

where λ is the specific lagrangian for the MHD case, whose functionality is expressed in the relation (46.2). Let $\boldsymbol{\pi}(\boldsymbol{x}, \boldsymbol{v}, t)$ be the generalized momentum whose i-component fulfill the following definition

$$\pi_i \equiv \frac{\partial \lambda}{\partial v^i} \,, \qquad (69.2)$$

which is the canonical momentum conjugate of the spatial coordinate $x^i(t)$. In this case in (69.1) we have that

$$\lambda = \pi_i v^i - h \,. \qquad (69.3)$$

The action integral is defined as usual so that in (14.1) and taking into account the condition (14.2), the result (16.4) is reached again; where the relation (16.5) was also used. On the other hand and because the specific hamiltonian has the following functionality $h(\boldsymbol{x}, \pi, \tilde{u}, \boldsymbol{B}, t)$, the calculus of variations applied to the relation (69.3) has as a direct consequence the following result

$$\rho \delta \lambda = \left[\rho \left(v^i - \frac{\partial h}{\partial \pi_i} \right) \right] \delta \pi_i + \rho \pi_i \delta v^i - \rho \frac{\partial h}{\partial x^i} \delta x^i - \rho \frac{\partial h}{\partial u_{ij}} \delta u_{ij} \qquad (69.4)$$

only, due to the fact that $\delta\boldsymbol{B}=0$ as it is easy to see from equation (46.4). On the other hand, from the relations (69.1) and (69.3) it is evident that $\partial h/\partial \pi_i = v^i$ so that the coefficient of the $\delta \pi_i$ becomes null, as it is easily seen in (69.4). In other words, an arbitrary variation of the π_i has no influence at all on the variation of the specific lagrangian and then it has no influence either on the temporary-space integral of $\rho\delta\lambda$. This gives the following important result: originally we required the vanishing of the first variation of the action integral, even when the π_i are not free variables, but certain given functions of the spatial coordinates and the velocity field. Thus, the fact that the variation in the π_i is in some way determined by the variation of x^i is fulfilled. However and since the local variation of π_i has no influence at all on the variation of the action integral, we can enlarge the validity of the original Hamilton-Type Variational Principle and state again that the action integral is an invariant even if the π_i can be varied arbitrarily. In other words this fundamental postulate is still valid in spite of the fact that the π_i are considered as a second set of independent variables. Thus, with the help of the results (18.4) and (33.4) substituted in (69.4) the following is obtained

$$\rho\delta\lambda = \frac{d}{dt}\left[\rho\pi_i\delta x^i\right] - \frac{d}{dt}(\rho\pi_i)\delta x^i - \rho\frac{\partial h}{\partial x^i}\delta x^i - \frac{\partial}{\partial x^j}\left[\rho\frac{\partial h}{\partial u_{ij}}u_{ik}\delta x^k\right]$$
$$+\frac{\partial}{\partial x^j}\left[\rho\frac{\partial h}{\partial u_{ij}}\right]u_{ik}\delta x^k;$$

where two integrations by parts were carried out. The fourth term of the right hand side of the former result is of the same form as the one we have in the expression (33.7), so that it also becomes null when it is integrated and Green´s theorem is used. Moreover, if on the other hand $\delta x^k = \delta^i_k\delta x^i$ it is evident that

$$\frac{\partial}{\partial x^j}\left[\rho\frac{\partial h}{\partial u_{ij}}\right]u_{ik}\delta x^k = \frac{\partial}{\partial x^j}\left[\frac{\partial h}{\partial u_{ij}}\right]\delta x^i; \qquad (69.5)$$

where the definition (33.5) was used and previously an integration by parts was made. In this case,

$$\rho\delta\lambda = \frac{d}{dt}\left[\rho\pi_i\delta x^i\right] - \frac{d}{dt}(\rho\pi_i)\delta x^i - \left[\rho\frac{\partial h}{\partial x^i} - \frac{\partial}{\partial x^j}\left(\frac{\partial h}{\partial u_{ij}}\right)\right]\delta x^i \qquad (69.6)$$

Let us consider now the first term of the right hand side of (69.6). When it is integrated with respect to time it is clear that a result equal in form to that obtained in the expression (33.8) is obtained. In consequence and with all the latter results substituted in (16.4) the following is obtained

$$\int_{t_1}^{t_2}\int_{R}\left[\mathcal{D}(\rho\pi_i)+\rho\frac{\partial h}{\partial x^i}-\frac{\partial}{\partial x^j}\left(\frac{\partial h}{\partial u_{ij}}\right)\right]dV dt\,\delta x^i=0. \quad (69.7)$$

According to the general conditions imposed on this type of integrals, the former relation is only satisfied if the integrand becomes null; that is to say if

$$\mathcal{D}(\rho\pi_i)+\rho\frac{\partial h}{\partial x^i}-\frac{\partial}{\partial x^j}\left(\frac{\partial h}{\partial u_{ij}}\right)=0. \quad (69.8)$$

Moreover, according to (7.1) and (9.2) it is clear that

$$\mathcal{D}(\rho\pi_i)=\rho\dot{\pi}_i. \quad (69.9)$$

In this case, (69.8) is reduced to the following relationship

$$\dot{\pi}_i+\frac{\partial h}{\partial x^i}-\frac{1}{\rho}\frac{\partial}{\partial x^j}\left(\frac{\partial h}{\partial u_{ij}}\right)=0. \quad (69.10)$$

On the other hand and according to (48.2), from (69.3) we have that

$$\frac{\partial h}{\partial u_{ij}}=\sigma_{ij}^{o}, \quad (69.11)$$

with $\tilde{\sigma}^{o}(\boldsymbol{x}, t)$ the generalized stress tensor whose general form is given by the relationship (44.4). Then, from (69.10) the following is finally obtained

$$\frac{\partial h}{\partial x^i}=-\dot{\pi}_i+\frac{1}{\rho}\frac{\partial\sigma_{ij}^{o}}{\partial x^j}. \quad (69.12)$$

According to (69.3) it is evident that

$$\frac{\partial h}{\partial \pi_i} = \dot{x}^i . \tag{69.13}$$

Now, if the relation (69.1) is taken into account and because λ is a cuadratic function in the velocity field as it is easy to observe from the equation (48.2), it is possible to write that $h+\lambda = 2t$, with $t = 1/2\, v^2$ the specific kinetic energy. From the energy conservation principle it is clear that $h+\lambda=2(e-\square)$, with e the total energy per unit mass and $\square$ the specific internal energy. As e is a constant and the specific internal energy $\square$ is not an explicit function of time, it is fulfilled that

$$\frac{\partial h}{\partial t} = -\frac{\partial \lambda}{\partial t} . \tag{69.14}$$

The relationships (69.12) and (69.13) are Hamilton´s canonical equations sought. They are a non-homogeneous differential equations set which in Phase Space describe the motion state of a viscous, compressible and conducting fluid which flows in an external magnetic field. This general case corresponds to the dynamical problem of the MHD. If the external magnetic field is zero, the situation is reduced to the study of flow of newtonian viscous fluid. The application of the calculus of variations and the Hamilton-Type Principle to such mechanical system implies the repetition of the former analysis. The only difference lies in the relation (69.12) where we have the mechanical stress tensor instead of the generalized stress tensor. That is so because in (48.3) only the term corresponding to the specific internal energy of the viscous fluid would be preserved.For the case of perfect fluid, the mechanical stress tensor is reduced to minus the hydrostatic pressure multiplied by the components of Kronecker´s delta. The calculus of variations and Hamilton-Type Principle applied to that system directly leads to a relation which has the same form as the result (69.12), has. Nevertheless, for the case of ideal fluids we have that instead of the divergence of a tensor we would have the gradient of the hydrostatic pressure which is a scalar. In consequence, the former analysis contains all the cases of fluids studied in this book so that also the equations (69.12), (69.13), and the condition (69.14) are Hamilton´s canonical equations for Fluid Dynamics in their non-homogeneous form.

§70. The energy balance equation

To obtain the energy balance equation for any fluid from the set of Hamilton´s canonical equations for the general case of the former paragraph, we use the same methodology that is used in the Theoretical Mechanics scheme. In fact let us consider that the specific hamiltonian for the MHD has the functional form given in the latter paragraph; that is to say

$$h = h(\boldsymbol{x}, \pi, \tilde{u}, \boldsymbol{B}, t).$$

Its total derivative with respect to time is

$$\frac{dh}{dt} = \frac{\partial h}{\partial x^i}\dot{x}^i + \frac{\partial h}{\partial \pi_i}\dot{\pi}_i + \frac{\partial h}{\partial u_{ij}}\frac{du_{ij}}{dt} + \frac{\partial h}{\partial B_i}\frac{dB_i}{dt} + \frac{\partial h}{\partial t}; \qquad (70.1)$$

which with the aid of the set of Hamilton´s canonical equations and from the condition (69.14) is transformed as follows

$$\frac{dh}{dt} = \frac{\dot{x}^i}{\rho}\frac{\partial \sigma^o_{ij}}{\partial x^j} + \frac{\partial h}{\partial u_{ij}}\frac{du_{ij}}{dt} + \frac{\partial h}{\partial B_i}\frac{dB_i}{dt} - \frac{\partial \lambda}{\partial t}. \qquad (70.2)$$

On the other hand

$$\frac{du_{ij}}{dt} = \frac{\partial}{\partial x^j}\left(\frac{du_i}{dt}\right) = \frac{\partial}{\partial x^j}\left(v^\circ \frac{\partial u_i}{\partial x^\circ}\right) = \frac{\partial}{\partial x^j}\left(u_{i\circ} v^\circ\right)$$

due to the fact that the displacement vector only depends on position. In this case

$$\frac{\partial h}{\partial u_{ij}}\frac{du_{ij}}{dt} = \frac{1}{\rho}\frac{\partial}{\partial x^j}\left[\frac{\partial h}{\partial u_{ij}} v^i\right] - \frac{v^i}{\rho}\frac{\partial}{\partial x^j}\left(\frac{\partial h}{\partial u_{ij}}\right); \qquad (70.3)$$

where and in order to reach the former result, some integrations by parts were made. Moreover and given that the mass density does not depend on the magnetic induction, in (70.2) the following result is obtained

$$\rho \frac{dh}{dt} = \frac{\partial}{\partial x^j}\left[\frac{\partial h}{\partial u_{ij}} v^i\right] + \frac{\partial \mathcal{H}}{\partial B_i}\frac{dB_i}{dt} - \rho \frac{\partial \lambda}{\partial t}. \tag{70.4}$$

According to definitions (19.14), (47.3) and (47.6) it is easy to see that

$$\frac{\partial \mathcal{H}}{\partial \boldsymbol{B}} \cdot \frac{\partial \boldsymbol{B}}{\partial t} = -div\ \boldsymbol{S}; \tag{70.5}$$

where clearly, $\boldsymbol{S}$ is Pointing´s vector. Joining all of the former results and considering that if the specific lagrangian of the system is not an explicit function of time so that $\partial\lambda/\partial t=0$, in (70.4) is finally obtained that

$$\frac{\partial \mathcal{H}}{\partial t} + \frac{\partial}{\partial x^j}\left[v^i\left(\mathcal{H}\,\delta_{ij} - \frac{\partial h}{\partial u_{ij}}\right) + S_j\right] = 0; \tag{70.6}$$

and this is the energy balance equation of MHD. Thus, in the scope of Fluid Dynamics, equation (70.6) is equivalent to the energy theorem of conservative dynamical systems of Theoretical Mechanics of particles.

It can be demonstrated that when the former calculus is made for the case of the viscous fluid it is fulfilled that

$$\frac{\partial \mathcal{H}}{\partial t} + \frac{\partial}{\partial x^j}\left[v^i\left(\mathcal{H}\,\delta_{ij} - \frac{\partial h}{\partial u_{ij}}\right)\right] = 0; \tag{70.7}$$

relationship which corresponds to the physical situation in which the external magnetic field is zero; Pointing´s vector becomes null and $\partial h/\partial u_{ij}=\sigma_{ij}$. It is evident that $\tilde{\sigma}(\boldsymbol{x},t)$ is the mechanical stress tensor and the hamiltonian density as well as the specific hamiltonian are functions referring to viscous fuid. For the case of perfect fluid and given that the viscosity as well as the thermal conduction are irrelevant, it is fulfilled that $\sigma_{ij}=-p\,\delta_{ij}$ and the relationship (70.7) is reduced to the equation (21.4). Here $p(\boldsymbol{x},t)$ is again the hydrostatic pressure. The former analysis shows that in the Theoretical Mechanics scheme and by the use of the set of Hamilton´s canonical equations in their non-homogeneous form, the energy balance equation for any fluid is directly obtained.

§71. The field of the specific enthalpy

In order to make homogeneous the set of Hamilton's canonical equations from paragraph 69, it is convenient to define a physical field in terms of the specific hamiltonian and the internal energy, respectively. In consequence, be

$$\omega' = h + \int d\square + \mathcal{E}'_o \tag{71.1}$$

the field of specific enthalpy, with $\square$ a function which depends on the small deformations strain tensor, on the external magnetic field, and on the specific entropy as it is easy to see from the relationship (48.3).

$\mathcal{E}'_o$ is some arbitrary function with units of energy per unit mass which depends on the thermodynamic state of the system and on the external magnetic field. It is introduced so that the field ω' will be constructed from a certain thermodynamic state of reference. Therefore the following relation is fulfilled

$$\frac{\partial \omega'}{\partial t} = \frac{\partial h}{\partial t}. \tag{71.2}$$

Now and according to (45.7)

$$d\square = \left(\frac{\partial \square}{\partial s}\right) ds + \left(\frac{\partial \square}{\partial u_{ij}}\right) du_{ij} + \left(\frac{\partial \square}{\partial B_i}\right) dB_i \cdot$$

If the relation (33.4) is taken into account it is possible to write that

$$du_{ij} = \frac{\partial}{\partial x^j}\left(u_{ik} dx^k\right),$$

so that

$$d\square = T ds + \frac{\partial}{\partial x^j}\left(\sigma^o_{ij} u_{ik} dx^k\right) - \frac{\partial \sigma^o_{ij}}{\partial x^j} u_{ik} dx^k + d\left(\frac{H^2}{8\pi\rho_o}\right);$$

where an integration by parts was made and the definitions (32.16), (45.8) and (45.9) were considered. Moreover, it is clear that

$$T ds = d\left[Ts - \int \frac{dp}{\rho} \right].$$

In order to be able to write the former result, an integration by parts was made and Gibbs-Duhem´s relationship (25.8), was used. On the other hand, it can be demonstrated that

$$\frac{\partial}{\partial x^j}\left(\sigma^o_{ij} u_{ik} dx^k\right) = \frac{\partial}{\partial x^j}\left[d\left(\sigma^o_{ij} u_{ik} x^k\right) - x^k d\left(\sigma^o_{ij} u_{ik}\right)\right]$$

$$= d\left[\frac{\partial}{\partial x^j}\left(\sigma^o_{ij} u_{ik} x^k\right) - \sigma^o_{ij} u_{ij} \right] = d\left[-\frac{x^i}{\rho_o}\frac{\partial \sigma^o_{ij}}{\partial x^j}\right];$$

where, an integration by parts was made, it was considered that $x^k = \delta^i_k x^i$, and the relation (32.14) was used. In this case

$$d\square' = d\left[Ts - \int \frac{dp}{\rho} + \frac{H^2}{8\pi\rho_o} - \frac{x^i}{\rho_o}\frac{\partial \sigma^o_{ij}}{\partial x^j}\right] - \frac{\partial \sigma^o_{ij}}{\partial x^j} u_{ik} dx^k .$$

On the other hand and again according to the relation (32.14) it is obvious that

$$-\frac{x^i}{\rho_o}\frac{\partial \sigma^o_{ij}}{\partial x^j} - \frac{\partial \sigma^o_{ij}}{\partial x^j} u_{ik} x^k = -\frac{x^i}{\rho_o}\frac{\partial \sigma^o_{ij}}{\partial x_j}\left(1 + \rho_o u_{oo}\right) = -\frac{x^i}{\rho}\frac{\partial \sigma^o_{ij}}{\partial x^j}.$$

In consequence, the field of the specific enthalpy has the following explicit form

$$\omega' = h - \frac{x^i}{\rho}\frac{\partial \sigma^o_{ij}}{\partial x^j} \tag{71.3}$$

only, due to the fact that by definition it is proposed that

$$\mathcal{E}'_o = \int \frac{dp}{\rho} - Ts - \frac{H^2}{8\pi\rho_o} = -\left[\frac{H^2}{8\pi\rho_o} + \int T ds\right]; \qquad (71.4)$$

where Gibbs-Duhem´s relationship was used again and an integration by parts was carried out. For the case of the simple viscous fluid and in absence of the external magnetic field it is fulfilled that

$$\mathcal{E}_o = -\int T ds \qquad (71.5)$$

whereas for the perfect fluid and because the specific entropy is a constant, it is evident that

$$\mathcal{E}_o^o = 0. \qquad (71.6)$$

According to the definitions (69.3) and (69.11), from (71.3) and taking into account the fact that λ is not an explicit function of the coordinates we have that

$$\frac{\partial \omega'}{\partial x^i} = -\dot{\pi}_i$$

and (71.7)

$$\frac{\partial \omega'}{\partial \pi_i} = \dot{x}^i .$$

The relationships (71.7) are the MHD homogeneous form of Hamilton´s canonical equations. Moreover, from (69.14) and (71.2) the following condition is obtained

$$\frac{\partial \omega'}{\partial t} = -\frac{\partial \lambda}{\partial t}. \qquad (71.8)$$

If the external magnetic field is zero, the case which is studied is the one of viscous fluid for which it is fulfilled that *d□=d□*; so that the corresponding specific hamiltonian is only $h(\boldsymbol{x}, \boldsymbol{\pi}, \tilde{u}, t)$. In consequence, instead of (71.3) and according to (32.17) we have the following relationship

$$\omega = h - \frac{x^i}{\rho} \frac{\partial \sigma_{ij}}{\partial x^j} \cdot \tag{71.9}$$

It is easy to prove that the application of calculus of variations and the Hamilton-Type Principle for to the case of viscous fluid, has as a direct consequence that instead of the equation (69.12) the following result is obtained

$$\frac{\partial h}{\partial x^i} = -\dot{\pi}_i + \frac{1}{\rho} \frac{\partial \sigma_{ij}}{\partial x^j}; \tag{71.10}$$

so that the set of homogeneous Hamilton's canonical equations for this case, turns to be again of the same form as the relationships (71.7).

If the continuous system that is being studied is a perfect fluid, the theoretical scheme of paragraph 69 directly leads to the following result

$$\frac{\partial h}{\partial x^i} = -\dot{\pi}_i + \frac{1}{\rho} \frac{\partial p}{\partial x^i}; \tag{71.11}$$

where $h(\boldsymbol{x}, \boldsymbol{\pi}, \rho, t)$ is the corresponding specific hamiltonian and clearly, $p(\boldsymbol{x}, t)$ is again the hydrostatic pressure. For that case, the definition of the field of the specific enthalpy turns out to be the following

$$\omega_o = h + \frac{x^i}{\rho} \frac{\partial p}{\partial x^i}, \tag{71.12}$$

due to the fact that the specific internal energy only depends on the mass density and obviously satisfies the relation (32.18). Thus, the homogenized Hamilton's canonical equations for the perfect fluid are going to have the same form which the set (71.7) has. Consequently, these relationships in Phase Space represent the homogenized Hamilton's canonical equations for Fluid Dynamics. It can be demonstrated by direct calculus that the proposed general definition of that physical field and with the homogenized Hamilton´s canonical equations have again as a consequence the energy balance equation for any fluid.

§72. Nature of interactions in fluid dynamics

In the mathematical Phase Space it is possible to define a physical field which has the property of transmitting the continuous medium the dynamic information which contains that space, in such a way that the field turns out to be responsible of the flow process. Thus, in Phase Space it is possible to define in a mathematical way certain regions or hamiltonian surfaces which have the property of containing all the dynamic information. It can be stated that it is about those regions for which the action integral is an invariant. Any fluid found in one of those dynamical surfaces will receive information from they through the field, which at every instant will indicate the fluid towards where and in what way it must move. The system respects the instructions received and displaces or flows until it reaches another hamiltonian region that will transmit it by means of the field, new dynamic information. Phase Space does not behave anymore as a simple scenario where the physical events occur but rather as the receptacle of dynamical information that transmits at the moment via the field, to the continuous medium. This transforms it right away into the corresponding process of flow. Evidently, that new space is not the same as the physical space. Now we have a scenario richer in qualities. A continuous medium inmerse in it feels the necessity to move urged by the information imposed through the field, by each one of the hamiltonian surfaces in which it moves. As it is easy to see from the general definition (71.8), the flux process is realized in such a way that the total energy balance is warranted. On the other hand, Phase Space obtains the dynamical information from the field sources, which are shown in the way of the divergence of a rank two tensor proposed in paragraph 71. It can be stated that Phase Space has a field whose sources not only give it the dynamic information but also act over the continuous medium to determine the nature of the flux. Thus, in the scope of Fluid Dynamics, the nature of interactions can be seen in terms of a primary physical entity which is the field of specific enthalpy. Its performance as a mediator agent between the continuous medium and the hamiltonian regions of Phase Space is an invitation to consider it as the essence of the force in Hydrodynamics as well as in MHD.

Selected Topics

The specific hamiltonian

Hamilton's generalization of Legendre's transformation between the specific lagrangian and hamiltonian functions, can be obtained from Euler-Lagrange's equations of Fluid Dynamics and from a specific lagrangian which only depends on the coordinates and the velocity field; according to the following analytical procedure. Let

$$\lambda = \lambda\big(\boldsymbol{x}(t); \boldsymbol{v}(\boldsymbol{x},t)\big)$$

be the specific lagrangian. Its total differential is

$$d\lambda = \frac{\partial \lambda}{\partial x^i} dx^i + \pi_i d\dot{x}^i ;$$

where the definitions (2.1) and (69.2) were used.

From Euler-Lagrange's equation (46.6) the following result is obtained

$$\frac{\partial \lambda}{\partial x^i} = \dot{\pi}_i + \frac{1}{\rho}\frac{\partial}{\partial x^j}\left(\frac{\partial \lambda}{\partial u_{ij}}\right);$$

in such a way that

$$d\lambda = \left[\dot{\pi}_i + \frac{1}{\rho}\frac{\partial}{\partial x^j}\left(\frac{\partial \lambda}{\partial u_{ij}}\right)\right] dx^i + d\big(\pi_i \dot{x}^i\big) - \dot{x}^i d\pi_i ;$$

where an integration by parts was made. In that case and according to definitions (45.8) and (49.2), from the former relation the following result is obtained

$$d\big(\pi_i \dot{x}^i - \lambda\big) = \dot{x}^i d\pi_i - \left(\dot{\pi}_i - \frac{1}{\rho}\frac{\partial \sigma^o_{ij}}{\partial x^j}\right) dx^i ,$$

with $\tilde{\sigma}^o(\boldsymbol{x},t)$ the generalized stress tensor and $\rho(\boldsymbol{x},t)$ the mass density. The argument of total differential is the specific hamiltonian as it can easily be seen from (69.3). From the relationship between differentials

$$dh = \dot{x}^i d\pi_i - \left(\dot{\pi}_i - \frac{1}{\rho}\frac{\partial \sigma_{ij}^o}{\partial x^j}\right)dx^i;$$

in which the spatial coordinates and the generalized momenta are the independent variables, the following relationships are obtained

$$\frac{\partial h}{\partial x^i} = -\dot{\pi}_i + \frac{1}{\rho}\frac{\partial \sigma_{ij}^o}{\partial x^j}$$

and

$$\frac{\partial h}{\partial \pi_i} = \dot{x}^i .$$

These are Hamilton´s canonical equations for Magnetohydrodynamics in their non-homogeneous form. Notice that for the transformation effects no dependence of the specific lagrangian on time was considered. It is easy to see that when time is included as another independent variable the following general condition is obtained

$$\frac{\partial h}{\partial t} = -\frac{\partial \lambda}{\partial t}.$$

The canonical integral

Because of the duality of Legendre´s transformation, it is possible to start with the specific hamiltonian and later construct the specific lagrangian and then obtain the expression (69.3). Next the action integral must be formed as follows

$$W = \int_{t_1}^{t_2}\int_R \left(\pi_i \dot{x}^i - h\right)\rho dV dt, \qquad (1)$$

and demand that it is an invariant under an arbitrary local variation of the spatial coordinates as well as the generalized momenta. The independent variation of these variables gives origin to a set of $2N$ first order differential equations. The new variational principle which we now have is equivalent to the original Hamilton-Type Variational Principle but if we take into account the simple structure of the resulting differential equations, it is even more powerfull. The new action integral is known by the name of canonical integral. The integrand has again the classical form of kinetical energy minus potential energy. In fact, even when $h=h(\boldsymbol{x}, \pi, \tilde{u}, \boldsymbol{B}, t)$, the second term of it is in general a function of the coordinates, whereas the first term has a lineal dependence with the field velocity.

It can be demonstrated by direct calculus that the invariance condition applied to the action integral (1) has the following result

$$\int_{t_1}^{t_2}\int_R \delta\left(\pi_i \dot{x}^i - h\right)\rho dV dt = 0\,, \qquad (2)$$

where the relationships (5.5) and (16.5) were used. Now,

$$\delta\left(\pi_i \dot{x}^i - h\right) = \frac{d}{dt}\left(\pi_i \delta x^i\right) - \dot{\pi}_i \delta x^i - \frac{\partial h}{\partial x^i}\delta x^i - \frac{\partial h}{\partial \pi_i}\delta \pi_i$$
$$- \frac{\partial h}{\partial u_{ij}}\delta u_{ij} + \dot{x}^i \delta \pi_i\,;$$

only, since according to (46.4) $\delta\boldsymbol{B}=0$. Moreover an integration by parts was made. The first term of the right hand side of the former result is zero when it is integrated and the boundary condition (14.3) is used. On the other hand and according to (33.4)

$$\rho\frac{\partial h}{\partial u_{ij}}\delta u_{ij} = \rho\frac{\partial h}{\partial u_{ij}}\frac{\partial}{\partial x^j}\left(u_{ik}\delta x^k\right) = \frac{\partial}{\partial x^j}\left(\rho\frac{\partial h}{\partial u_{ij}}u_{ik}\delta x^k\right)$$
$$- \frac{\partial}{\partial x^j}\left(\rho\frac{\partial h}{\partial u_{ij}}\right)u_{ik}\delta x^k\,.$$

The first term of the former result is zero when it is integrated and Green's theorem is used, whereas according to (33.5) and (33.6)

$$\rho \frac{\partial h}{\partial u_{ij}} \delta u_{ij} = -\frac{\partial}{\partial x^j}\left(\frac{\partial h}{\partial u_{ij}}\right)\delta x^i .$$

With all the last results substituted in (2) the following is obtained

$$\int_{t_1}^{t_2}\int_R \left[\left(\dot{\pi}_i + \frac{\partial h}{\partial x^i} - \frac{1}{\rho}\frac{\partial \sigma^o_{ij}}{\partial x^j}\right)\delta x^i + \left(\frac{\partial h}{\partial \pi_i} - \dot{x}^i\right)\delta \pi_i\right]\rho dV dt = 0 .$$

Due to the fact that the variations $\delta\pi_i$ and δx^i are independent and arbitrary, the former expression is only zero if the two integrands disappear separately. The former means that

$$\frac{\partial h}{\partial x^i} = -\dot{\pi}_i + \frac{1}{\rho}\frac{\partial \sigma^o_{ij}}{\partial x^j} ,$$

and

$$\frac{\partial h}{\partial \pi_i} = \dot{x}^i .$$

These are again the MHD Hamilton's canonical equations in their non-homogeneous form. It is easy to see that in the former relationships the cases of newtonian viscous fluid and of perfect fluid are contained in such a way that the realized calculus is in general valid for Fluid Dynamics. The term that makes non-homogeneous the first of the equations emerges from the fact that the specific hamiltonian as well as the specific lagrangian also depend on the small deformations strain tensor, and on the external magnetic field.

Energy theorem

In Fluid Dynamics the energy balance equation is equivalent to the energy conservation principle of particle mechanics. It is obtained from the action invariance under continuous and infinitesimal temporary variations and as a consequence of time uniformity. According to definition (16.2), it is easy to demonstrate by direct calculus that

$$\int_{t_1}^{t_2}\int_R \left(\delta^+\circ - \frac{d\circ}{dt}\delta^+ t\right)dVdt = \int_{t_1}^{t_2}\int_R \left[\rho\left(\delta^+\lambda - \frac{d\lambda}{dt}\delta^+ t\right)\right]dVdt\,.$$

It is convenient to express the integral of the right hand side of the former result in terms of the specific hamiltonian

$$h = \pi_i \dot{x}^i - \lambda\,;$$

where π_i is the i-component of generalized momentum according to definition (69.2). In this case

$$\delta^+\lambda - \frac{d\lambda}{dt}\delta^+ t = \frac{dh}{dt}\delta^+ t - \delta^+ h\,;$$

in such a way that the action invariance under temporary transformations can be expressed as follows

$$\int_{t_1}^{t_2}\int_R \left[\rho\left(\delta^+ h - \frac{dh}{dt}\delta^+ t\right)\right]dVdt = 0\,. \tag{1}$$

According to the functionality of h and with the results of calculus of variations,

$$\delta^+ h = \left[\frac{\partial h}{\partial x^i}\dot{x}^i + \frac{\partial h}{\partial \pi_i}\dot{\pi}_i + \frac{\partial h}{\partial u_{ij}}\frac{du_{ij}}{dt} + \frac{\partial h}{\partial B_i}\frac{\partial B_i}{\partial t} + \frac{\partial h}{\partial t}\right]\delta^+ t\,;$$

so that with the help of the canonical equations (69.12) and (69.13) we have that

$$\delta^+ h = \left[\frac{\dot{x}^i}{\rho} \frac{\partial \sigma^o_{ij}}{\partial x^j} + \frac{\partial h}{\partial u_{ij}} \frac{du_{ij}}{dt} + \frac{\partial h}{\partial B_i} \frac{\partial B_i}{\partial t} + \frac{\partial h}{\partial t} \right] \delta^+ t \,.$$

It can be demonstrated by direct calculus that

$$\frac{\partial h}{\partial u_{ij}} \frac{du_{ij}}{dt} = \frac{1}{\rho} \frac{\partial}{\partial x^j} \left(\frac{\partial h}{\partial u_{ij}} \dot{x}^i \right) - \frac{\dot{x}^i}{\rho} \frac{\partial}{\partial x^j} \left(\frac{\partial h}{\partial u_{ij}} \right).$$

In this case and with the aid of the result (69.11) and given that

$$\frac{1}{\rho} \frac{\partial \mathcal{H}}{\partial \boldsymbol{B}} \cdot \frac{\partial \boldsymbol{B}}{\partial t} = -\frac{1}{\rho} \, div \, \boldsymbol{S} \,;$$

with $\boldsymbol{S}$ Pointing´s vector, the following is obtained

$$\rho \delta^+ h - \rho \frac{dh}{dt} \delta^+ t = -\left[\frac{\partial}{\partial x^j} \left(S_j - \dot{x}^i \sigma^o_{ij} \right) + \rho \frac{\partial \lambda}{\partial t} + \rho \frac{dh}{dt} \right] \delta^+ t \,.$$

In consequence in (1) we have that

$$\int_{t_1}^{t_2} \int_R \left[\frac{\partial}{\partial x^j} \left(S_j - \dot{x}^i \sigma^o_{ij} \right) + \rho \frac{\partial \lambda}{\partial t} + \rho \frac{dh}{dt} \right] \delta^+ t dV dt = 0 \,;$$

result that is only fulfilled if

$$\rho \frac{dh}{dt} + \frac{\partial}{\partial x^j} \left(S_j - \dot{x}^i \sigma^o_{ij} \right) + \rho \frac{\partial \lambda}{\partial t} = 0 \,. \qquad (2)$$

On the other hand

$$\rho \frac{dh}{dt} = \frac{\partial \mathcal{H}}{\partial t} + div(\mathcal{H}\dot{\boldsymbol{x}});$$

where $\mathcal{H}$ is the hamiltonian density, an integration by parts was made, and the continuity equation (9.2) was used. In this case in (2) the following result is obtained

$$\frac{\partial \mathcal{H}}{\partial t} + \frac{\partial}{\partial x^j}\left[\dot{x}^i\left(\mathcal{H}\delta_{ij} - \sigma^o_{ij}\right) + S_j\right] + \rho\frac{\partial \lambda}{\partial t} = 0\,;$$

where the general condition (69.14) was used. According to Theoretical Mechanics, time uniformity implies that the specific lagrangian is not an explicit function of time, in such a way that $\partial \lambda / \partial t = 0$ is fulfilled. The condition above mentioned has as a consequence the MHD energy balance equation

$$\frac{\partial \mathcal{H}}{\partial t} + \frac{\partial}{\partial x^j}\left[\dot{x}^i\left(\mathcal{H}\delta_{ij} - \sigma^o_{ij}\right) + S_j\right] = 0\,.$$

The former calculus is equally valid for the newtonian viscous fluid as well as for the perfect fluid, in such a way that it is valid for Fluid Dynamics.

The variation of generalized momentum

In the calculus of variations it is easy to demonstrate that the variation of the generalized momentum can be expressed in terms of the variations of the spatial coordinates, so long as the π_i are not free variables but rather functions of time, of spatial coordinates, and the field velocity; that is to say $\pi_i = \pi_i(\boldsymbol{x}, \boldsymbol{v}, t)$. In that case

$$\delta \pi_i = \frac{\partial \pi_i}{\partial x^j}\delta x^j + \frac{\partial \pi_i}{\partial v^j}\delta v^j$$

only, because $\delta t=0$ for every t. According to (69.2) and (18.4)

$$\delta\pi_i = \left[\frac{\partial\pi_i}{\partial x^j} - \frac{d}{dt}\left(\frac{\partial\pi_i}{\partial v^j}\right)\right]\delta x^j + \frac{d}{dt}\left[\frac{\partial\pi_i}{\partial v^j}\delta x^j\right];$$

where an integration by parts was made. Now,

$$\frac{\partial\pi_i}{\partial x^j} = \frac{\partial}{\partial x^j}\left(\frac{\partial\lambda}{\partial v^i}\right) = \frac{\partial}{\partial v^i}\left(\frac{\partial\lambda}{\partial x^j}\right);$$

because $\partial/\partial\boldsymbol{x}$ and $\partial/\partial\boldsymbol{v}$ are differential operators independent between them, in such a way that they can be interchanged. The last term is zero for the following. Let

$$\frac{\partial\lambda}{\partial\boldsymbol{x}} = -\boldsymbol{f}(\boldsymbol{x},t)$$

be the external force per unit mass and since it does not depend explicitly on the velocity, $\partial\boldsymbol{f}/\partial\boldsymbol{v}=0$. On the other hand and given that λ is a cuadratic function in the velocities

$$\frac{\partial}{\partial v^j}\left(\frac{\partial\lambda}{\partial v^i}\right) = \frac{\partial v^i}{\partial v^j} = \delta^i_j;$$

with δ^i_j the components of Kronecker´s delta. In consequence

$$\frac{d}{dt}\left(\frac{\partial\pi_i}{\partial v^j}\right) = \frac{d}{dt}\left(\delta^i_j\right) = 0.$$

In the same way

$$\frac{d}{dt}\left[\frac{\partial\pi_i}{\partial v^j}\delta x^j\right] = \frac{d}{dt}\left(\delta^i_j\delta x^j\right).$$

Therefore it is fulfilled that

$$\delta\pi_i = \frac{d}{dt}\left(\delta x^i\right);$$

which is what we wanted to demonstrate.

The homogeneous form of the canonical equations

By definition we have that

$$\omega' = h - \frac{x^i}{\rho}\frac{\partial\sigma_{ij}^o}{\partial x^j}$$

with ω' the field of generalized specific entalphy. Now and given that

$$\frac{\partial h}{\partial x^i} = -\dot{\pi}_i + \frac{1}{\rho}\frac{\partial\sigma_{ij}^o}{\partial x^j},$$

we have that

$$\frac{\partial\omega'}{\partial x^n} = -\dot{\pi}_n + \frac{x^i}{\rho^2}\frac{\partial\rho}{\partial x^n}\frac{\partial\sigma_{ij}^o}{\partial x^j} - \frac{x^i}{\rho}\frac{\partial}{\partial x^n}\left(\frac{\partial\sigma_{ij}^o}{\partial x^j}\right). \qquad (1)$$

Nevertheless and according to (69.3) and (69.11)

$$\sigma_{ij}^o = -\frac{\partial\lambda}{\partial u_{ij}},$$

in such a way that the last two terms of the right hand side of (1) are separately zero. In fact, the divergence of the former expression can be written as follows

$$\frac{\partial \sigma_{ij}^{o}}{\partial x^{j}} = -\frac{\partial}{\partial u_{ij}}\left(\frac{\partial \lambda}{\partial x^{j}}\right); \tag{2}$$

because the differential operators $\partial/\partial u_{ij}$ and $\partial/\partial x$ are independent between them and can be interchanged. If in general we have that $\partial\lambda/\partial \boldsymbol{x}=-\boldsymbol{f}$, with $\boldsymbol{f}(\boldsymbol{x},t)$ the external force per unit mass which certaintly does not depend on the small deformations strain tensor, the terms before quoted become null. In this case in (1) we have that

$$\frac{\partial \omega'}{\partial x^{n}} = -\dot{\pi}_{n}$$

which is the homogeneus form of the first equation (71.7). The other canonical equation has that form already.

Cyclic variables

Although it is true that Hamilton's canonical equations have a more simple structure than those of Lagrange's, there is not any general method for their integration. However, Hamilton's scheme is specially configured for the treatment of problems that implies cyclic variables or ignorable coordinates†.

The ignorable variables are very important in the integration process of the motion equations in both theoretical schemes. Whatever the variable may be, it is always possible to make a partial integration in order to simplify the given mechanical problem. The simplification process is much simpler within Hamilton's form of mechanics that it is in Lagrange's scheme. If the lagrangian does not contain a certain variable, the hamiltonian will not contain it either. If the generalized coordinate does not appear in the lagrangian and due to the form of Euler-Lagrange's equations, its conjugate momentum is a constant. In fact, if q_n is ignorable, we have that

$$\frac{d}{dt}\left(\frac{\partial L}{\partial \dot{q}_{n}}\right) = \dot{p}_{n} = 0$$

† *If Lagrange's function does not contain any generalized coordinate even when it does have the corresponding generalized velocity, then it is said that the coordinate is cyclic or ignorable.*

and $p_n = constant$. In the same way if a conjugate variable does not appear in the hamiltonian, its conjugate canonical momentum is conserved. In fact, from canonical equations we have that if q_n does not appear in the hamiltonian, $\partial H / \partial q_n = 0$, so that p_n is constant again. This is also true in the scope of Fluid Dynamics when the homogeneous form of the canonical equations is used.

References

1. Aris, R. "Vector, Tensors, and the Basic Equations of Fluid Mechanics". Prentice Hall, Inc. (1966).
2. Fierros Palacios, A. "Las ecuaciones canónicas de Hamilton en la dinámica de los fluidos". Rev. Mex. de Fís. 46 (4) 314-390 (2000).
3. Goldstein, H. "Classical Mechanics", Addison-Wesley Publishing Co. (1959).
4. Landau, L.D. and Lifshitz, E.M., "Fluid Mechanics". Addison-Wesley Publishing Co. (1960).
5. Landau, L.D. and Lifshitz, E.M., "Electrodynamics of Continuous Media". Addison-Wesley Publishing Co. (1960).
6. Landau, L.D. and Lifshitz, E.M., "Theory of Elasticity". Addison-Wesley Publishing Co. (1959).
7. Lanczos, C. "The Variational Principles of Mechanics". University of Toronto Press, 4th. Edition (1970).
8. Jackson, J.P. "Classical Electrodynamics". John Wiley & Sons, Inc. New York, London (1962).

Chapter XIII

Thermodynamics

§73. The system of a single phase

Classical Thermodynamics is a part of physics which studies the properties of matter concerning temperature changes. It is a phenomenological science. The systems studied are macroscopic in such a way that no model concerning the intimate structure of matter is considered. It is a physical theory of universal contents settled on an small set of basic postulates called the thermodynamic laws. Its formal structure is elegant and the scope of application of its fundamental concepts is very wide.

The simplest thermodynamic system consists of a single continuous and homogeneous medium. Its structure founds the hydrodynamics of compressible fluids in such a way that its mathermatical description can be appropriately made with the Theoretical Mechanics methods within the Hamilton-Type Variational Principle. One single phase is described by certain thermodynamic variables of state as the volume V, the total internal energy E, the pressure p, the total entropy S, and the absolute temperature T. Some of them as the energy and the volume are well defined macroscopic concepts which have the same significance in Mechanics as well as in Thermodynamics. Other variables as the entropy and the temperature are the result of purely statistical regularities, which only have meaning for macroscopic systems. The structure of the phase is specified by means of certain relationships between the thermodynamic variables. The combinations which can be established between them are known by the name of equations of state. It is well known that when two any variables of state are fixed the remainders are determined.

As it will be seen next, the problem of Classical Thermodynamics as a branch of Lagrange´s Analytical Mechanics may be tackled by means of the theoretical formalism of Hamilton-Type Variational Principle.

§74. Total internal energy

Let us consider a simple thermodynamic system which consists of a single homogeneous continuous medium, at rest every moment, in some region R of three-dimensional Euclidean Space. It is assumed that this system experiments some process which changes its thermodynamic state without suffering any motion as a whole. In order to describe that physical situation a very simple lagrangian density is proposed which has the following functional form

$$\circ = \circ(\rho, t); \tag{74.1}$$

where $\rho(\boldsymbol{x}, t)$ is the mass density and t the time. From (16.2) the corresponding specific lagrangian is

$$\lambda = \lambda(\rho, t). \tag{74.2}$$

According to continuity condition (16.5) that governs the local variation of mass density

$$\rho \delta \lambda = -\rho^2 \frac{\partial \lambda}{\partial \rho} div(\delta \boldsymbol{x}).$$

With this result substituted in the equation (16.4) the following is obtained

$$\int_{t_1}^{t_2}\int_R \left[-\frac{\partial}{\partial x^i}\left(\rho^2 \frac{\partial \lambda}{\partial \rho} \delta x^i \right) + \frac{\partial}{\partial x^i}\left(\rho^2 \frac{\partial \lambda}{\partial \rho} \right) \delta x^i \right] dV dt = 0; \tag{74.3}$$

where an integration by parts was made. The application of Green´s theorem to the first term of former integral produces the following result

$$-\int_{t_1}^{t_2}\oint_{\Sigma} \rho^2 \frac{\partial \lambda}{\partial \rho} \delta x^i da_i dt;$$

where Σ is the surface that limits the region R and da the differential of area. The surface integral is zero by the same arguments which were before given for this type of integrals, in such a way that in (74.3) only the following remains

$$\int_{t_1}^{t_2}\int_{R}\left[\frac{\partial}{\partial x^i}\left(\rho^2\frac{\partial\lambda}{\partial\rho}\right)\right]\delta x^i dV dt = 0\,, \tag{74.4}$$

result which is only fulfilled if the integrand becomes null; that is to say if

$$\frac{\partial}{\partial x^i}\left(\rho^2\frac{\partial\lambda}{\partial\rho}\right) = 0\,. \tag{74.5}$$

These are the field differential equations. When the former relationship is integrated the following result is obtained

$$\rho^2\frac{\partial\lambda}{\partial\rho} = f(V,t)\,; \tag{74.6}$$

with *f (V,t)* some thermodynamic of state function whose gradient is zero. Be by definition

$$f(V,t) \equiv E(V,t)\,; \tag{74.7}$$

where E is the total internal energy contained in the volume V, which in this case is only a function of time and of the volume occupied by the system; so that in (74.6) the following is obtained

$$\rho^2\frac{\partial\lambda}{\partial\rho} = E\,. \tag{74.8}$$

Next it is proposed that the proper lagrangian density has the explicit following form

$$\circ = \rho(-\mathcal{L} + q)\,, \tag{74.9}$$

where $\square(\rho, t)$ is the specific internal energy and $q(t)$ the quantity of heat per unit mass; quantity which is only function of time. Evidently,

$$\lambda = -\square + q\,. \tag{74.10}$$

According to (17.1) it is fulfilled that

$$\rho^2 \frac{\partial \lambda}{\partial \rho} = -p\,,$$

with p the pressure which is exerted over the volume V of the region occupied by the system. In this case, in (74.8) the following result is obtained

$$E = -pV\,.$$

The total derivative with respect to time of the former expression is the analytical representation of the infinitesimal process before mentioned which changes the thermodynamic state of the system, that is to say

$$\frac{dE}{dt} = -p\frac{dV}{dt}\,.$$

This is the change on time of the specific internal energy if the system is in thermal isolation. In that situation it is said that the process is adiabatic and it is stated that the total entropy of the system remains constant. In consequence, the adiabatic work realized over the system is only due to a change in the volume occupied by the system, that is to say

$$\left(\frac{dR}{dt}\right)_{ad.} = -p\frac{dV}{dt}\,. \tag{74.13}$$

Here R is the mechanical work realized over the system. This last result is applied to reversible as well as irreversible processes and it is subjected to only one condition. During the whole process the system must stay in a mechanical equilibrium state; which means that at any moment the pressure must be a constant throught it. It can be stated that in the scope of

Classical Thermodynamics and according to Theoretical Mechanics, the existence of the internal energy as a thermodynamic function of state, is a consequence of homogenity of the space.

§75. The change in the total internal energy

According to Hamilton-Type Variational Principle, the invariance of the action under continuous and infinitesimal temporary transformations has as a result the energy balance equation. Next, the consequences of that invariance in the scope of Classical Thermodynamics will be determined. The temporary variation of the lagrangian density (74.1) is equal to

$$\delta^+ \circ = \frac{\partial \circ}{\partial \rho}\delta^+\rho + \frac{\partial \circ}{\partial t}\delta^+ t\,. \tag{75.1}$$

If the result (19.3) is used we have that

$$\delta^+ \circ = \left(-\circ\, div\, \mathbf{v} - \rho^2 \frac{\partial \lambda}{\partial \rho}\, div\, \mathbf{v} + \frac{\partial \circ}{\partial t} \right)\delta^+ t\,;$$

where the relationship (16.5) was used and the fact that

$$\rho \frac{\partial \circ}{\partial \rho}\, div\, \mathbf{v} = \circ\, div\, \mathbf{v} + \rho^2 \frac{\partial \lambda}{\partial \rho}\, div\, \mathbf{v}\,,$$

together with the definition (16.2). In this case

$$\delta^+ \circ - \frac{d\circ}{dt}\delta^+ t = \left(-\mathcal{D}\circ - \rho^2 \frac{\partial \lambda}{\partial \rho}\, div\, \mathbf{v} + \frac{\partial \circ}{\partial t} \right)\delta^+ t\,. \tag{75.2}$$

Substituting this result in (19.1), which is the resulting expression from the invariance of the action under temporary transformations, the following is obtained

$$\int_{t_1}^{t_2}\int_R \left(\mathcal{D}\circ + \rho^2 \frac{\partial \lambda}{\partial \rho} div\, \boldsymbol{v} - \frac{\partial \circ}{\partial t} \right) \delta^+ t\, dV dt = 0 . \qquad (75.3)$$

In the former expressions $\mathcal{D}$ is Reynolds′ differential operator. Again and for the same arguments for this type of integrals before used, the former relationship is only fulfilled if

$$\mathcal{D}\circ + \rho^2 \frac{\partial \lambda}{\partial \rho} div\, \boldsymbol{v} - \frac{\partial \circ}{\partial t} = 0 . \qquad (75.4)$$

This is the energy balance differential equation. According to Theoretical Mechanics the time uniformity has as a consequence the energy conservation law. It is said then that the lagrangian function of the system must not be an explicit function of time. For Classical Thermodynamics the condition $\partial \circ / \partial t = 0$ gives as a result the change in the total energy of the system. In fact, from definition (7.1) of Reynolds′ differential operator and from the explicit form (74.9) for the lagrangian density we have that

$$\mathcal{D}\circ = -\frac{dE}{dt} + \frac{dQ}{dt} ; \qquad (75.5)$$

where the continuity equation (9.2) was used and the fact that the continuous medium remains at rest as a whole. In this case in (75.4) we have the following

$$\frac{dE}{dt} = \frac{dQ}{dt} - p\, div\, \boldsymbol{v} . \qquad (75.6)$$

Here the definition (17.1) was used. On the other hand and according to (9.2)

$$p\, div\, \boldsymbol{v} = p \frac{dV}{dt} \qquad (75.7)$$

where $V = 1/\rho$ is the specific volume and p the pressure over the volume V. In consequence, it is fulfilled that

$$\frac{dE}{dt} = \frac{dQ}{dt} - p\frac{dV}{dt}, \tag{75.8}$$

where dV is the volume element in the region occupied by the system. This is the change in the total internal energy per unit time if there is no thermal isolation. In this case the energy of the system is acquired or granted from or toward the environs or other objects in thermal contact with it by direct transmission throughout its boundaries, as well as by mechanical work too. The part of the change in energy which is not identified with some mechanical process is called the quantity of heat received or granted by the thermodynamic system.

§76. The second law of thermodynamics

Let us consider now a single phase thermodynamic system which experiments some process. In any given instant it is asumed that the system will be in a thermal state of equilibrium represented by the values of its energy and volume in that instant; values which correspond to what was expressed by the relationship (75.8). The above mentioned does not mean that the process must be necessarily reversible, as it does not mean either that the system is not necessarily in equilibrium with its evirons. To describe such situation a lagrangian density with the following functional form is proposed

$$\circ = \circ(\rho, s, t), \tag{76.1}$$

where $s(\rho,\square t)$ is the entropy per unit mass and

$$\lambda = \lambda(\rho, s, t) \tag{76.2}$$

is the functional form of the specific lagrangian. Therefore,

$$\rho\delta\lambda = \rho\frac{\partial\lambda}{\partial\rho}\delta\rho + \rho\frac{\partial\lambda}{\partial s}\delta s.$$

Be by definition

$$\frac{\partial \lambda}{\partial s} \equiv -T \,, \tag{76.3}$$

with T the absolute temperature. On the other hand, the total entropy of any thermodynamic system is defined as a function of time, of the energy, and of the volume; that is to say $S = S(E,V,t)$. That function is introduced into Classical Thermodynamics proposing the following relationship

$$S \equiv k\sigma \,; \tag{76.4}$$

where σ is the entropy as it is defined in the scope of Statistical Physics and $k = 1.372 \times 10^{-16}$ ergs·grade^{-1} the Boltzmann's constant which indicates the number of ergs that we have in a grade and has the mission of dimensionally balance the former relationship. The total entropy in this way defined turns out to be independent from the geometrical parameters in such a way that

$$\delta S = \frac{\partial S}{\partial \alpha}\,\delta\alpha = 0 \,.$$

However and from Classical Thermodynamics view point it is also true that $S = \rho s$ so that the following relation is fullfiled

$$\delta s = s\,div(\delta \mathbf{x}) \,; \tag{76.5}$$

where the equation (16.5) was used. It is easy to prove by direct calculus that

$$\rho\delta\lambda = -\frac{\partial}{\partial x^i}\left[\left(\rho^2\frac{\partial\lambda}{\partial\rho} + \rho Ts\right)\delta x^i\right] + \frac{\partial}{\partial x^i}\left(\rho^2\frac{\partial\lambda}{\partial\rho} + \rho Ts\right)\delta x^i \,. \tag{76.6}$$

In order to obtain the former relation an integration by parts was made. Let us consider the first term of the right hand side of (76.6). Integrating and using Green's theorem the following result is obtained

$$\int_{t_1}^{t_2}\oint_{\Sigma}\rho^2\frac{\partial\lambda}{\partial\rho}\,\delta x^i da_i dt - \int_{t_1}^{t_2}\oint_{\Sigma}\rho Ts\,\delta x^i da_i dt \,;$$

where again, Σ is the surface which encloses the region R and $d\boldsymbol{a}$ is the element of area. Both surface integrals are zero. The first one by the same arguments which are used related to this type of integrals, and the second one because over the integration surface it is assumed that $\delta\boldsymbol{x}$ and $d\boldsymbol{a}$ are perpendicular between them so that their scalar product becomes null.

Therefore, in (16.4) only the following is kept

$$\int_{t_1}^{t_2}\int_R \left[\frac{\partial}{\partial x^i}\left(\rho^2\frac{\partial \lambda}{\partial \rho} + \rho T s\right)\right]\delta x^i \, dV dt = 0\,; \tag{76.7}$$

relationship which is only fulfilled if

$$\frac{\partial}{\partial x^i}\left(\rho^2\frac{\partial \lambda}{\partial \rho} + \rho T s\right) = 0\,. \tag{76.8}$$

These are the field differential equations for this case. The same assumption made before can be made again writing that

$$\rho^2\frac{\partial \lambda}{\partial \rho} + \rho T s = E\,; \tag{76.9}$$

where now $E = E(S,V,t)$. For the present situation it is proposed that the lagrangian density has the following explicit form

$$\circ = -\rho\square \tag{76.10}$$

because the quantity of heat in the state of thermal equilibrium initially considered is a constant. The corresponding specific lagrangian is

$$\lambda = -\sqsubset \tag{76.11}$$

and of course the definition (17.1) is fulfilled, so that in (76.9) the following result is obtained

$$E = TS - pV\,. \tag{76.12}$$

This is Euler´s form of the first and second laws of thermodynamics. It can be stated again that the total internal energy $E(S,V,t)$ emerges from the theoretical formalism used as a consequence of the homogeneity of space.

§77. Entropy and quantity of heat

Let us consider now the consequences which result from the action invariance under temporary transformations. According to the lagrangian functionality given in (76.1) the temporary variaton of $\circ$ is the following

$$\delta^+\circ = \frac{\partial\circ}{\partial\rho}\delta^+\rho + \frac{\partial\circ}{\partial s}\delta^+ s + \frac{\partial\circ}{\partial t}\delta^+ t\,. \tag{77.1}$$

From the definition of temporary variation the following result is obtained

$$\delta^+ s = \frac{ds}{dt}\delta^+ t\,. \tag{77.2}$$

On the other hand and according to the definition (76.3) it is possible to write that

$$\frac{\partial\circ}{\partial s} = -\rho T\,. \tag{77.3}$$

Thus and with the aid of (16.5) we have that

$$\delta^+\circ = \left(-\circ\, div\, \boldsymbol{v} - \rho^2\frac{\partial\lambda}{\partial\rho} div\, \boldsymbol{v} - \rho T\frac{ds}{dt} + \frac{\partial\circ}{\partial t}\right)\delta^+ t\,,$$

so that

$$\delta^+\circ - \frac{d\circ}{dt}\delta^+ t = \left(-\mathcal{D}\circ - \rho^2\frac{\partial\lambda}{\partial\rho} div\, \boldsymbol{v} - \rho T\frac{ds}{dt} + \frac{\partial\circ}{\partial t}\right)\delta^+ t\,.$$

With this last result substituted in (19.1) the following is obtained

$$\int_{t_1}^{t_2}\int_R \left(\mathcal{D}\circ + \rho^2 \frac{\partial \lambda}{\partial \rho} \operatorname{div} \boldsymbol{v} + \rho T \frac{ds}{dt} - \frac{\partial \circ}{\partial t} \right) \delta^+ t dV dt = 0 . \quad (77.4)$$

That relationship is only fulfilled if the integrand becomes null; that is to say if

$$\mathcal{D}\circ + \rho^2 \frac{\partial \lambda}{\partial \rho} \operatorname{div} \boldsymbol{v} + \rho T \frac{ds}{dt} - \frac{\partial \circ}{\partial t} = 0 . \quad (77.5)$$

This is the corresponding energy balance differential equation. If again it is considered that $\partial \circ / \partial t = 0$ and because the system stays at rest as a whole, it is fulfilled that

$$\frac{dE}{dt} = T \frac{dS}{dt} - p \frac{dV}{dt} ; \quad (77.6)$$

or also

$$dE = TdS - pdV . \quad (77.7)$$

This is the so called thermodynamic identity. It is one of the most important thermodynamic relations. It defines the total differential of the function $E(S,V)$, which is the internal energy of the system in the state of equilibrium. According to Gibbs′ formulation, the variables p and T are defined as follows

$$p = -\left(\frac{\partial E}{\partial V} \right)_S , \quad \text{and} \quad T = \left(\frac{\partial E}{\partial S} \right)_V .$$

If the results (77.6) and (75.8) are compared the following relationship is obtained

$$\frac{dQ}{dt} = T\frac{dS}{dt}, \tag{77.9}$$

which is the expression for the quantity of heat received by the system. It is important to mention that the mechanical work *dR* and the quantity of heat *dQ* gained by the system in an infinitesimal change of its state are inexact differentials because they can not be precisely related to any particular thermodynamic state. Only the sum *dR*+*dQ*; that is to say, the change in the total energy *dE* has a precise physical sense as a total or exact differential. Hence, one can speak of the energy *E* in a given thermodynamic state but one can not make reference to the quantity of heat or the mechanical energy possessed by the system in that particular state. That means that the energy of a thermodynamic system can not be split into heat energy and mechanical energy. However, the change in energy due to transition of the system from one state to another can be divided into the quantity of heat gained or lost by the system, and in the work done on it or by it on another systems. This subdivision is not uniquely determined by the initial and final states of the system but also depends on the actual nature of the process. In other words the mechanical work and the quantity of heat are functions of the process undergone by the system, and not only of its initial and final states. On the other hand, it is stated that if a closed thermodynamic system is found in any instant of its history in a macroscopic state of non-equilibrium, the most possible consequence is that its entropy increases in successive instants. This is the so called law of the entropy increase or second law of thermodynamics. It was initially postulated by R. Clausius and later statistically founded by L. Boltzmann.

§78. Temperature

Finally, let us consider a physical system which is found in a thermodynamic state such that the volume which occupies does not change. In order to describe that situation a lagrangian density whose functional form is the following is proposed

$$\circ = \circ(s,t); \tag{78.1}$$

where again, $s(\square\rho,t)$ is the entropy per unit mass, so that

$$\lambda = \lambda(s,t) \tag{78.2}$$

is the corresponding specific lagrangian and

$$\rho\delta\lambda = -\rho Ts\ div(\delta\boldsymbol{x}).$$

In this case,

$$\rho\delta\lambda = -\frac{\partial}{\partial x^i}\left(\rho Ts\,\delta x^i\right) + \frac{\partial}{\partial x^i}(\rho Ts)\,\delta x^i\ ;$$

where an integration by parts was made. The first term of the right hand side is zero according to the result that was obtained in paragraph 76 to integrate the first term of the relation (76.6). In consequence, in (16.4) only the following is kept

$$\int_{t_1}^{t_2}\int_R \left[\frac{\partial}{\partial x^i}(\rho Ts)\right]\delta x^i\, dVdt = 0\ ;$$

relation which is only fulfilled if

$$\frac{\partial}{\partial x^i}(\rho Ts) = 0\ .$$

These are the field differential equations for this case. Since T as well as ρ are constants, the same asumption made before can be accomplished and write the following

$$TS = E\ ;$$

with $E(S,t)$ the total internal energy. This thermodynamic function of state rises again from the theoretical scheme as a consequence of the homogeneity of space.

With respect to temporary variaton of the lagrangian density and taking into account the relations (77.2) and (77.3) we have that

$$\delta^{+}\circ=\left(-T\frac{dS}{dt}+\frac{\partial\circ}{\partial t}\right)\delta^{+}t\,;$$

in such a way that

$$\delta^{+}\circ-\frac{d\circ}{dt}\delta^{+}t=\left(-\frac{d\circ}{dt}-T\frac{dS}{dt}+\frac{\partial\circ}{\partial t}\right)\delta^{+}t\,.$$

In this case in (19.1) we have the following

$$\int_{t_1}^{t_2}\int_{R}\left[\frac{d\circ}{dt}+T\frac{dS}{dt}-\frac{\partial\circ}{\partial t}\right]\delta^{+}t\,dV\,dt=0\,;$$

result that is only fulfilled if

$$\frac{d\circ}{dt}+T\frac{dS}{dt}-\frac{\partial\circ}{\partial t}=0\,. \tag{78.3}$$

This is the energy balance equation for the present case. If again it is fulfilled that $\partial\circ/\partial t=0$ and the explicit form (76.10) for the lagrangian density is used, the following result is obtained

$$\frac{dE}{dt}=T\frac{dS}{dt}\,; \tag{78.4}$$

or also

$$dE=TdS\,. \tag{78.5}$$

From the former equation the following is obtained

$$\frac{dS}{dE}=\frac{1}{T}\,. \tag{78.6}$$

In Classical Thermodynamics it is stated that the inverse of the derivative of the total entropy S with respect to the total energy E is a constant for the system; so that the temperatures of two systems which are found in thermal equilibrium between them are equal. In other words, it is fulfilled that for that state of thermodynamic equilibrium

$$T_1 = T_2; \tag{78.7}$$

and this result is valid for any number of thermodynamic systems in thermal equilibrium between them. This is an expression of what is known by the name of the zero principle or the zero law of thermodynamics, whereas (78.6) is an equation which defines the macroscopic concept of temperature.

Moreover, the total derivative concerning time of Euler's form (76.12) for the first and second laws of thermodynamics, is the analytic representation of the process that the system experiments, mentioned in paragraph 76, that is to say

$$\frac{dE}{dt} = T\frac{dS}{dt} + S\frac{dT}{dt} - p\frac{dV}{dt} - V\frac{dp}{dt}. \tag{78.8}$$

If this result is compared to the relationship (77.6) we have that

$$S\frac{dT}{dt} - V\frac{dp}{dt} = 0 \tag{78.9}$$

and this is the well known relationship of Gibbs-Duhem.

Finally, a single phase thermodynamic system has two important heat capacities[†]. It is stated that the quantity of heat that raises the temperature of a thermodynamic system by one unit of temperature (for instance one degree) is its specific heat. Also, it is stated that the specific heat of a system depends on the process by means of which it is warmed up. It is usual to make the distinction between the specific heat c_V at a constant volume and the specific heat c_P at a constant pressure. They are defined as follows

† *When the phase is defined per unit mass we talk of specific heat*

$$c_V = T\left(\frac{\partial S}{\partial T}\right)_V , \qquad (78.10)$$

$$c_p = T\left(\frac{\partial S}{\partial T}\right)_p . \qquad (78.11)$$

Both specific heat are variables of state. For a single phase the experimental evidence determines that

$$c_p > c_V > 0 . \qquad (78.12)$$

§79. Hamilton´s formulation

The fundamental equation for the equilibrium state of a thermodynamic system can be formulated in terms of the entropy or the energy proposing any of them as the dependent variable. There is a remarkable symmetry between those functions in such a way that the role played by any of them in Thermodynamics can be interchanged without difficulty. In other words, the election that we make is totally arbitrary and is a consequence of the fact that the principles of maximum entropy and of minimum energy are wholly equivalent. The Maximum Entropy Principle characterizes the equilibrium state of a system as one having maximum entropy for a given total energy, whereas the Minimum Energy Principle can be used to describe the same equilibrium state as one having minimum energy for a fixed total entropy. From a mathematical point of view and in both representations the extensive parameters play the role of mathematically independent variables, whereas the intensive parameters arise in the theory as derived concepts. On the other hand, the parameters of the system which are more easly measured and controlled in the laboratory are the intensive ones, in such a way that it is usual to think, from a practical view point, of the intensive parameters as operationally independent variables and assign the character of derived quantities to extensive parameters. That difference between what normaly occurs in practice and the analytical scheme used has propitiated the search of a mathematical formalism in which the extensive properties were replaced by the intensive parameters in their role as independent variables.

That mathematical tool exists in Theoretical Mechanics and is known by the name of Hamilton´s formulation. In that theoretical frame a generalization of Legendre´s transformation is proposed as another alternative method to the one of Lagrange´s to formulate and solve the problems in Theoretical Physics.

§80. Generalized momenta and the hamiltonian density

In order to apply Hamilton´s scheme to thermodynamic systems, the existence of a general lagrangian density is assumed as a continuous function with continuous derivatives as far as third order in their arguments which have the following functional form

$$\circ = \circ(S, V, t), \tag{80.1}$$

where t is the time, $S(t)$ the total entropy, and $V(t)$ the volume. According to the functionality of $\circ$ two possible definitions for the generalized momentum can be proposed. One of them consists of defining it in terms of the dependence of $\circ$ with the total entropy in the following manner

$$\frac{\partial \circ}{\partial S} \equiv -T \tag{80.2}$$

with $T(t)$ the temperature. The other one is to consider the functionality of $\circ$ with the total volume in such a way that

$$\frac{\partial \circ}{\partial V} \equiv p\,, \tag{80.3}$$

with $p(t)$ the pressure. Therefore and according to the general definition of Hamilton´s function, which is made in Theoretical Mechanics, in the scope of Thermodynamics, the hamiltonian density can be defined in any of the two following forms

$$\mathcal{H} = \frac{\partial \circ}{\partial S} S - \circ\,, \tag{80.4}$$

or also as

$$\mathcal{H} = \frac{\partial \circ}{\partial V} V - \circ\,. \tag{80.5}$$

Every former representation analitically describes some specific thermodynamic process so that the theoretical scheme outlined gives the required mathematical tool in order to formulate and study several physical situations.

§81. Helmholtz free energy

Let us consider a single phase thermodynamic system which consists of a homogeneous continuous medium and some event that changes its state. Let us suppose that the change of state is due to an infinitesimal adiabatic process which maintains the entropy value of the system constant. For that physical situation the definition (80.2) and (80.4) will be used to obtain the following relationship

$$\mathcal{H} = -TS - \circ \,. \tag{81.1}$$

Given that $-T$ is the generalized momentum and V the generalized coordinate, the proposed hamiltonian density has the following functional form

$$\mathcal{H} = \mathcal{H}(T, V, t)\,. \tag{81.2}$$

From (81.1) it can be stated that T and S are the conjugate canonical variables. The total derivative with respect to time of (81.2) is

$$\frac{d\mathcal{H}}{dt} = \frac{\partial\mathcal{H}}{\partial T}\dot{T} + \frac{\partial\mathcal{H}}{\partial V}\dot{V} + \frac{\partial\mathcal{H}}{\partial t}\,,$$

whereas from (81.1) the following result is obtained

$$\frac{d\mathcal{H}}{dt} = -S\dot{T} - \frac{\partial\circ}{\partial V}\dot{V} - \frac{\partial\circ}{\partial t}\,;$$

where the definition (80.2) was used. Theoretical Mechanics demands that for the former derivatives to be compatible between them it is required that the following general condition must be satisfied

$$\frac{\partial\mathcal{H}}{\partial t} = -\frac{\partial\circ}{\partial t}\,, \tag{81.3}$$

and the following relationships are also fulfilled

$$\frac{\partial \mathcal{H}}{\partial T} = -S$$

and (81.4)

$$\frac{\partial \mathcal{H}}{\partial V} = -p\,;$$

where the definition (80.3) was used. The relationships (81.4) will be called Hamilton´s canonical equations for this case. In the thermodynamic state of equlibrium we have that $\mathcal{H} = \mathcal{H}(T,V)$ in such a way that the following relationship is fulfilled

$$d\mathcal{H} = -SdT - pdV\,.$$

That last result can be written in another way if we have the following integration by parts

$$-SdT - pdV = -SdT + Vdp + d(-pV)\,.$$

According to the relation (79.8)

$$-SdT + Vdp = 0\,.$$

From Euler´s form of the first and the second laws of thermodynamics (76.12) we have that

$$\mathcal{H} = E - TS\,; \tag{81.5}$$

where $E(t)$ is the total internal energy of the system. That new thermodynamic function of state is known by the name of Helmholtz free energy and it is labeled with the letter F. In terms of it the thermodynamic identity is

$$dF = -SdT - pdV\,,$$

whereas the condition (81.3) is transformed into

$$\frac{\partial F}{\partial t} = -\frac{\partial \circ}{\partial t} = 0\,;$$

while Hamilton's canonical equations (81.4) take the following form

$$\left(\frac{\partial F}{\partial T}\right)_V = -S$$

and (81.6)

$$\left(\frac{\partial F}{\partial V}\right)_T = -p\,.$$

§82. The heat function

Let us suppose that the change of state of the system is due to an isothermal, infinitesimal and reversible process, which is realized at a constant volume. For this new physical situation and according to the definitions (80.3) and (80.5) the proper hamiltonian density has the following form

$$\mathcal{H} = pV - \circ\,. \tag{82.1}$$

For this case, p and V are the conjugate canonical variables. Since p is the generalized momentum and S the generalized coordinate, the hamiltonian density has the following functional form $\mathcal{H} = \mathcal{H}(p,S,t)$.

If the procedure of calculus used in the former paragraph is repeated, it can be proven that again the general condition (81.3) is satisfied and the following relationships are fulfilled

$$\frac{\partial \mathcal{H}}{\partial p} = V$$

and (82.2)

$$\frac{\partial \mathcal{H}}{\partial S} = T\,.$$

These are the corresponding Hamilton's canonical equations. Given that for the thermodynamic state of equilibrium $\mathcal{H} = \mathcal{H}(p,S)$,

$$d\mathcal{H} = Vdp + TdS\,.$$

If in the second term of the right hand side of the last result an integration by parts is made and then the relations (76.12) and (78.9) are used, it can be proven that

$$\mathcal{H} = E + pV\,. \tag{82.3}$$

That new thermodynamic function of state is known as the heat function or heat content. Also, it is called the enthalpy of the system and it is labeled with the letter W. Therefore, the corresponding thermodynamic identity takes the following form

$$dW = Vdp + TdS\,,$$

whereas the general condition (81.3) is now

$$\frac{\partial W}{\partial t} = -\frac{\partial o}{\partial t} = 0\,;$$

and Hamilton´s canonical equations (81.6) are transformed into the following relationships

$$\left(\frac{\partial W}{\partial p}\right)_S = V$$

and (82.4)

$$\left(\frac{\partial W}{\partial S}\right)_p = T\,.$$

§83. Mechanical work and quantity of heat

In an infinitesimal adiabatic process the total entropy remains fixed. In order so that the system is always in a state of mechanical equilibrium, it is stated that also the pressure must be constant during the whole process. So from the thermodynamic identity (76.12), from Gibbs-Duhem´s relationship (78.9), and the equation (81.5) the following is obtained

$$d\mathcal{H} = -pdV\,. \tag{83.1}$$

According to this result the hamiltonian density is only a function of time t and of volume $V(t)$. In those circumstances it can be stated that if the change of state of the system is only due to a change in volume which occupies it, is legitimate to assure that

$$\frac{d\mathcal{H}}{dt} = \left(\frac{dR}{dt}\right)_{ad}, \tag{83.2}$$

where R is the mechanical work done over the system. Therefore, for an adiabatic process the relationship (74.13) is fulfilled.

From the relations (81.5) and (76.12) it can be demonstrated by direct calculus that for the present conditions $\mathcal{H} = E(V,t)$, with E the total internal energy of the system.

During an isothermal process the temperature as well as the volume occupied by the system remain constant so that the change of state is only due to the quantity of heat gained. In this case from the hamiltonian function (82.3) and from (76.12) we have that

$$d\mathcal{H} = TdS. \tag{83.3}$$

It can be proven by direct calculus that from the relationship (82.3) and (76.12) we have that $\mathcal{H} = E(S,t)$.

If the equation (83.3) is compared to the relationship (77.9) it is easy to see that

$$\frac{d\mathcal{H}}{dt} \equiv \frac{dQ}{dt}; \tag{83.4}$$

with Q the quantity of heat received by the system.

§84. The thermodynamic identity

In the general case of a single phase thermodynamic system which is not thermally isolated and experiments an infinitesimal process which changes its state, the gain or lose of energy is due to two mechanisms. One is the mechanical work made on or by the system and the other is the quantity of heat acquired or lost by it. For that physical situation the following functional form for the hamiltonian density is proposed

$$\mathcal{H}(S,V,t) = \frac{\partial \circ}{\partial S} S + \frac{\partial \circ}{\partial V} V - 2\circ.$$

In consequence, the generalized momenta are $-T(t)$ and $p(t)$ and the generalized coordinates are $V(t)$ and $S(t)$ respectively. It can be said that the couples (T,S) and (p,V) are the conjugate canonical variables. According to (80.12), (80.3) and (76.12)

$$\frac{\partial \circ}{\partial S}S+\frac{\partial \circ}{\partial V}V=-(TS-pV)=-E(S,V,t),$$

with $E(S,V,t)$ the total internal energy of the system which is equal to $-\circ(S,V,t)$. In this case the relation between the hamiltonian and lagrangian densities is the following

$$\mathcal{H}(S,V,t)=-\circ(S,V,t). \tag{84.1}$$

Their total derivative with respect to time is

$$\frac{\partial \mathcal{H}}{\partial S}\dot{S}+\frac{\partial \mathcal{H}}{\partial V}\dot{V}+\frac{\partial \mathcal{H}}{\partial t}=-\frac{\partial \circ}{\partial S}\dot{S}-\frac{\partial \circ}{\partial V}\dot{V}-\frac{\partial \circ}{\partial t}.$$

For this result to be consistent it is required that the general condition (81.3) is fulfilled and the following relationships are met

$$\frac{\partial \mathcal{H}}{\partial S}=T \tag{84.2}$$

and

$$\frac{\partial \mathcal{H}}{\partial V}=-p.$$

These are Hamilton´s canonical equations. The thermodynamic state of equilibrium should be defined by the functional relationship $H(V,S)$ and evidently

$$d\mathcal{H}=TdS-pdV.$$

If on the right hand side of the last result two integrations by parts are carried out and the relationships (76.12) and (78.9) are used, it can be seen that

$$\mathcal{H}(S,V)=E(S,V) \tag{84.3}$$

and of course it is fulfilled that $\circ(S,V) = -E(S,V)$. Therefore, for the equilibrium state the thermodynamic identity is the relationship (77.7), whereas Hamilton´s canonical equations are now,

$$\left(\frac{\partial E}{\partial S}\right)_V = T$$

and (84.4)

$$\left(\frac{\partial E}{\partial V}\right)_S = -p;$$

besides the general condition

$$\frac{\partial E}{\partial t} = -\frac{\partial \circ}{\partial t} = 0,$$

which is identically fulfilled.

§85. Gibbs´ free energy

The former results (84.4), (82.4), and (81.6) show that given any one of the thermodynamic variables of state E, W, and F as a function of the appropiate variables, it is possible to obtain all the other thermodynamic parameters by constructing their partial derivatives. Because of this property the above mentioned variables of state are sometimes called the thermodynamic potentials by analogy with the mechanical potential. They are also known by the name of characteristic functions; that is, the energy E, with respect to the variables V and S, the heat function W, with respect to S and p, and the Helmholtz free energy F, with respect to V and T.

Nevertheless, the theory requires a thermodynamic potential which relates the variables p and T. Let us suppose that T is the generalized momentum and p the generalized coordinate in such a way that the hamiltonian density has the following functional form

$$\mathcal{H} = \mathcal{H}(T, p, t). \tag{85.1}$$

In this case the lagrangian density is the following

$$\circ = \circ(p, S, t). \tag{85.2}$$

If this functional relationship is compared to the one which has $\mathcal{H}$ in the equation (82.1) it is easy to agree that for this case the lagrangian density has the same functional form as the enthalpy, so that it can be written that

$$\circ = E + pV\,. \tag{85.3}$$

Therefore, the following relationships are fulfilled

$$\frac{\partial \circ}{\partial S} = \frac{\partial E}{\partial S} = T$$

and (85.4)

$$\frac{\partial \circ}{\partial p} = V\,.$$

Consequently,

$$\mathcal{H} = TS - \circ \tag{85.5}$$

is the explicit form of the hamiltonian density with T and S the conjugate canonical variables. The procedure of calculus that has been used systematically in the former paragraphs allows to demonstrate that also the general condition (81.3) is fulfilled as well as the following relationships

$$\frac{\partial \mathcal{H}}{\partial T} = S$$

and (85.6)

$$\frac{\partial \mathcal{H}}{\partial p} = -V\,;$$

which are now the corresponding Hamilton's canonical equations. Since for the state of equilibrium we have that $\mathcal{H} = \mathcal{H}(T, p)$, the total differential of that function can be calculated to obtain the following result

$$d\mathcal{H} = SdT - Vdp\,.$$

On the other hand, from the relations (85.3) and (85.5) we can verify that $-\mathcal{H} = \Phi$, with

$$\Phi = E - TS + pV \qquad (85.7)$$

a new thermodynamic function of state known by the name of thermodynamic potential or Gibbs' free energy. In terms of it the thermodynamic identity takes the following form

$$d\Phi = -SdT + Vdp\,.$$

On the other hand, the general condition (81.3) is now

$$\frac{\partial \Phi}{\partial t} = \frac{\partial \circ}{\partial t} = 0\,,$$

whereas Hamilton's canonical equations are the following

$$\left(\frac{\partial \Phi}{\partial T}\right)_p = -S$$

and

$$\left(\frac{\partial \Phi}{\partial p}\right)_T = V\,.$$

§86. Maxwell relations

Due to the fact that in the double partial derivative the order in which the derivatives appear is irrelevant, from Hamilton's canonical equations (84.4) the following results can be obtained

$$\left(\frac{\partial T}{\partial V}\right)_S = \frac{\partial^2 E}{\partial V \partial S}$$

and

$$-\left(\frac{\partial p}{\partial S}\right)_V = \frac{\partial^2 E}{\partial S \partial V};$$

so that the following relationship is satisfied

$$\left(\frac{\partial T}{\partial V}\right)_S = -\left(\frac{\partial p}{\partial S}\right)_V. \tag{86.1}$$

in the same way, from (82.4) we have that

$$\left(\frac{\partial V}{\partial S}\right)_p = \frac{\partial^2 W}{\partial S \partial p}$$

and

$$\left(\frac{\partial T}{\partial p}\right)_S = \frac{\partial^2 W}{\partial p \partial S}.$$

In this case

$$\left(\frac{\partial V}{\partial S}\right)_p = \left(\frac{\partial T}{\partial p}\right)_S. \tag{86.2}$$

From the equations (81.6) it is easy to obtain the following results

$$\left(\frac{\partial S}{\partial V}\right)_T = -\frac{\partial^2 F}{\partial V \partial T}$$

and

$$\left(\frac{\partial p}{\partial T}\right)_V = -\frac{\partial^2 F}{\partial T \partial V};$$

so that the following relationship is fulfilled

$$\left(\frac{\partial S}{\partial V}\right)_T = \left(\frac{\partial p}{\partial T}\right)_V . \tag{86.3}$$

Finally from (85.8) we have that

$$-\left(\frac{\partial S}{\partial p}\right)_T = \frac{\partial^2 \Phi}{\partial p \partial T}$$

and

$$\left(\frac{\partial V}{\partial T}\right)_p = \frac{\partial^2 \Phi}{\partial T \partial p},$$

in such a way that the following equality is verified

$$\left(\frac{\partial S}{\partial p}\right)_T = -\left(\frac{\partial V}{\partial T}\right)_p . \tag{86.4}$$

The results (86.1) to (86.4) are the Maxwell relations. They form a set of mathematical equations which are very useful in many of the conventional applications of Classical Thermodynamics.

Selected Topics

Adiabatic processes and mean values

Let us consider a single phase thermodynamic system which consists of a homogeneous continuous medium. If we assume that the system is undergoing an adiabatic process and define the total derivative of its energy with respect to time dE/dt, a general expression to calculate mean values by the use of purely thermodynamic analitycal methods can be obtained.

From the definition of energy made in Thermodynamics it can be assured that $E=\mathcal{H}(S,V;\lambda)$, where $\mathcal{H}(S,V;\lambda)$ is the hamiltonian density of the system as a function of the total entropy $S(t)$, the volume $V(t)$, and of the external parameter λ. The dash above that thermodynamic variable indicates an average over the statistical distribution of equilibrium. The calculus of the partial time derivative of the total internal energy gives the following result

$$\frac{\partial E(S,V;\lambda)}{\partial t}=\frac{\partial E}{\partial S}\dot{S}+\frac{\partial E}{\partial V}\dot{V};$$

where the dot means the total time derivative. According to relations (84.2) and (77.6) in the former result we have that

$$\frac{\partial E(S,V;\lambda)}{\partial t}=\frac{dE(S,V;\lambda)}{dt}.$$

As $E(S,V;\lambda)$ depends explicitly on time and $\lambda=\lambda(t)$ we can write that

$$\frac{dE(S,V;\lambda)}{dt}=\frac{\partial E(S,V;\lambda)}{\partial\lambda}\frac{d\lambda}{dt}.$$

On the other hand and since the totally arbitrary order in which the operations of averaging over the statistical distribution and the total time derivative is performed, it can be written that

$$\frac{dE}{dt}=\frac{dE(S,V;\lambda)}{dt}=\frac{\partial E(S,V;\lambda)}{\partial\lambda}\frac{d\lambda}{dt};\qquad(1)$$

where $d\lambda/dt$ is a given function of time and may be taken out from underneath the averaging sign.

A very important fact which is necessary to detach is that due to the fact that the process is considered adiabatic, the mean value of the derivative $\partial E(S,V;\lambda)/\partial\lambda$ in (1) can be interpreted as a mean over a statistical distribution corresponding to the equilibrium state[†] for a given value of the exterternal parameter. That means that the mean value is determined for certain external conditions which are established at some given instant. The total derivative dE/dt can also be expessed in another form if the energy is considered as a function of the entropy and the external parameter only. Since the total entropy remains constant in an adiabatic process we can write that

$$\frac{dE}{dt}=\left(\frac{\partial E}{\partial\lambda}\right)_S\frac{d\lambda}{dt}; \qquad (2)$$

where the subscript indicates that the derivative is taken for constant S. Comparing the results (1) and (2) we see that

$$\overline{\frac{\partial E(S,V;\lambda)}{\partial\lambda}}=\left(\frac{\partial E}{\partial\lambda}\right)_S. \qquad (3)$$

This is the required formula by means of which we can calculate only by thermodynamic methods and over statistical distribution of equlibrium, the mean values of expressions of the type $\partial E(S,V;\lambda)/\partial\lambda$. Those expressions often occur when one studies the average properties of macroscopic objects. In those cases, the formula (3) is of great advantage from a statistic viewpoint.

Entropy and temperature

Let us consider a thermal equlibrium state between two thermodynamic systems each of a single phase, which consist of two homogeneous continuous media which together form a closed system[*]. It can be assumed that

---

[†] *The statistical equilibrum is also called thermodynamic or thermal equlibrium.*
[*] *It is said that a system is closed if it does not interact in any way with another system.*

the total entropy of the system reaches its maximum possible value for a given value of its energy. As the energy is an additive quantity, the total energy E of the system is equal to the sum of the energies E_1 and E_2 of the separate systems. On the other hand, the entropy of each system is a function of its energy, so that the total entropy of the closed system is equal to the sum of the entropies of each isolated system. In other words, σ is also an additve quantity in such a way that

$$\sigma(E) = \sigma_1(E_1) + \sigma_2(E_2).$$

Because the total energy is a constant, the total entropy can be expressed as a function of a single independent variable writing that $E_2 = E - E_1$. The necessary condition so that σ has its maximum possible value can be expressed as follows[†]

$$\frac{d\sigma}{dE_1} = \frac{d\sigma_1}{dE_1} + \frac{d\sigma_2}{dE_2}\frac{dE_2}{dE_1} = \frac{d\sigma_1}{dE_1} - \frac{d\sigma_2}{dE_2} = 0 .$$

In this case it is fulfilled that

$$\frac{d\sigma_1}{dE_1} = \frac{d\sigma_2}{dE_2}$$

result which is valid for any number of thermodynamic systems in thermal equilibrium between them. For such situation it is stated that

$$\frac{d\sigma}{dE} = \frac{1}{\Theta};$$

where Θ is a constant known as the absolute temperature of the system. Therefore, the absolute temperature of two thermodynamic systems which are found in thermal equilibrium between them is equal; that is to say it is fulfilled that

---

† *Evidently,* $\frac{dE_2}{dE_1} = \frac{dE}{dE_1} - \frac{dE_1}{dE_1} = -1$; *because* E = *constant.*

$$\Theta_1 = \Theta_2 .$$

As the total entropy is a dimensionless quantity and E has the dimensions of energy, it is usual to propose the following definition

$$\Theta = kT ,$$

where T is the temperature in degrees and k is Boltzmann´s constant which has the units of *ergs·grade*$^{-1}$. The former discussion allows to consider the case of a closed system formed by two systems which are not in thermal equilibrium between them in such a way that their temperature Θ_1, and Θ_2 are different. Experience states that as time goes by the equilibrium will be set up so that temperatures will gradually equalize. The total entropy of the closed system will increase while this process takes place, giving as a consequence that its total time derivative is positive; that is to say

$$\frac{d\sigma}{dt} = \frac{d\sigma_1}{dt} + \frac{d\sigma_2}{dt} = \frac{d\sigma_1}{dE_1}\frac{dE_1}{dt} + \frac{d\sigma_2}{dE_2}\frac{dE_2}{dt} > 0 .$$

Since the total enegy is kept

$$\frac{dE_1}{dt} + \frac{dE_2}{dt} = 0 ;$$

in such a way that

$$\frac{d\sigma}{dt} = \left(\frac{d\sigma_1}{dE_1} - \frac{d\sigma_2}{dE_2}\right)\frac{dE_1}{dt} = \left(\frac{1}{\Theta_1} - \frac{1}{\Theta_2}\right)\frac{dE_1}{dt} > 0 .$$

If the temperature of the second system is higher than that of the first; that is to say if $\Theta_2 > \Theta_1$, the following condition is fulfilled

$$\frac{1}{\Theta_1} - \frac{1}{\Theta_2} > 0 ;$$

and of course,

$$\frac{dE_1}{dt} > 0.$$

That means that whereas the energy of the first system increases the energy of the second one decreases. In other words, it is fulfilled that

$$\frac{dE_2}{dt} < 0.$$

This property of temperature indicates that the energy passes from the system which has a higher temperature to that which has a lower temperature.

Temperature and energy

Let us consider a closed system which is observed for a very long temporary interval in comparison to its relaxation time†. This fact implies that the system is in a thermodynamic state of equilibrium. Its temperature can be determined only using its thermodynamic parameters if it is assumed that the volume it occupies remains constant. Under those conditions it is possible to propose that

$$\Theta = \frac{\partial E(S, V; \sigma)}{\partial \sigma};$$

where Θ is a constant known as the absolute temperature of the system and $\sigma(E)$ is the entropy of Statistical Mechanics. As the system is closed, it can be assumed that it is in a total state of statistic equilibrium, so that its entropy can be expressed only as a function of its total energy. Thus, the function $\sigma(E)$ determines the density of the energy levels in the spectrum of the macroscopic system. Let σ be the external parameter in such a way that in the former relation the following result is obtained

† *The relaxation time is the temporary period in which the transition to the state of statistical equilibrium occurs.*

$$\Theta = \left(\frac{\partial E}{\partial S} \right)_V \frac{\partial S}{\partial \sigma}.$$

Because by definition $S = k\sigma$ with k Boltzmann's constant, it is fulfilled that $\partial S / \partial \sigma = k = constant$ and then $\Theta = kT$, with T the temperature in grades such that

$$T = \left(\frac{\partial E}{\partial S} \right)_V.$$

Pressure

The additivity property that the energy as well as the entropy have, gives as a consequence the following very important conclusion. If a system is in thermal equlibrium, then it can be asserted that its entropy, for a given value of its energy or its energy for a given value of its entropy, depends only on its volume. The macroscopic state of a stationary system in thermal equilibrium can be completely defined by only two thermodynamic quantities, for example, its energy and volume considered as independent variables. The other thermodynamic parameters can be expressed as functions of these two variables. That is true for any couple of thermodynamic parameters of state.

Now, and in order to obtain the force with which a thermodynamic system acts on the surface limiting the volume it occupies, the well known analytic methods of Theoretical Mechanics will be used. Let

$$\boldsymbol{F} = -\boldsymbol{grad}\, E(S, V; \boldsymbol{x})$$

be the force which acts over a given element of surface $d\boldsymbol{s}$. The thermodynamic potential $E(S,V;\boldsymbol{x})$ is the total internal energy of the system as a function of entropy, volume, and the vector radius $\boldsymbol{x}$ which in this case plays the role of an external parameter. From the last expression it can be stated that the mean force is given by

$$\boldsymbol{F} = -\overline{\frac{\partial E(S, V; \boldsymbol{x})}{\partial \boldsymbol{x}}};$$

where the dash indicates the average. Since the entropy is not an explicit function of $\boldsymbol{x}$, from the former result the following is obtained

$$\bar{\boldsymbol{F}} = -\left(\frac{\partial E}{\partial V}\right)_S \frac{\partial V}{\partial \boldsymbol{x}} .$$

Since the change in the volume dV is equal to a $d\boldsymbol{s} \cdot d\boldsymbol{x}$, it can be assured that $\partial V/\partial \boldsymbol{x} = d\boldsymbol{s}$, so that

$$\bar{\boldsymbol{F}} = -\left(\frac{\partial E}{\partial V}\right)_S d\boldsymbol{s} ,$$

result which indicates that the mean force acting on a surface element is directed along the normal to it and is proportional to its area. This is known as Pascal's law. The absolute value of the force acting on the unit element of surface is equal to

$$p = -\left(\frac{\partial E}{\partial V}\right)_S .$$

This quantity is known by the name of pressure. It can be stated that the systems which are found in thermodynamic equilibrium between them have equal pressures. This follows directly from the fact that thermal equilibrium itself always presupposes the mechanical equilibrium. In fact, the set of forces acting between two systems over the surface of their common boundary must balance, in such a way that their absolute values must be the same and their directions opposite.

The equality of pressure between systems which are found in equilibrium between them can also be deduced from the condition that the entropy should be a maximum. To do this, let us consider two parts of a closed system which are in contact between them. One of the necessary conditions for the entropy to be a maximum is that it should be a maximum with respect to a change in the volumes V_1, and V_2 of these two parts, whereas the state of all the other parts remain unchanged. That means that the sum V_1+V_2 must remain fixed. If S_1 and S_2 are the entropies of the parts considered we have that

$$\frac{\partial S}{\partial V_1} = \frac{\partial S_1}{\partial V_1} + \frac{\partial S_2}{\partial V_2}\frac{\partial V_2}{\partial V_1} = \frac{\partial S_1}{\partial V_1} - \frac{\partial S_2}{\partial V_2} = 0 .$$

From the thermodynamic identity the following result is obtained

$$dS = \frac{1}{T}dE + \frac{p}{T}dV ;$$

in such a way that

$$\frac{\partial S}{\partial V} = \frac{p}{T},$$

because the total energy remains constant. In consequence it is fulfilled that

$$\frac{p_1}{T_1} = \frac{p_2}{T_2} .$$

Since the temperatures T_1, and T_2 are equal for thermal state of equilibrium, it can be said that the pressures in the state of mechanical equilibrium are also equal.

The Helmholtz free energy as a mechanical potential

Let us consider a continuous medium formed by a large number of equal particles each one of them with mass *m*, which are contained in a volume *V* and found in a thermodynamic state of equilibrium characterized by a temperature *T=constant*. If the form (81.2) for the hamiltonian density is used to describe the proposed physical situation, an analytical expression for the mean kinetic energy of the particles can be calculated. That function can be written as follows

$$\mathcal{H}(T,V) = E_c(T) + U(V);$$

where $E_c(T)$ is the mean kinetic energy and $U(V)$ the potential energy of interactions of the particles. According to the elemental Statistic Mechanics the mean kinetic energy of the particles and the temperature of the system have the following relationship $T=\alpha E_c$, with α a constant and $E_c = p^2/2m$. Here p is the magnitude of the momentum of each particle, so that the mean kinetic energy is a cuadratic function of the momenta and inversely proportional to the mass of the particles. In this case the mass as the external parameter can be considered and then write that

$$\frac{\partial \mathcal{H}(T,V;m)}{\partial m} = -\frac{1}{m}\left(\frac{p^2}{2m}\right) = -\frac{E_c(T)}{m} .$$

From the definition (81.5) of Helmholtz free energy we have that

$$\frac{\partial \mathcal{H}(T,V;m)}{\partial m} = \left(\frac{\partial F}{\partial m}\right)_{T,V} ;$$

so that the expression searched is the following

$$E_c(T) = -m\left(\frac{\partial F}{\partial m}\right)_{T,V} .$$

The generalized thermodynamic potentials

Let us consider a single phase conducting continuous medium confined in a region R of three-dimensional Euclidean Space where a magnetic field exists. It is assumed that the system experiments some process which changes its thermodynamic state. If the change is due to an infinitesimal adiabatic process, the entropy remains constant. In the state of thermodynamic equilibrium, the hamiltonian density in its argument must contain the external magnetic field, in such a way that $\mathcal{H}=\mathcal{H}(T, V, B_i)$. The total differential of this function is

$$d\mathcal{H} = \frac{\partial \mathcal{H}}{\partial T} dT + \frac{\partial \mathcal{H}}{\partial V} dV + \frac{\partial \mathcal{H}}{\partial B_i} dB_i .$$

For conducting thermodynamic systems in a magnetic field, the general lagrangian density (80.1) must have the following functional dependence

$$\circ = \circ\left(S, V, B_i, t\right).$$

In this case and according to (81.1) and (47.3) it is fulfilled that

$$\frac{\partial \mathcal{H}}{\partial B_i} = -\frac{\partial \circ}{\partial B_i} \equiv \frac{H_i}{4\pi}; \qquad (1)$$

where $\boldsymbol{B}(\boldsymbol{x},t)$ is the magnetic induction and $\boldsymbol{H}(\boldsymbol{x},t)$ the external magnetic field. In this case and according to the relation (81.4),

$$d\mathcal{H} = -SdT - pdV + \frac{H_i}{4\pi} dB_i .$$

If the second and third terms of the right hand side are integrated by parts and Gibbs-Duhem´s equation (78.9) is used, the following is obtained

$$d\left(\mathcal{H} + TS - E - \frac{B^2}{8\pi}\right) = 0;$$

where the relationship (76.12) was used. Then and given that $\mathcal{H} = F$ it is fulfilled that

$$\hat{F} = F + \frac{B^2}{8\pi},$$

with $\hat{F}$ the generalized Helmholtz free energy.

Suppose that the change of state is reached by means of an isothermal, infinitesimal and reversible process which maintains constant the volume occupied by the system. For the thermodynamic state of equlibrium, the hamiltonian density has the following functional form $\mathcal{H} = \mathcal{H}(p, S, B_i)$. In this case

$$d\mathcal{H} = Vdp + d\left(TS + \frac{B^2}{8\pi}\right) - SdT ;$$

where two integrations by parts were made. According to the relations (78.9) and (76.12)

$$d\left(\mathcal{H} - E - pV - \frac{B^2}{8\pi}\right) = 0 .$$

Since for this case it is fulfilled that $\mathcal{H} = W$, from the former result we have that

$$\hat{W} = W + \frac{B^2}{8\pi} ,$$

with $\hat{W}$ the generalized heat function or generalized enthalpy.

If the change of state is accomplished when the mechanical work is done over the system and a certain amount of heat is adquaired by it[†], the hamiltonian density for the state of thermodynamic equilibrium is such that $\mathcal{H} = \mathcal{H}(S, V, B_i)$, in such a way that

$$d\mathcal{H} = TdS - pdV + \frac{H_i}{4\pi} dB_i ;$$

where the relationships (84.2) were used. It can be demonstrated by direct calculus that

$$\hat{E} = E + \frac{B^2}{8\pi} ;$$

where the relationship (84.3) was used and $\hat{E}$ is the generalized total internal energy.

Finally, let $\mathcal{H} = \mathcal{H}(T, p, B_i)$ be in such a way that

---

[†] *The system can also make mechanical work and give a certain amount of heat.*

$$d\mathcal{H} = SdT - Vdp + \frac{B^2}{8\pi}.$$

According to what is stated in paragraph 85, $\mathcal{H}$=–Φ in such a way that following the same analytical procedure used in this section it can be demonstrated that

$$\hat{\Phi} = \Phi - \frac{B^2}{8\pi},$$

with $\hat{\Phi}$ the generalized Gibbs′ free energy, also refered to as the generalized thermodynamic potential when the external magnetic field is taken into account.

References

1. Callen, H.B. "Thermodynamics". John Wiley & Sons, Inc. New York. London Sydney. 5th Printing (1965).
2. Fierros Palacios, Angel. "Termodinámica. Una formulación lagrangiana" Rev. Mex. de Fís. Vol. 45 No.3, 308-314 (1999).
3. Fierros Palacios, Angel. "El esquema de Hamilton en la termodinámica". Rev. Mex. de Fís. 45 (4) 414-417 (1999).
4. Goldstein, H. "Classical Mechanics". Addison-Wesley Publishing Co. (1960).
5. Lanczos, C. "The Variational Principles of Mechanics" University of Toronto Press, 4th Edition (1970).
6. Landau, L.D. and Lifshitz, E.M. "Statistical Physics". Addison-Wesley Publishing Co. (1958).
7. Landau, L.D. and Lifshitz, E.M. "Electrodynamics of Continuous Media". Addison-Wesley Publishing Co. (1960).
8. Pippard, A.B. "Elements of Classical Thermodynamics". Cambridge at the University Press (1961).
9. Serrin, J."Mathematical Principles of Classical Fluid Mechanics". Handbuch der Physic, VIII/1, Springer-Verlag, Berlin (1959)

Chapter XIV

The Magnetic Field in the Stability of the Stars

§87. The self-generated magnetic field

The theory developed by Eddington considers the case of a star composed by a fluid which from the thermodynamic viewpoint behaves as an ideal gas. To determine the state of the material, two ordinary differential equations are proposed in order to mathematically express the following conditions:

The mechanical equilibrium of the star. In order to fulfil this condition it is necessary that at any internal region of the star the pressure has the right value in order to support the weight of the material above it.

The existence of thermal equilibrium. This condition requires that temperature distribution in the star is capable of maintaining itself automatically not withstanding the continual transfer of heat from one part of the star to another.

The proposed equations are integrated and conditions of the material at any point are determined. This way the distribution of pressure, mass density and temperature are obtained. As work hypothesis it is considered that at first approach, the scheme distribution is *homologous* from star to star. This means that all gaseous stars copy the same stellar model within their appropriate scale of mass, length, temperature, etc. In order to, once and for all, simplify the task, a general solution to the problem is formulated and then the question is reduced to adapting it to the scale of the particular star being studied. It is important to notice that Eddington abandons J. Homer Lane's hypothesis when he does not consider that the thermal equilibrium in a gaseous star is due to the existence of internal convective streams. The hypothesis of convective equilibrium is replaced by the radiative equilibrium and this last basic principle is applied to the internal conditions of the Sun and the stars.

In the theoretical scheme which will be developed next, a magnetic field as the fundamental element for the stars equilibrium, will be introduced. It is proposed as a basic hypothesis, that all gaseous stars generate an intense magnetic field at an early stage of their evolution. Thus, consider a large mass of compressible gaseous fluid, viscous and conductor, isolated in space and at very high temperature and pressure conditions, that remains together by its own gravitational attraction and at dynamic equilibrium with the force produced by the sum of the pressures of radiation and the hot gases. The source of power generation lies on the central part of this huge mass of gas and a little closer to the surface the so called *convective zone*. The convective streams generated there are responsible of removing the heat produced and the ashes of the thermonuclear combustion, feeding at the same time the power generating source with new nuclear fuel. It is assumed that these convective streams are made up by neutral atoms and a lot of electrically charged particles both positive and negative. It is also assumed that these convective streams contain a steady-state current distribution localized in some region inside the convective zone, produced by a process of maximum ionization, which by hypothesis is the generator of the magnetic field which in a first approach, may be considered similar to the bipolar magnetic field produced by a magnetized bar. Irregularities observed in the magnetic field should be attributed to the fact that the structure of the localized steady-state current distribution is not always the same; this is because the condition of the convective streams is not always the same either and can be dependent on the magnitude of the thermonuclear explosions produced in the oven. It is significant to emphasize the importance that the self-generated magnetic field has for the equilibrium and the stability of this object. Additionally, this enormous concentration of matter is found distributed in a configuration which has spherical symmetry. A heavenly object with the above mentioned characteristics is a *star*.

It can be assured that at any point of the stellar fluid in equilibrium there is a hydrostatic pressure which is the same in every direction. If a closed surface within this configuration is drawn, the reaction of the external fluid to that surface over the internal fluid consists of a force per unit area exerted on the surface and along the inward normal. So that for the dynamic equilibrium to exist, it is necessary for these surface forces to be balanced by body forces such as the gravitational which acts towards the inside of the star. It is also necessary to consider the contribution of the magnetic field. Its contribution to the state of equilibrium must be included in the theory in the way of a force per unit area known as the *magnetic hydrostatic pressure*. As it is well known, a magnetic field has the property of generating in a conducting fluid, a magnetic viscosity that confers to it a certain rigidity. That rigidity in the conducting stellar fluid may be interpreted as if the star were supported by a superstructure made up by the magnetic lines of force. It will be assumed that this magnetic superstructure has the mission of keeping the shape of the star even when this object should be animated by a rotational movement, since the magnetic lines of force are frozen in the stellar fluid and move along with it.

§88. The internal structure and the stability of a gaseous star

In order to make an appropriate theoretical treatment of the problem of the stability of a star and of its internal structure, it is necessary to include the self-generated magnetic field in the fundamental equation that governs the state of equilibrium which in this case is magneto mechanical; that is to say

$$\frac{d}{dx}\left(p-\frac{H^2}{8\pi}\right)=-\rho g\,; \tag{88.1}$$

where $p(x,t)$ is the whole pressure, $\rho(x,t)$ the mass density, g the constant value of gravity acceleration, and $H^2/8\pi$ the hydrostatic magnetic pressure, with $\boldsymbol{H}(x,t)$ being the magnetic field self-generated by the star. The previous functions depend on time t and on an x distance measured from any internal region of the star to the center of it. The other relationship that must be considered is the equation for the radiative equilibrium

$$\frac{dp_r}{dx}=-\frac{k\rho\square}{c}. \tag{88.2}$$

In the previous formula we have that $p_r\,(x,t)$ is the pressure of radiation, c the velocity of light in the empty space, □the radiation energy by cm^2 and by second, and k the coefficient which determines radiation absorption by the stellar fluid. From (88.1) and (88.2) it is easy to see that

$$dp_r=\left(\frac{k\square}{cg}\right)d\left(p-\frac{H^2}{8\pi}\right). \tag{88.3}$$

Let us consider a fluid mass distribution within the star contained in a radius sphere equal to x. In a stationary state , the amount of energy released per second within the sphere is equal to $L(x)$ and it is such that

$$\square = \frac{L(x)}{4\pi x^2}; \tag{88.4}$$

where $4\pi\square x^2$ is the whole of the radiation that flows per second through the surface of the sphere and it is clearly equal to the amount of energy released by the central thermonuclear oven. On the other hand, the gravitational force at the position x is only due to the mass $M(x)$ which is found in the inside region to x, so that

$$g = \frac{GM(x)}{x^2}; \tag{88.5}$$

with G the universal gravitational constant. From (88.4) and (88.5) the following result is obtained

$$\frac{\square}{g} = \frac{1}{4\pi G}\frac{L(x)}{M(x)}; \tag{88.6}$$

where $L(x)/M(x)$ is the average rate of energy release per gram from the inner region to x. It is assumed that this energy release is greater in the dense center of the star than it is in its external parts, in such way that this reason decreases as x is increased when succeeding layers of colder material are added to the average.

Be M the total mass of the star and L the total emission of energy per second from its surface in such a way that L/M is the boundary value that reaches the reason $L(x)/M(x)$. For all of the above mentioned it is possible to write the following relation

$$\frac{L(x)}{M(x)} = \eta\frac{L}{M}. \tag{88.7}$$

There are reasons to state that η is a magnitude which is increased from 1 on the star surface up to an unknown value although not very big in the center of it. Its analytical form depends on the law of energy release by the thermonuclear oven. Nevertheless, it is possible to assume that this law is approximately of the same kind for all the gaseous stars since presumably the nuclear energy releasing mechanism is basically the same for all of them. If (88.6), and (88.7) are substituted in (88.3) the following result is obtained

$$dp_r = \frac{L\eta k}{4\pi GcM} d\left(p - \frac{H^2}{8\pi}\right). \tag{88.8}$$

This last relation is an exact equation which allows to establish an upper limit for the value of the opacity for any star for which magnitudes L and M are known as a result of observation, regardless of whether the star is constituted or not by a fluid obeying the thermal equation of perfect gas. In any case it is expected that the values of temperature and mass density are increased towards the inner part of the star. Consider that the whole pressure p is such that

$$p = p_g + p_r; \tag{88.9}$$

where $p_g(x,t)$ is the pressure of hot gases. As on the other hand it is clear that

$$d\left(p - \frac{H^2}{8\pi}\right) = dp_r + d\left(p_g - \frac{H^2}{8\pi}\right), \tag{88.10}$$

so that in (88.8) the following fundamental relation is obtained

$$dp_r = \frac{L\eta k}{4\pi GcM - L\eta k} d\left(p_g - \frac{H^2}{8\pi}\right). \tag{88.11}$$

Since the tendency of temperature and mass density is that of increasing its values towards the inner part of the star, it can be stated that the pressure of the material p_g and the magnetic hydrostatic pressure $H^2/8\pi$ should increase their intensities so that the stability of the star is kept. Under these conditions the following inequality is expected to be fulfilled

$$dp_r < d\left(p_g - \frac{H^2}{8\pi}\right). \tag{88.12}$$

According to (88.11)

$$\frac{L\eta k}{4\pi GcM - L\eta k} < 1. \tag{88.13}$$

Since η is always positive we have that $\eta>1$, in such a way that from the previous relation the following is obtained

$$k < 2\pi Gc\, M / L, \tag{88.14}$$

which with the constant numerical values is transformed into

$$k < 12{,}500 \frac{M}{L}. \tag{88.15}$$

The previous numerical factor is only half of what is reported in the specialized literature. For the most brilliant component of Capella's binary system we have the following basic data

$$M = 8.3 \times 10^{33} gr$$
$$L = 4.8 \times 10^{35} ergs \cdot sec^{-1}$$

In that case

$$k < 216\ cm^2 \cdot gr^{-1}.$$

For the Sun we have that $k<6{,}564\ cm^2 \cdot gr^{-1}$. The physical reason for the existence of that upper limit known by the name of *Eddington limit* is the following: so that the radiation is observed it has to be emitted reaching its way through the star, and if there were too much obstruction it would blow up the star. However, the existence of the self-generated magnetic field is a determining factor for the star to keep its stability. In fact, the superstructure built by the magnetic lines of force is a dynamic obstacle which prevents the star from collapsing because of gravitational compression or exploding because of the added effect of the radiation pressures and of the hot gases. Ultimately, the role played by the magnetic hydrostatic pressure in this part of the theory is that of diminishing to a half the limit value of k.

This fact is important since it indicates that in general, gaseous stars must be more luminous than it is believed and therefore that their age can be less than that which has been determined by Eddington's theory. In specialized literature it is stated that the age of stars is inversely proportional to luminosity. Since luminosity is inversely proportional to opacity as it will be seen further on, if the value of the Eddington limit of the opacity is in fact smaller than the one calculated so far, luminosity will be greater and therefore the age of the stars will be smaller.

Incidentally, an increase in luminosity is indicative of the fact that the star will live less since it is burning its fuel at a greater rate. *With this result the apparent paradox of the stars which are older than the Universe itself, can be solved.* In conclusion and within the present theoretical scheme, *it can be stated that the age of gaseous stars is ruled by the modified Eddington limit of the coefficient of opacity.*

§89. The magnetic field on the surface of a star

On the surface of any gaseous star the value of the self-generated magnetic field follows a very simple law as we shall see next. Its absolute value can be determined from the condition of magneto mechanical equilibrium (88.1) and considering that $\rho=\rho(R,t)$, with R the star's radius. When that relation is integrated it is easy to obtain the following result

$$p-\frac{H^2}{8\pi}=\rho \boldsymbol{g}\cdot\boldsymbol{x}+constant\,. \tag{89.1}$$

On the other hand, it is assumed that the pressure of hot gases $p_g(\boldsymbol{x},t)$ satisfies the hydrostatic equation, that is to say

$$p_g(\boldsymbol{x},t)=\rho \boldsymbol{g}\cdot\boldsymbol{x}+constant\,. \tag{89.2}$$

In that case and with the help of the relation (88.9) it can be proved that

$$p_r = \frac{H^2}{8\pi}. \tag{89.3}$$

The preceding formula must be considered as a *relation of equivalence* more than an equality. From it and with the help of the expression for the pressure of the radiation, the absolute value of the magnetic field can be calculated. Since for a perfect gas it is fulfilled that

$$p_r = \frac{1}{3}aT^4; \tag{89.4}$$

where $a = 7.64\times10^{-15}$ is Stefan's constant. From (89.3) it can be proved that

$$H = mT_e^2, \tag{89.5}$$

with

$$m = \left[\frac{8\pi a}{3}\right]^{1/2} \tag{89.6}$$

a universal constant which has the following numerical value

$$m = 2.53\times10^{-7}\ gauss\cdot K^{-2}. \tag{89.7}$$

The relation (89.5) allows calculating the absolute value of the average magnetic field on the surface of a any gaseous star if its effective temperature T_e is known. The effective temperature can be calculated from the radiation theory of the black body. If E is the energy density radiated by the black body, it can be shown that

$$\mathsf{E} = aT^4. \tag{89.8}$$

The former result is known by the name of Stefan´s law. Since for a perfect gas $\mathsf{E}=3p_r$, the relation (89.4) is satisfied. On the other hand, the radiation emitted by a radius sphere r every second is

$$L = \left(4\pi r^2\right)\left(\frac{acT_e^4}{4}\right) = \pi acr^2 T_e^4 . \tag{89.9}$$

Therefore, the effective temperature of a star is defined by the following relation

$$T_e = \left[\frac{L}{\pi acR^2}\right]^{1/4} ; \tag{89.10}$$

since it gives the black body temperature which produces the same amount of radiation emitted by the star. In (89.9) and (89.10) L is the *luminosity*. It is important to clarify that the effective temperature is a conventional measurement that specifies the reason of radiant flux heat per unit area. It must not be considered as the temperature at any particularly significant level in the star. Finally, it is known that the temperature of the photosphere of the Sun is equal to $5.741\times10^3 K$. Consequently $H_\odot$=8.4 *gauss* on its surface.

Another example is that of the most brilliant component of Capella's binary system for which it is known that its effective temperature is equal to $5.2\times10^3 K$. In that case H_c= 6.8 *gauss*.

§90. The mass-luminosity relation and the coefficient of opacity

Let us suppose that in the equation (88.11) the product ηk is a constant through the star. In order to support that hypothesis it is required that the absorption coefficient k be practically constant decreasing a little towards the center to counterbalance the increase of η, in that way assuring that the product of these two quantities remains constant. It is said that there are very good reasons to believe that in general, k behaves in this manner, wherefore it is possible to state that the constancy of this product is a good approximation. Be therefore

$$\eta k = k_o , \tag{90.1}$$

with k_o a constant that somehow represents the boundary value of k. It is important to emphasize that the value of k in the stellar photosphere could be very different from the one that k_o would have there. Using the approximation (90.1) in (88.11) and integrating, the following result is obtained

$$p_r = \frac{Lk_o}{4\pi GcM - Lk_o}\left(p_g - \frac{H^2}{8\pi}\right); \tag{90.2}$$

that with the help of the relation (89.3) is transformed into what follows

$$p_r = \frac{Lk_o}{4\pi GcM} p_g. \tag{90.3}$$

Clearly, it fulfils the fact that for a particular star the ratio p_r/p_g always maintains the same relation. Consequently

$$\frac{L}{M} = \frac{4\pi Gc}{k_o}\left(\frac{p_r}{p_g}\right). \tag{90.4}$$

Now, it is important to introduce a constant ε defined as follows

$$\begin{aligned} p_r &= (1-\varepsilon)p \\ p_g &= \varepsilon\, p; \end{aligned} \tag{90.5}$$

where p is the whole pressure. ε represents the ratio between hot gas pressure and whole pressure, whilst (1-ε) is the ratio between radiation pressure and whole pressure. Therefore, the ratio (1-ε)/ε can be considered as a measurement of the degree of stability of a star. Substituting (90.5) in (90.4) we have that

$$\frac{L}{M} = \frac{4\pi Gc}{k_o}\left(\frac{1-\varepsilon}{\varepsilon}\right). \tag{90.6}$$

For gaseous stars the value of ε is determined from the following quadric equation

$$1-\varepsilon = 3.09\times 10^{-3}(M/\odot)^2 \mu^4 \varepsilon^4; \tag{90.7}$$

where the symbol ⊙ indicates the mass of the Sun and μ the average molecular weight. The relation (90.7) can be solved for various values of the mass M in terms of the Sun's mass, e. g. M=1/4, 1/2, 1,... times ⊙; and also for some average value of μ. It is important to point out that ε only depends on the mass and on the average molecular weight of the material forming the star and it is independent of its radius and its opacity. Once obtained the value of ε by solving the equation (90.7) it is possible to determine the opacity k_o from the relation (90.6). In specialized literature there are tables which can be consulted to use in numerical calculations which contain values from (1-ε) for different masses and molecular weights.

§91. Luminosity and opacity

If the possible small changes of μ with the temperature and density are neglected, it is possible to see from (90.7) that (1-ε) is a function only of the mass of the star. In that case and according to (90.6) it can be stated that *for gaseous stars of the same mass, the luminosity L is inversely proportional to the opacity* k_o.

However the approximate constancy of k_o from star to star should be distinguished from the approximate constancy of the product ηk within a single star. According to theoretical analysis carried out by other researchers, it can be stated that for a particular star the following relation is satisfied

$$\eta T^{-1/2} = \frac{\eta k}{k_c} = \frac{k_o}{k_c}; \tag{91.1}$$

where k_c refers to density and temperature conditions in the center of the star. On the other hand, the law of absorption of the radiation commonly accepted has the following analytical form

$$k = k_1 \rho T^{-7/2}, \tag{91.2}$$

with k_1 a constant. In that case it is customary to write that

$$k_o = \alpha k_c; \tag{91.3}$$

where α is a constant that according to (91.1) is equal to

$$\alpha = \eta T^{-1/2}. \tag{91.4}$$

The constancy of $\eta T^{-1/2}$ depends on the ratio of energy release; release which is likewise related to the distribution of energy sources in the inner part of the star. What it normally does is to consider various degrees of concentration of those sources and to examine how the ratio of energy released per gram, □ for different temperature powers varies. However, even when the constancy of that factor is better for □~ T , it is customary to explore other possibilities. For cases in which $\eta T^{-1/2}$ is not absolutely constant, the best approximation to the ratio k_o/k_c may be obtained from the average. From specialized literature we have that

	α=1.32	1.74	2.12	2.75	
for					(91.5)
	□~ *constant*	T	T^2	T^4,	

respectively. It is said that when the sources are very concentrated in the center of the star, its brightness diminishes. In the case in which numerical calculations are required, the value α =2.5 is adopted, which is indicative of an intense and uniform concentration of sources of energy release in the center of the star. Consequently in (90.6) the *modified mass-luminosity relation* is obtained

$$L = \frac{4\pi GcM}{\alpha k_c} \cdot \left(\frac{1-\varepsilon}{\varepsilon} \right). \tag{91.6}$$

When numerical calculations are made it is customary to assume that k_c=k.

§92. The central temperature

In order to calculate the central temperature in any gaseous star, relations (89.1) and (89.2) are used as well as the fact that the *relation of equivalence* (89.3)is always fulfilled as much on the surface as in any inner region of the star. Since for gaseous stars the thermal equation of ideal gas is fulfilled

$$p_g = \frac{\mathcal{R} \rho T}{\mu}; \tag{92.1}$$

where $\mathcal{R}=\mathcal{R}_o/m_H$ is the gas universal constant, $\mathcal{R}_o$ is Boltzmann's constant, m_H hydrogen atom mass, and μ the molecular weight which is generally taken as a constant numerically equal to 2.11, from (88.9) and from relations (90.5) the following result is obtained

$$\frac{T^3}{\rho}=\frac{3\mathcal{R}(1-\varepsilon)}{a\mu\varepsilon}. \tag{92.2}$$

The previous relation is equivalent to the equation

$$T^3/\rho=1.53\times10^{18}k_cL/M; \tag{92.3}$$

where results (91.6) and (92.2) were used.

In general, the ratio T^3/ρ which is a constant through any star is also the same for all the stars which have the same mass; whenever it is possible small differences in the average molecular weight μ that may exist among them, should be ignored. Consequently in stars of the same mass, *the temperature at homologous points in the interior varies as the cube root of mean density measured at these points does*. As it is easy to see from paragraph 89, effective temperature is subjected to a different law.

§93. The problem of variable stars of the cepheid type

The theory commonly accepted about the variable stars of the Cepheid type attributes the variation in their brightness to a regular pulsation they experiment. To make an adequate theoretical analysis of the problem, it will be assumed that they are gaseous objects where the self-generated magnetic field has lost much of its original intensity but, that in the position of maximum compression keeps enough of it, so as to reduce the oscillation preventing the final collapse. Next, and with the help of radiation pressure and that of hot gases, that diminished magnetic field bounces, starting the subsequent expansion. If we assume that the Cepheids are in a stage of their evolution such that the fluid which forms them is no longer totally supported by the superstructure formed by the magnetic lines of force, it can be stated that the important dynamic agents acting on them are the huge gravitational force as well as the combined pressures of radiation and hot gases. Subjected to these dynamic conditions, the Cepheids variables oscillate around some equilibrium position losing and recovering brightness alternately, as they expand and contract themselves with a noticeable regularity and with a perfectly determined period. In accordance to the laws of gases, when the star collapses because of the effect of its huge weight and reaching even an extreme position or minimum size, the gaseous fluid heats up and the brightness of the Cepheid is increased. Right after that the hot gases, the radiation pressure, and the residual magnetic field which behaves as a spring, act against gravity distending the star as far as another extreme position of maximum amplitude, causing the stellar fluid to cool and the Cepheid to lose brightness. The cycle is repeated once and once again with a regularity very much alike that observed in the movement of a simple harmonic oscillator. To obtain the differential equation governing the phenomenon and also an expression for the period of harmonic oscillation, equations (88.11), (89.3), and (90.1) must be considered in order to obtain the following relation

$$dp_r = \frac{Lk_o}{4\pi GcM}\, dp_g \,. \tag{93.1}$$

Integrating the above equation the following is obtained

$$p_r + p_r^o = \frac{Lk_o}{4\pi GcM}\left(p_g + p_g^o\right); \tag{93.2}$$

where $p^o{}_r$ and $p^o{}_g$ are integration constants with unities of force per unit area. Besides p_g satisfies the hydrostatic equation (89.2). Let us now consider a radius sphere r concentric with the star and with its surface almost coinciding with that of the star in such a way that the amount of mass $M(r)$ contained in it is practically equal to the total mass M of the star. Under these conditions, the gravity acceleration on the surface of the star is

$$g \approx \frac{GM}{r^2}\,.$$

It is proposed that as much on the surface of the sphere as on that of the star, $p_g{}^o$ and the constant in (89.2) be equal to zero. However, there $p_r{}^o \neq 0$ because the pressure of radiation does not disappear on the surface. According to the first of the definitions (90.5), in (93.2) the following result is obtained

$$(1-\varepsilon)p + p_r^o = -\frac{Lk_o\rho}{4\pi cr}; \tag{93.3}$$

where p is the whole pressure in such a way that

$$p = \frac{f}{4\pi R^2}, \tag{93.4}$$

with f the magnitude of total force, $4\pi R^2$ the area of the surface of the star, and R its radius. Be

$$p_r^o = \frac{f}{4\pi r^2}, \tag{93.5}$$

in such a way that in (93.3) the following is obtained

$$f\left[1-\varepsilon+\frac{R^2}{r^2}\right] = -\frac{Lk_o\rho}{c}\frac{R^2}{r}.$$

In $r = R$ we have that

$$f = -\frac{Lk_o\rho}{c(2-\varepsilon)}r. \tag{93.6}$$

According to Newton's second law, from (93.6) the following result is obtained

$$M\ddot{r} + Kr = 0; \tag{93.7}$$

where $\ddot{r} = d^2r/dt^2$. The above relation is the differential equation that governs the harmonic oscillations observed on the Cepheid variable stars. In it,

$$K = \frac{Lk_o\rho}{c(2-\varepsilon)} \tag{93.8}$$

is a constant which depends on some basic parameters of the particular star being studied. The oscillations period is given by the relation $\tau=2\pi(M/K)^{1/2}$, namely

$$\tau = 6.28 \times \left[\frac{Mc(2-\varepsilon)}{L\alpha k_c \rho_m}\right]^{1/2}; \qquad (93.9)$$

where the relation (91.3) was used.

Since M and L are the total mass and lumiminosity respectively, in (93.9) the average density ρ_m must be used. From the *mass-luminosity relation* (91.6) and given that $4\pi G$=8.3×10^{-7}*ergs.cm·gr*$^{-2}$, it is easy to see that

$$\tau = 6.87\times10^3\left[\frac{\varepsilon(2-\varepsilon)}{(1-\varepsilon)\rho_m}\right]^{1/2}. \qquad (93.10)$$

As it was to be expected , *the product of the period and the square root of the mean density is equal to a constant* which can be calculated from the theory. Next the periods of three known Cepheids are calculated and each result is compared to the one measured by direct observation.

1. *δ Cephei*.

For this intrinsic variable there are the following data

$$\rho_m = 3.5\times10^{-4}\, gr\cdot cm^{-3}$$
$$\varepsilon = 0.545$$
$$1-\varepsilon = 0.455$$
$$2-\varepsilon = 1.455$$

then,

$$\tau = 5.61\, days\,.$$

The period directly measured is of 5.366 days. As it is easy to see, theoretical calculation and direct measuring are practically equal.

2. Polaris

For this case we have that

$$\begin{aligned}\rho_m &= 0.61\times10^{-3} gr\cdot cm^{-3}\\ \varepsilon &= 0.545\\ 1-\varepsilon &= 0.455\\ 2-\varepsilon &= 1.455\end{aligned}$$

Therefore

$$\tau = 5\,days\,.$$

The period measured for this variable is equal to 3.968 days.

3. ε Cephei

The data we have are as follows

$$\begin{aligned}\rho_m &= 0.7 gr\cdot cm^{-3}\\ \varepsilon &= 0.67\\ 1-\varepsilon &= 0.33\\ 2-\varepsilon &= 1.33\end{aligned}$$

Consequently, $\tau = 0.156$ days; while τ (measured))=0.190 days. It is possible that discrepancies observed in the two last mentioned cases are due to the fact that there are not more accurate data for these stars. However, the magnitude order is adequate.

§94. The magnetic field in the inner part of a gaseous star

Just as temperature follows different laws on the surface and in the inner part of a star, something similar happens to the self-generated magnetic field. On the surface of the star its behavior is ruled by the relation (89.5) whereas in their inner part it follows another different law as it will next be seen. In order to calculate the magnitude of the magnetic field at the center of the star as well as at any other inner point of it, it is necessary to use *the polytropic gas sphere theory*. In terms of the gravitational potential $\phi(x)$, the acceleration of gravity is by definition

$$g = -\frac{d\phi}{dx}. \tag{94.1}$$

Be $P = p - H^2/8\pi$, in such a way that in (88.1) the following result is obtained

$$dP = -g\rho dx ,$$

that with the aid of (94.1) is transformed into what follows

$$dP = \rho d\phi . \tag{94.2}$$

From Poisson's equation for the gravitational potential

$$\nabla^2 \phi = -4\pi G\rho ;$$

we have that for spherical symmetry the above relation takes the following form,

$$\frac{d^2\phi}{dx^2} + \frac{2}{x}\frac{d\phi}{dx} = -4\pi G\rho . \tag{94.3}$$

Now we have the relations (94.2) and (94.3) in order to be able to determine the three following unknown functions of x: P, ρ and ϕ. Then, a third equation is also required in order to take into account the thermodynamic state of the star. In general and regardless of whether the stellar gas is perfect or not, it is always possible to make any value of P correspond to a given mass density if temperature is adequately fixed. What is usual is to use as a third relation the following formula

$$P = \kappa\rho^{\gamma} ; \tag{94.4}$$

where κ and γ are disposable constants. Thus and for different values of κ and γ it is possible to investigate a variety of temperature distributions. It is said that distribution is *polytropic* if it obeys an equation such as (94.4). The problem is reduced to redoing the analysis carried out by Eddington, task that will not be repeated in this paragraph. However, and following that methodology, we have that the whole pressure p in terms of gravitational potential has the following form

$$p = \frac{\rho\phi}{(n+1)\varepsilon}; \tag{94.5}$$

where n is a positive integer number. Given that the thermal equation of the state of the ideal gas (92.1) is fulfilled, from (94.5) we have that

$$T = \frac{\mu\phi}{\mathcal{R}(n+1)}. \tag{94.6}$$

As it is easy to see, T=*constant* ϕ. Be ϕ_o the gravitational potential in the center of the sphere and T_o the central temperature so that T_o=*constant* ϕ_o ; where the constant is the same as before. It can be seen that according to (94.6) and with the results of specialized literature, it is easy to see that for a particular star

$$T_o = \frac{\mu}{\mathcal{R}(n+1)} \cdot \frac{GM}{M'} \cdot \frac{R'}{R}; \tag{94.7}$$

where M' y R' are parameters calculated from the polytropic gas sphere theory and

$$\phi_o = \frac{GM}{M'}\frac{R'}{R}. \tag{94.8}$$

The relation (94.7) is used to calculate the central temperature of any gaseous star for which its radius and mass are known. Apart from this and according to the first of the relations (90.5), from (94.5) the following result is obtained

$$p_r = \frac{\rho\phi(1-\varepsilon)}{(n+1)\varepsilon}. \tag{94.9}$$

For gaseous stars the relation (94.4) takes the following form

$$P \approx \rho^{4/3}; \tag{94.10}$$

in such a way that $n = 3$. With this result and with the *relation of equivalence* (89.3), it is easy to see from (94.9) that for a particular star

$$\frac{H^2}{\rho\phi} = constant; \tag{94.11}$$

where

$$constant = 2\pi\left(\frac{1-\varepsilon}{\varepsilon}\right). \tag{94.12}$$

Consequently, for stars of the same mass as well as *for homologous points in the inner part of a given star, the self-generated magnetic field varies like the square root of the product of the mass density and the gravitational potential*; both calculated at those points.

For any inner region of the star, the magnitude of the self-generated magnetic field can be calculated from (94.11), namely

$$H = \left[2\pi\rho\phi\left(\frac{1-\varepsilon}{\varepsilon}\right)\right]^{1/2}. \tag{94.13}$$

It is customary for numerical calculations to use the values given in the following table

TABLE
($n = 3, \gamma = 1.3333$)

z	u	u^n	u^{n+1}	$-du/dz$	$-zdz/3du$	$-z^2du/dz$
0.00	1.00000	1.00000	1.00000	.00000	1.0000	.0000
0.25	.98975	.96960	.95966	.08204	1.0158	.0051
0.50	.95987	.88436	.84886	.15495	1.0756	.0387
0.75	.91355	.76242	.69650	.21270	1.1754	.1196
1.00	.85505	.62513	.53451	.25219	1.3218	.2522
1.25	.78897	.49111	.38747	.27370	1.5224	.4276
1.50	.71948	.37244	.26797	.27993	1.7862	.6298
1.75	.64996	.27458	.17847	.27460	2.1213	.8110
2.00	.58282	.19796	.11538	.26149	2.5195	1.0450
2.50	.46109	.09803	.04520	.22396	3.7210	1.3994
3.00	.35921	.04635	.01665	.18393	5.4370	1.6553
3.50	.27629	.02109	.005828	.14859	7.8697	1.8203
4.00	.20942	.009185	.001923	.11998	11.113	1.9197
4.50	.15529	.003746	.000582	.09748	15..387	1.9740
5.00	.11110	.001371	.000152	.08003	20.826	2.0007
6.00	.04411	.000086	.000004	.05599	35.720	2.0156
6.80	.00471	.000001	.000000	.04360	51.987	2.0161
6.9011	.00000	.000000	.000000	.04231	54.360	2.0150

Usually the problem is reduced to finding the internal distribution for the density as well as for the pressure, in a star for which its mass and its radius or its mean density are known. The expressions used are the following

$$R' = (z)_{u=0} \quad ; M' = \left(-z^2 \frac{du}{dz} \right)_{u=0} . \tag{94.14}$$

The condition $u = 0$ indicates the boundary of the star, whereas the numerical values of the parameters R' and M' may be consulted in the last line of the table. Another important relation is the one for the ratio between the mean and central densities

$$\frac{\rho_m}{\rho_o} = \left(-\frac{3}{z}\frac{du}{dz} \right)_{u=0} . \tag{94.15}$$

Its numerical value can be found at the bottom of the sixth column. On the other hand and since $\phi/\phi_o=u$, it is easy to see that

$$\frac{T}{T_o}=u; \tag{94.16}$$

where T_o is the central temperature which can be calculated from the relation (94.6) for a constant value of the average molecular weight μ equal to 2.2. The numerical value of u can be looked up on the second column of the table for different points within the star. For density we have the following relation

$$\frac{\rho}{\rho_o}=u^n; \tag{94.17}$$

with ρ the density at some inner region of the star. It is calculated from ρ_o and using the numerical values of u^n for different points within the star which are recorded on the third column. In order to show how the formulae above and the table should be used, consider as the first example that of Capella's brightest component.

Amongst the many data known of that star, we have that $M=8.3\times10^{33}\ gr$ and $R=9.55\times10^{11}\ cm$. Its mean density is ρ_m=0.00227 and in column sixth it can be seen that the ratio ρ_o/ρ_m is equal to 54.36; so that in the center of the star

$$\rho_o=0.1234 gr\cdot cm^{-3}.$$

Also from the last line of the table the following values are obtained

$$\begin{aligned} M' &= 2.015 \\ R' &= 6.901 \end{aligned}.$$

According to (94.8) the gravitational potential in the center is

$$\phi_o=\frac{6.66\times10^{-8}\times8.3\times10^{33}\times6.901}{2.015\times9.55\times10^{11}}=1.982\times10^{15}.$$

On the other hand it is known that

$$1-\varepsilon = 0.283$$
$$\varepsilon = 0.717;$$

in such a way that the magnitude of the self-generated magnetic field in the center of the star is according to (94.13)

$$H_o = \left[\frac{2\times 3.14\times 0.2446\times 10^{15}\times 0.283}{0.717}\right]^{1/2} = (6.063)^{1/2}\times 10^7;$$

in such a way that

$$H_o = 2.46\times 10^7\ gauss.$$

Consider another point inside the star, like for example the line $z = 3.5$ from the table. According to (94.14) it is easy to see that

$$\frac{z}{R'} = \frac{3.5}{6.901} = 0.507.$$

The point considered is found placed a little further from half the center of the star. From column sixth of the table we have that

$$\frac{M(r)}{M} = \frac{1.8203}{2.0150} = 0.90.$$

In other words, with that choice 90% of star´s mass is being taken into account. The gravitational potential at that point is

$$\phi = \frac{6.66\times 10^{-8}\times 8.3\times 10^{33}\times 3.5}{1.8203\times 9.55\times 10^{11}} = 1.113\times 10^{15}.$$

It is obtained from the third column of the table that

$$\rho = 0.02109\rho_o = 0.00260 gr\cdot cm^{-3},$$

and

$$\rho\phi = 2.6\times 10^{-3}\times 1.113\times 10^{15} = 2.8938\times 10^{12}.$$

The magnetic field at that point has the following absolute value

$$H = \left[\frac{6.28 \times 2.8938 \times 10^{12} \times 0.283}{0.717}\right]^{1/2} = (7.1729)^{1/2} \times 10^{6};$$

namely

$$H = 2.68 \times 10^{6}\, gauss\, .$$

If H and H_o are compared, it can be seen that

$$H = 0.11\, H_o;$$

which means that the magnitude of the magnetic field at the middle part of the star has been reduced to 11% of the value that it has in the center. Since due to boundary condition $\phi_o = 0$ is always taken on the surface, in that region the magnetic field must be calculated with the formula (89.5). That way it is obtained that on the surface

$$H_s = 6.8\, gauss\, .$$

With these three points it is possible to build the following graphic showing the general behaviour of the field.

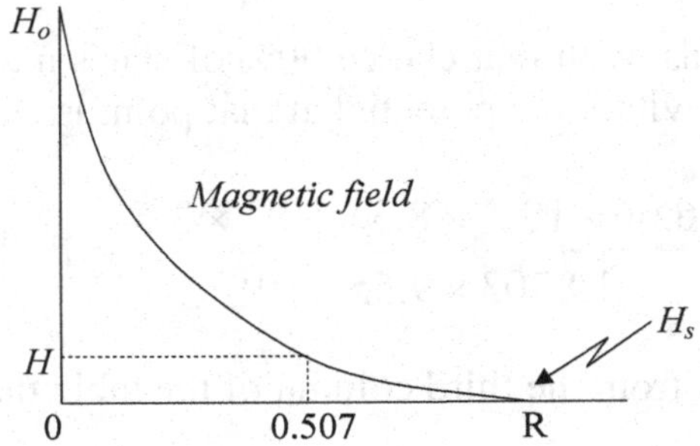

Consider the case of the Sun, for which we have the following data

$$M_{\odot} = 1.985 \times 10^{33}\, gr$$
$$R_{\odot} = 6.951 \times 10^{10}\, cm$$

$$1-\varepsilon=0.0498$$
$$\varepsilon=0.9502$$
$$\rho_m=1.414\ gr\cdot cm^{-3}$$
$$\rho_o=76.9\ gr\cdot cm^{-3}$$

In the center of the Sun the gravitational potential has the following value

$$\phi_o=\frac{6.66\times10^{-8}\times1.985\times10^{33}\times6.901}{2.015\times6.951\times10^{10}}=6.51\times10^{15};$$

and then

$$\phi_o\rho_o=5.01\times10^{17}.$$

The central magnetic field has a magnitude equal to

$$H_o=4.06\times10^{8}\ gauss.$$

Consider the same point used in Capella´s case, in such a way that

$$\rho=1.62\ gr\cdot cm^{-3}.$$

In that position the gravitational potential has the following value

$$\phi=3.66\times10^{15},$$

and of course

$$\rho\phi=5.93\times10^{15}.$$

Besides,

$$\frac{1-\varepsilon}{\varepsilon}=0.0524.$$

In consequence

$$H=4.4\times10^{7}\ gauss.$$

Comparing H to H_o we have that for the case of the Sun

$$H = 0.11\,H_o.$$

As it was expected, the magnitude of the solar magnetic field diminishes in the same proportion as Capella's magnetic field. Indeed, its value is again equal to the 11% of the one it has in the center. Consequently the corresponding graphic is equal to that of the previous example, as it is easy to see in the following figure.

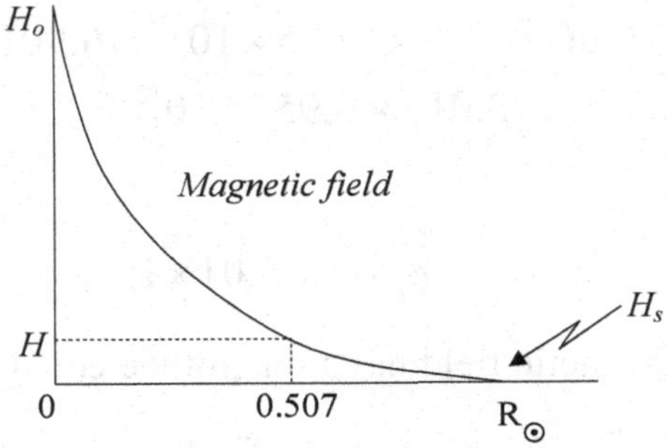

The behavior of the magnetic field is in general very similar to the one observed in the density of the stellar mass in *Eddington's theory*.
From the theoretical scheme developed in the above paragraph, it is possible to derive an expression to calculate the average value of the residual magnetic field which acts in the Cepheid type variable stars, in the maximum gravitational compression as well as in that of maximum expansion. If the absolute value of total force is considered, from the equations (93.4) and (93.6) as well as from the *mass-luminosity relation* (91.6), it is possible to demonstrate that in $r = R$

$$\overline{H} = [8\pi GM(1-\varepsilon)]^{1/2} \cdot \left[\frac{(1-\varepsilon)\overline{\rho}}{\varepsilon(2-\varepsilon)}\right]^{1/2} \cdot \left(\frac{1}{R}\right)^{1/2}; \qquad (94.18)$$

where the bar indicates the average values of those quantities; H is the residual magnetic field absolute value, ρ the mass density, and R star's radius. Besides, the first of the definitions (90.5) was used. On the other hand, from the result (93.10) it is easy to see that

$$\left[\frac{(1-\varepsilon)\overline{\rho}}{\varepsilon(2-\varepsilon)}\right]^{1/2} = \frac{6.87\times 10^{3}}{\tau}, \qquad (94.19)$$

in such a way that in (94.18) the following is obtained

$$\overline{H} = \frac{\theta}{\tau R^{1/2}}; \tag{94.20}$$

where

$$\theta = 8.89 \times [M(1-\varepsilon)]^{1/2} \tag{94.21}$$

is a constant which depends on the basic parameters of the particular star being studied.

Therefore, it is possible to state that for each particular Cepheid, *the residual magnetic field is inversely proportional to the product of the period and the square root of the radius*; so that for the situation of maximum compression we have that

$$\overline{H}_{\text{máx.}} = \frac{\theta}{\tau R_{\text{mín.}}^{1/2}}; \tag{94.22}$$

whereas for the maximum expansion

$$\overline{H}_{\text{mín.}} = \frac{\theta}{\tau R_{\text{máx.}}^{1/2}}. \tag{94.23}$$

The maximum and minimum values of the radius reached by the star in its extreme volumes, can be calculated from (89.10) and from the observational data obtained for the luminosity and the effective temperature at the position of maximum and minimum amplitude of the harmonic oscillations carried out by the particular Cepheid being observed.

Finally, suppose that the gravitational collapse suffered by a star which is in the last stage of its evolution is so fast and vigorous that the magnetic field noticeably weakened does not have anymore an intensity powerful enough so as to stop it initiating the rebound. Under these conditions, the star collapses and due to the combined effect of material and radiation pressures, as well as the sudden hydrogen combustion still present in large amounts in the regions of the star atmosphere, a huge explosion is produced which can send into the space up to 9/10 parts of the star total mass. In that case it is said that a *nova* or a *supernova* has risen depending on the magnitude of the sidereal catastrophe. Whether one thing or another should occur possibly depends on how massive the star is. The final outcome of the catastrophe will also depend on that last condition. It is possible to have a *recurrent nova*, a *white dwarf*, a *pulsar*, or even a *black hole*.

Selected Topics

The scale of stellar magnitudes

The scale of stellar magnitudes which is now generally adopted is based on the fact that we recive about 100 times as much light from a star of the first magnitude as from one of the sixth. This means that one magnitude is equivalent in brightness to a factor of 2.512, because it is the fifth root of 100. Thus, a fifth magnitude star has a brightness 2.512 greater than another of the sixth magnitude; whereas one of the fourth magnitude is 2.512 times brighter than one of the fifth or 2.512^2 times brighter than a sixth-magnitude star. In consequence, from stars of third, second, and first magnitude, we receive 2.512^3, 2.512^4, and 2.512^5 more light than from a sixth-magnitude star, respectively. It is important to note that the numerically smaller magnitudes are associated with the brighter stars, whereas a numerically large magnitude refers to a faint star; in such a way that for stars of maximun brightness it is necessary to give them negative magnitudes.

If all the stars had the same intrinsec brightness or luminosity without taking into account the fact that the interstellar space is not entirely free of opaque matter, we would have to assume that the difference in the apparent brightness is due only and exclusively to the distance. Let us assume that this is the case. In order to be able to compare the amount of light which the stars emit into space, it is necessary to calculate how much light we would receive from each star if all stars were at the same distance from us. Fortunately, if we know a star´s distance, like the Sun for example, we can make such a calculation.

The Sun is the most brilliant star which can be observed because is the one closer to us. Its apparent magnitude is equal to –26.5. To compare the intrinsec luminosity of the Sun with that of other stars, it is necessary to determine what magnitude the Sun would have if it were at a given distance typical of the distances at which the other stars are found. Suppose we choose 10 parsecs as a more or less representative distance of the nearest stars. Since 1 parsec is about 2×10^5 astronomical units (one astronomical unit $=150\times10^6$ km), the Sun would be 2×10^6 times as distant as it is now if it were removed to a distance of 10 parsecs. Consequently and given that the amount of light received by a star is inversely proportional to the square of the distance, the amount of light that would deliver to us would be $(1/2\times10^6)=1/4\times10^{12}$ of the light it now sends. In this case and in order to know how many magnitudes correspond to a factor of 4×10^{12}, the following calculus will be made

$$(2.512)^n = 4\times10^{12}. \tag{1}$$

Taking logarithms, it is easy to see that $n = 31.5$ magnitudes. Therefore, if the Sun were at a distance of 10 parsecs it would appear fainter by some 31.5 magnitudes than its present magnitude of –26.5; that is, its absolute magnitude would be +5 and it would appear as a fifth magnitude star. A more exact estimation indicates that the absolute or bolometric magnitude of the Sun is equal to +4.9. For all the previous, the absolute magnitude of a star is defined as the magnitude that the star would have if it were at the standard distance of 10 parsecs (1 parsec = 3.26 light-years).

Luminosity and stellar radius

The calculus of the luminosity of any gaseous star is made from the relationship (89.8) of the text and using the Sun as a measure unit, to obtain the following result

$$\frac{L}{L_\odot} = \left(\frac{T_e}{T_{e\odot}}\right)^4 \cdot \left(\frac{R}{R_\odot}\right)^2; \tag{1}$$

where T_e is the effective temperature and R star´s radius. Taking logarithms in (1) the following fundamental relationship is obtained

$$\log\frac{L}{L_\odot} = 4\log\frac{T_e}{T_{e\odot}} + 2\log\frac{R}{R_\odot}. \tag{2}$$

In the former formula, $T_{e\odot} = 5740\ K$ and $R_\odot = 6.951 \times 10^{10}$ cm.

In order to know the bolometric magnitude of any star, it is necessary to consider the change of brightness with the distance. If we have this observational datum, it can be used to calculate the luminosity of the star. What is usual to do is to use the Sun as intermediary and the following basic equation

$$\log\frac{L}{L_\odot}=\frac{1}{2.512}\left(M_{bol\odot}-M_{bol}\right); \tag{3}$$

where $M_{bol_\odot}=+4.9$ and $L_\odot=3.78\times10^{33}$ ergs.sec^{-1}.

As a first example let us consider Sirius which is the brightest star of the sky, from which we know that $M_{bol}=+1.4$. Then

$$\log\frac{L}{L_\odot}=\frac{1}{2.512}(4.9-1.4)=\frac{3.5}{2.512}=1.3933;$$

so, taking antilogarithms we see that

$$L(Sirus)=25L_\odot.$$

In other words, Sirius is 25 times brighter than the Sun and in consequence

$$L(Sirus)=9.45\times10^{34}\,ergs.sec^{-1}.$$

For the brightest component of Capella, it is known that its bolometric magnitude is equal to –0.40, so that

$$L(Capella)=129L_\odot,$$

that is, Capella is almost 130 times brighter than the Sun, which means that

$$L(Capella)=4.87\times10^{35}\,ergs.sec^{-1}.$$

On the other hand, for any gaseous star the relationship (89.5) is fulfilled; so that substituting that formula in (1) the following result is obtained

$$\log\frac{L}{L_\odot}=2\log\left(\frac{RH}{R_\odot H_\odot}\right); \tag{4}$$

with H the value of the self-generated magnetic field which we have on the surface of the star. Comparing the relations (3) and (4) we have that

$$\log\left(\frac{RH}{R_{\odot}H_{\odot}}\right) = \frac{1}{5.024}\left(M_{bol\odot} - M_{bol}\right). \quad (5)$$

The last formula can be used to obtain the radius of any gaseous star for which its effective temperature and bolometric magnitude are known as a result of the observation, all of it in terms of the relevant Sun's parameters. With the data that we have on Capella, it is easy to calculate that its radius is equal to 9.3×10^{11} cm.

Let us consider the case of Betelgeuse. It is said that it is a red giant which has an effective temperature of about 3×10^{3} *K*, so that the value on its surface of the self-generated magnetic field is of 2.3 gauss. Also it is known that its bolometric magnitude is equal to –5.5. Consequently, from (3) it is easy to obtain that

$$L(Betelgeuse) = 13{,}808 L_{\odot}.$$

In other words, Betelgeuse is almost 14,000 times more luminous than the Sun. However, its effective temperature is a little more than half of that of the Sun; in such a way that in order to be so luminous it is necessary to be gigantic. From the equation (5) we have that

$$R(Betelgeuse) = 298\times10^{6}\,km\,.$$

The radius of Betelgeuse is greater than the radius of Mars´orbit and certainly, it is a red giant.

Stellar distances

Since the absolute magnitude of a star is the magnitude which the star would have at the standard distance of 10 parsecs, it is about an actual measure of the rate of emission of visible light energy by the star; emission which, of course is independent of the star´s actual distance. On the other hand, the apparent magnitude is a measure of how bright the star appears to be and consequently, that measure depends on both star´s actual rate of light output and its distance. The difference between star´s apparent magnitude M_{vis} and its absolute or bolometric magnitude M_{bol}, can be calculated from the inverse-square law of light and from a knowledge of how much greater or less that 10 parsecs star´s distance actually is. This way, the difference

(M_{vis} - M_{bol}) depends only on star′s distance. That difference is known as the distance modulus and it is measured in parsecs. For effect of calculus, let $\circ(r)$ be the observed light of a star at its actual distance r from the observer and $\circ(10)$ the amount of light we would receive from it if it were at a distance of 10 parsecs.

From the definition of stellar magnitudes, it can be demonstrated that

$$M_{vis} - M_{bol} = 2.512 \log \frac{\circ(10)}{\circ(r)}\ ; \tag{1}$$

whereas from the inverse-square law of light, we have the following result

$$\frac{\circ(10)}{\circ(r)} = \left(\frac{r}{10}\right)^2 .$$

Combining both equations the following is obtained

$$M_{vis} - M_{bol} = 5.024 \log \frac{r}{10} .$$

The quantity 5.024 log(r/10) is the distance modulus. Let us consider the following examples

Sirius A

$$M_{vis} = -1.42; M_{bol} = +1.4$$

$$r = 2.75\, parsecs = 8.95\, light - years$$

Procyon A

$$M_{vis} = +0.38; M_{bol} = +2.7$$

$$r = 3.45\, parsecs = 11.3\, light - years$$

Capella

$$M_{vis} = -0.26;\ M_{bol} = -0.4$$

$$r = 10.7\, parsecs = 35\ light - years$$

Betelgeuse

$$M_{vis} = +0.41; M_{bol} = -5.5$$

$$r = 150\, parsecs = 500\ light - years .$$

Sirius and *Procyon* are binary systems. Both stars have a very small and very massive companion known as *Sirius B* and *Procyon B*. They are known by the name of *white dwarfs*.

Time scale of stellar evolution

When the star's evolution problem is studied, it is usual to fix one's attention on the rate of change of hydrogen content. Therefore, be $\bar{X}$ the average hydrogen content in a gaseous star, defined as follows

$$\bar{X} = \int_0^1 X dq \,; \tag{1}$$

where X represents the abundance of hydrogen in the star and dq the differential of effective cross section for collitions; in such a way that the total amount of hydrogen in a star of mass M is equal to $M\bar{X}$. The total time derivative of this quantity gives the rate of hydrogen burning in the thermonuclear oven, that is $Md\bar{X}/dt$. If we multiply this rate by the following conversion factor

$$E_{cc}^{*} = \frac{E_{cc}}{4h} = 6.0 \times 10^{18}\, ergs.gr^{-1}; \tag{2}$$

factor which turns the number of grams burned into ergs, the total luminosity of the star is obtained

$$L = -ME_{cc}^{*}\frac{d\bar{X}}{dt}. \tag{3}$$

In the relationship (2)

$$E_{cc} = 4.0 \times 10^{-5}\, ergs \tag{4}$$

is the total energy delivered for each helium atom formed and h represents the hydrogen atom; whereas in (3) the sign minus indicates that the quantity of hydrogen fuel disposable is diminished because it is burning.

The age of a star can be defined as the time elapsed since the star first started to burn its hydrogen fuel until the evolutive phase in which it is at that time. It can be determined by integrating the equation (2) over the time, from the initial evolutive phase to the evolutionary phase which will be considered, that is to say

$$\tau = \int_{X}^{X_o} \frac{ME_{cc}^*}{L} d\bar{X}; \tag{5}$$

where the limits of the integral were inversed in order to eliminate the sign minus and the zero subscript refers to the initial state. Be

$$\tau = \tau_o \tau^*; \tag{6}$$

where

$$\tau_o = \frac{ME_{cc}^*}{L_o} \tag{7}$$

is the expected life of a star, since this quantity represents the time a star would live if it started with one hundred per cent of hydrogen and it were capable of completely burning up maintaining its luminosity equal to its initial luminosity throughout all of its life; whereas

$$\tau^* = \int_{X}^{X_o} \frac{L_o}{L} d\bar{X} \tag{8}$$

is the relative age, since this term gives the fraction of the expected life which a star has actually lived.

If in the initial phase the modified mass-luminosity relationship is taken into account, it is easy to see that

$$L_o = \frac{z_o M}{k_c}, \tag{9}$$

with

$$z_o = \frac{4\pi Gc}{\alpha} \cdot \left(\frac{1-\varepsilon}{\varepsilon} \right) \tag{10}$$

a constant and k_c the opacity coefficient. In this case,

$$\tau_o = constant\ k_c \tag{11}$$

where

$$constant = \frac{E_{cc}^*}{z_o}. \tag{12}$$

As it is easy to see, the age of a gaseous star is ruled by the value of its coefficient of opacity. To calculate the Sun's expected life the following data will be used

$$1-\varepsilon = 0.0498$$
$$\varepsilon = 0.9502$$

so that z_o=5.26 × 10^2; whereas E_{cc}^*/z_o = 1.14 × 10^{16}. Since it is known that k_c = 2.76 × 10^2 and 1 sec ≈10^{-7} years,

$$\tau_o \approx 10^{11} years.$$

For Capella's case we have that $1-\varepsilon$ =0.283, in such a way that ε =0.717. With those data it is clear that z_o=3.96×10^2. Likewise, E_{cc}^*/z_o =1.52 ×10^{15}, k_c=6.9×10 and therefore

$$\tau_o \approx 10^{10} years.$$

A simple model to estimate p_c, T_c, and H_c

The equation (88.1) which governs the magneto mechanical equilibrium can be used to calculate some formulae by means of which it is possible to estimate pressure, temperature, and the magnetic field magnitude orders in the center of any gaseous star in terms of its relevant parameters. Consider then that it is possible to draw within the star, a radius sphere r which contains an amount of mass equal to $M(r)$. When the result (88.5) is used in (88.1), that relation is transformed into the following

$$d\left(p - \frac{H^2}{8\pi}\right) = -\rho \frac{GM(r)}{r^2} dr. \tag{1}$$

From the *relation of equivalence* (89.3) and from the first of the definitions (90.5), it is easy to obtain in (1) that

$$dp = -\rho \frac{GM(r)}{\varepsilon} \frac{dr}{r^2}. \tag{2}$$

In order to simplify the calculation, consider that the surface of the sphere is located in the middle region between the center of the star and its boundary, in such way that $M(r) \approx M/2$, with M the total mass of the star. For dp, the difference between the central whole pressure and the pressure on the surface will be taken, and the latter will be considered as zero.
Be $-dr/r^2 \approx 2/R$ with R star´s radius. In that case from (2) the following result is obtained

$$p_c = \frac{\rho_c \phi_c}{\varepsilon}; \tag{3}$$

where subscript c refers to the star´s center and

$$\phi_c = \frac{GM}{R} \tag{4}$$

is the gravitational potential. Central mass density is obtained from the following ratio

$$\frac{\rho_c}{\rho_m} = 54.36\,, \tag{5}$$

and according to the last column of the table of the text.

From the relation (3) and from the second of the definitions (90.5) the following result is obtained

$$T_c = \frac{\mu\phi_c}{\mathcal{R}}\,, \tag{6}$$

where μ is the average molecular weight, and $\mathcal{R}$ the gas universal constant, and the thermal equation of the ideal gas (92.1) was used.

Finally, to obtain the formula which allows to estimate the magnitude of the magnetic field, the *relation of equivalence* (89.3), and the first of the definitions (90.5) are used in the formula (3). Thus, for each particular case

$$H_c = constant(\rho_c\phi_c)^{1/2}; \tag{7}$$

where

$$constant = 2\times\left[\frac{2\pi(1-\varepsilon)}{\varepsilon}\right]^{1/2}. \tag{8}$$

The brightest component of Capella's binary system will be considered as a first example. Its relevant data are the following

$$\begin{aligned} M &= 8.30\times 10^{33}\,gr \\ R &= 9.55\times 10^{11}\,cm \\ \mu &= 2.2 \\ \varepsilon &= 0.717 \\ 1-\varepsilon &= 0.283 \end{aligned}$$

Besides we have that

$$\begin{aligned} G &= 6.66\times 10^{-8}\,ergs.\,cm\cdot gr^{-2} \\ \mathcal{R} &= 8.26\times 10^{7}\,ergs.\,mol\cdot gr^{-1}K^{-1} \end{aligned}$$

$$2\times\left[\frac{2\pi(1-\varepsilon)}{\varepsilon}\right]^{1/2}=3.15$$

$$\phi_c=5.8\times10^{14}\,ergs\cdot gr^{-1}.$$

Since it is also known that $\rho_m = 0.00227$, we have that

$$\rho_c=0.1234\;gr\cdot cm^{-3}.$$

With all previous numerical data it is easy to obtain the following results

$$\begin{aligned}p_c&=9.96\times10^{13}\;dyne\cdot cm^{-2}\\T_c&=15.4\times10^{6}K\\H_c&=2.7\times10^{7}\,gauss.\end{aligned}$$

On the other hand and since

$$\begin{aligned}&L=4.8\times10^{35}\;ergs\cdot sec^{-1}\\&4\pi Gc=2.51\times10^{4}ergs.cm^{2}\cdot gr^{-2}sec^{-1}.\\&\alpha=2.5\end{aligned}$$

we have that

$$k_c=69\;cm^{2}\cdot gr^{-1}.$$

As a second example consider the case of the Sun for which we have the following data

$$\begin{aligned}M_\odot&=1.985\times10^{33}gr\\L_\odot&=3.78\times10^{33}ergs\cdot sec^{-1}\\R_\odot&=6.951\times10^{10}cm\\\varepsilon&=0.9502\\1-\varepsilon&=0.0498\\\phi_c&=2.0\times10^{15}ergs\cdot gr^{-1}.\\\mu&=1\\\rho_m&=1.414\,gr\cdot cm^{-3}\\\rho_c&=77gr\cdot cm^{-3}.\end{aligned}$$

Consequently

$$\begin{aligned} p_c &= 1.6\times10^{17}\, dyne\cdot cm^{-2} \\ T_c &= 24.2\times10^{6} K \\ H_c &= 4.5\times10^{8}\, gauss \\ k_c &= 276\ \ cm^{2}\cdot gr^{-1}. \end{aligned}$$

Polytropic gas sphere

When the equation (94.4) is fulfilled it is stated that the fluid stellar distribution is polytropic. If that formula is derived, the following is obtained

$$dP = \gamma\kappa\rho^{\gamma-1} d\rho\,; \tag{1}$$

which with the help of (94.2) is transformed into the following relation

$$\gamma\kappa\rho^{\gamma-2} d\rho = d\phi\,. \tag{2}$$

Integrating , the following result

$$\frac{\gamma}{\gamma-1}\kappa\rho^{\gamma-1} = \phi + constant\,, \tag{3}$$

can be obtained.

The point of space where the gravitational potential is zero is elected in a totally arbitrary way. The usual convention is to make ϕ vanish at an infinite distance from all matter. Nevertheless, in Astrophysics it is usual to take the zero of ϕ at the boundary of the star where the mass density is practically zero, in such a way that the integration constant in (3) is equal to zero. Be by definition

$$\gamma = 1 + \frac{1}{n}, \tag{4}$$

with n a positive integer. In this case in (3) we have that

$$\rho = \left[\frac{\phi}{(n+1)\kappa} \right]^n . \tag{5}$$

Substituting this last result in (94.4) the following is obtained

$$P = \frac{\rho\phi}{n+1} . \tag{6}$$

It can be demonstrated that the relationship (94.5) of the text is obtained from (6) and according to (89.1) and the second one of the definitions (90.5). Now, if in the equation (94.3) of the text, the result (5) is substituted, the following is obtained

$$\frac{d^2\phi}{dx^2} + \frac{2}{x}\frac{d\phi}{dx} + \alpha^2 \phi^n = 0 ; \tag{7}$$

where

$$\alpha^2 = \frac{4\pi G}{[(n+1)\kappa]^n} . \tag{8}$$

What is usually done next is introducing two new variables u and z which are proportional respectively to ϕ and x; in such a way that if ϕ_o is the value of ϕ at the center of the distribution, the following definitions are fulfilled

$$\phi = \phi_o u$$

and

$$x = \frac{z}{\alpha \phi_o^{1/2(n-1)}} ; \tag{9}$$

which when substituted in (7) the following differential ordinary equation is obtained

$$\frac{d^2u}{dz^2} + \frac{2}{z}\frac{du}{dz} + u^n = 0 ; \tag{10}$$

equation which is subjected to the following conditions in the center of the distribution: u=1 and du/dz=0, if z=0. The proposed substitution only means a change of units which is introduced in order to bring the differential equation (7) and its limiting conditions have the usual form of the specialized literature. The condition du/dz=0 at the central part of the distribution, originates from the fact that at that point the relationship $g=-d\phi/dx$ disappears. On the other hand, the solution of (10) for a large number of values of n has been calculated by *R.* Emden. Its result can be seen in very complete tables, as the one included in the text. Given that u = 0 indicates the frontier of the star, it is fulfilled that

$$R=(x)_{u=0}; \tag{11}$$

and

$$GM=\left(-x^2\frac{d\phi}{dx}\right)_{u=0}. \tag{12}$$

In this case and according to (9) and (94.14)

$$\frac{R}{R'}=\frac{1}{\alpha\phi^{1/2(n-1)}}, \tag{13}$$

and

$$\frac{GM}{M'}=\frac{1}{\alpha\phi^{1/2(n-3)}}; \tag{14}$$

in such a way that the relationship (94.8) is fulfilled. On the other hand,

$$\left(\frac{GM}{M'}\right)^{n-1}\cdot\left(\frac{R'}{R}\right)^{n-3}=\frac{1}{\alpha^2}. \tag{15}$$

In the last formula, R is star´s radius. Those are all the relations required in order to calculate the relevant parameters of any gaseous star. For instance, the mass density ρ_o in the center of the distribution can be calculated in terms of the mean mass density ρ_m, , as follows

$$\rho_m = \frac{M}{4/3\pi R^3} = \frac{1}{G}\left[-\frac{1}{4/3\pi x^3}\cdot x^2 \frac{d\phi}{dx}\right]_{u=0};$$

where the relation (12) was used.

According to the second of the definitions (9)

$$\frac{1}{x}\frac{d\phi}{dx} = \alpha^2\phi_o^n \cdot \frac{1}{z}\frac{du}{dz};$$

so that

$$\rho_m = \frac{3\alpha^2\phi_o^n}{4\pi G}\left(-\frac{1}{z}\frac{du}{dz}\right)_{u=0}. \tag{16}$$

Nevertheless and according to (5) and (8)

$$\alpha^2\phi_o^n = 4\pi G\rho_o, \tag{17}$$

in such a way that substituting (17) in (16) the relation(94.15) is obtained.

Gaseous stars

Let us consider a star composed by a fluid which from the thermodynamics view point behaves as an ideal gas, so that it obeys the thermal equation of state (92.1) as well as the relation between the radiation pressure and the temperature (89.4). According to definitions (90.5), the total pressure fulfils the following relationships

$$p = \frac{\mathcal{R}}{\mu\varepsilon}\rho T$$

and

$$p = \frac{aT^4}{3(1-\varepsilon)}.$$

Eliminating T between them, the following result is obtained

$$p = \kappa \rho^{4/3}; \tag{1}$$

where

$$\kappa = \left[\frac{3\mathcal{R}^4 (1-\varepsilon)}{a\mu^4 \varepsilon^4} \right]^{1/3}. \tag{2}$$

Consequently, the distribution is polytropic with γ=1.3333 and n = 3. In the former relations a is Stefan´s coefficient.

From the formula (15) of the former paragraph we have that for $n = 3$,

$$\left(\frac{GM}{M'} \right)^2 = \frac{(4\kappa)^3}{4\pi G} = \frac{48\mathcal{R}^4 (1-\varepsilon)}{\pi G a \mu^4 \varepsilon^4}. \tag{3}$$

In this case,

$$1 - \varepsilon = CM^2 \mu^4 \varepsilon^4; \tag{4}$$

where

$$C = \frac{\pi G^3 a}{48\mathcal{R}^4 M'^2} \tag{5}$$

is a numerical constant equal to 7.83×10^{-70} $\text{mol}^{-4}.\text{gr}^{-2}$. In the relation (4) the mass of the star is expressed in grams and the molecular weight in terms of the hydrogen atom. It is more convenient to write it using the mass of the Sun as a measure unit. Since ⊙ = 1.985×10^{33} gr is the solar mass, from the formula (4) the quadric equation (90.7) is obtained.

A numerical equation for ε

The mass of any gaseous star can be calculated from the mass-luminosity relation (91.6). The procedure is the following

$$L=\frac{4\pi cGM(1-\varepsilon)}{\alpha k_c\,\varepsilon}=\frac{4\pi cGM(1-\varepsilon)}{\alpha k_1\,\varepsilon}\frac{T^3}{\rho}T_c^{1/2}$$
$$=\frac{4\pi cG}{\alpha k_1}\cdot\frac{3\mathcal{R}}{a\mu}\cdot\left(\frac{G}{4\mathcal{R}}\cdot\frac{R'}{M'}\frac{\mu}{R}\right)^{1/2}M^{3/2}\frac{(1-\varepsilon)^2}{\varepsilon^2}; \tag{1}$$

where the relationships (91.2), (92.2) and (94.7) were used. From the equations (4) and (5) of the former paragraph, the mass M can be eliminated to obtain that

$$M^{3/2}=\left(\frac{48\mathcal{R}^4M'^2}{\pi G^3a}\right)^{3/4}\frac{(1-\varepsilon)^{3/4}}{\mu^3\varepsilon^3}, \tag{2}$$

so that

$$L=\frac{4\pi cG}{\alpha k_1}\cdot\frac{3\mathcal{R}}{a}\cdot\left(\frac{G}{4\mathcal{R}}\cdot\frac{R'}{M'}\right)^{1/2}\cdot\left(\frac{48\mathcal{R}^4M'^2}{\pi G^3a}\right)^{3/4}\frac{(1-\varepsilon)^{11/4}}{\mu^{7/2}R^{1/2}\varepsilon^5}. \tag{3}$$

With the value of the constants we have that

$$L=\frac{1.443\times10^{71}(1-\varepsilon)^{11/4}}{\alpha k_1R^{1/2}\mu^{7/2}\varepsilon^5}. \tag{4}$$

For k_1, it is usual to utilize the value of that factor calculated for the brightest component of Capella's binary system, as follows. From the mass-luminosity relation (91.6) and given that for Capella we have the following data

$$L=4.8\times10^{35}\,ergs.sec^{-1}$$
$$M=8.3\times10^{33}\,gr$$
$$\varepsilon=0.717$$
$$1-\varepsilon=0.283$$

the following results are obtained

$$k_c = 69cm^2 \cdot gr^{-1}$$
$$\rho_c = 0.1234 gr \cdot cm^{-3}$$
$$T_c = 9.09 \times 10^6 K.$$

In order to obtain T_c the relation (92.3) was used. From (91.2) it is easy to see by direct calculus that

$$k_1 = 1.26 \times 10^{27}.$$

On the other hand, $\alpha = 2.5$ and $\mu = 2.11$ in such a way that in (4) the following result is obtained

$$LR^{1/2} = 0.34 \times 10^{43} \frac{(1-\varepsilon)^{11/4}}{\varepsilon^5}. \tag{5}$$

In that case, for a particular gaseous star we have that

$$(1-\varepsilon)^{11/4} = \varepsilon_o \varepsilon^5; \tag{6}$$

where

$$\varepsilon_o = 2.94 \times 10^{-43} LR^{1/2}. \tag{7}$$

Consider the case of the intrinsic variable *δ Cephei*. From its absolute magnitude the following data are obtained

$$L = 2.81 \times 10^{36} ergs.sec^{-1}$$
$$R = 2.32 \times 10^{12} cm;$$

so that ε_o=1.3 and in (6) an equation is obtained whose only unknown quantity is ε, that is to say

$$(1-\varepsilon)^{11/4} = 1.3\varepsilon^5. \tag{8}$$

Using Newton´s method the following numerical solution can be obtained

$$\varepsilon = 0.542$$
$$1 - \varepsilon = 0.458$$
$$2 - \varepsilon = 1.458.$$

Substituting the two first values in the quadric equation (90.7) the mass of δ *Cephei* is

$$M = 1.82 \times 10^{34}\ gr\ .$$

On the other hand, since its volume is V=4.58×10^{37}cm^3, its mean mass density is ρ_m=4.0×10^{-4}gr·cm^{-3}.

The magnetic field in homologous inner points

If the relation of equivalence (89.3) and also the definitions (90.5) and the formula (92.1) are used, the following can be obtained

$$H = \left[\frac{8\pi \mathcal{R}\ \rho T(1-\varepsilon)}{\mu\varepsilon}\right]^{1/2}; \qquad (1)$$

by means of which it is also possible to calculate the magnitude of the self-generated magnetic field, in both, the center as well as the inner homologous points of any gaseous star.

From the results of paragraph 94 and with the numerical data from the table, from relations (94.7) and (94.15) it is obtained for Capella's case that

$$T_c = 13.2 \times 10^6 K$$
$$\rho_c = 0.1234\ gr \cdot cm^{-3}.$$

Thus, in the center of the star, the magnitude of the self-generated magnetic field is again

$$H_c = 2.46 \times 10^7\ gauss\ .$$

From the second column of the table we have that for z=3.5, u=0.27629 in such a way that at this point

$$T = 3.65 \times 10^6\, K$$
$$\rho = 2.6 \times 10^{-3}\, gr \cdot cm^{-3}$$
$$H = 2.0 \times 10^6\, gauss.$$

Likewise, for the center of the Sun the following results are obtained

$$T_c = 19.7 \times 10^6\, K$$
$$\rho_c = 77\, gr \cdot cm^{-3}$$
$$H_c = 4.06 \times 10^8\, gauss.$$

It can be proven that in the point z=3.5,

$$T = 5.4 \times 10^6\, K$$
$$\rho = 1.62\, gr \cdot cm^{-3}$$
$$H = 3.1 \times 10^7\, gauss.$$

It is easy to see that for the calculated values of H and H_c with this last method, it is also fulfilled that in both examples the ratio H/H_c always keeps the same proportion.

In fact, let us consider the equation (1) for two inner homologous points. One of them in the center of the star and the other one at any inner point for which H_c = *constant* $(\rho_c T_c)^{1/2}$ and H = *constant* $(\rho T)^{1/2}$. In this case, it is fulfilled that

$$H = \left(u^{n+1}\right)^{1/2} H_c. \qquad (2)$$

The numerical values of u^{n+1} can be looked up on the fourth column on the table for different points within the star.

The mass and the luminosity

From the equation (89.5), (89.7), (89.10) and taking into account the mass-luminosity relation (91.6), the following expression to calculate the mass for any gaseous star, using some of their basic parameters is obtained; that is to say

$$M = \left(\frac{\varepsilon k_c}{1-\varepsilon}\right) \cdot \left(\frac{\alpha a}{4m^2 G}\right) \cdot (RH)^2 . \tag{1}$$

With the numerical values of the constants, the following is obtained

$$\left(\frac{\alpha a}{4m^2 G}\right) = 1.12 \times 10^6 \, gr^2 \cdot cm^{-4} \cdot gauss^{-2} . \tag{2}$$

Consequently, for a particular star it is fulfilled that

$$M = constant \times (RH)^2 \tag{3}$$

where

$$constant = 1.12 \times 10^6 \left(\frac{\varepsilon k_c}{1-\varepsilon}\right) . \tag{4}$$

With the data on the Sun, it is easy to prove by direct calculus that $M_{\odot}$=2.0×10^{33}gr; whereas for Capella we have that M=8.3×10^{33}gr.

From the equations (89.5) and (89.9) from the text, it can be demonstrated that also

$$L = constant \times (RH)^2 , \tag{5}$$

where now

$$constant = \pi ac/m^2 = 1.1\times10^{10} ergs\cdot sec^{-1}\cdot cm^{-2}\cdot gauss^{-2}; \tag{6}$$

in such a way that for the Sun, it can be verified that

$$L_\odot = 3.82\times10^{33} ergs\cdot sec^{-1}. \tag{7}$$

For Capella the following result is obtained

$$L = 4.72\times10^{35} ergs\cdot sec^{-1}. \tag{8}$$

Consequently, *the mass as well as the luminosity of a gaseous star, are proportional to the square of the product of its radio and the magnitude which the self-generated magnetic field has on its surface.* Clearly, the constant of proportionality is different for both parameters.

The effective temperature and the absolute magnitude

In this *Selected Topic* a theoretical procedure to calculate the values of fundamental parameters for any gaseous star will be presented if as a result of the observation, its effective temperature and its absolute magnitude are known.

As an illustrative example the case of the brightest component of the binary system *V Puppis* will be considered. According to the observational data, that star is an eclipsing variable of the spectral class BI. Its effective temperature is equal to 1.9×10^4 K and its bolometric magnitude is –4.75 and its visual magnitude is –3.2.

Using the relations (89.5) and (89.7) and the data of effective temperature, the magnitude that the self-generated magnetic field has on its surface can be calculated. In this way it is obtained that

$$H_* = 91.3\, gauss. \tag{1}$$

With this result, with the bolometric magnitude, and taking the Sun as a measure unit, the radius is calculated from the following relation

$$log\left[\frac{R_* H_*}{R_\odot H_\odot}\right] = \frac{M_{bol_\odot} - M_{bol*}}{5.024};$$

where $M_{bol_\odot} = +4.9$. From the direct calculus the following result is obtained

$$R_* = 5.33 \times 10^{11}\, cm. \tag{2}$$

With the effective temperature and the value of the radius, it is easy to calculate the luminosity from (89.9) and thus obtain that

$$L_* = 2.67 \times 10^{37}\, ergs \cdot sec^{-1}. \tag{3}$$

To calculate the mass it is necessary to know beforehand the magnitude of the parameter ε_*; value which can be determined from the relation (4) obtained on page 376 in the *Selected Topic* to develop a numerical equation for ε. Once those calculus are accomplished it is usual to use, as a measure unit, the brightest component of Capella's binary system. This way and given that the equation before mentioned is valid for any gaseous star, the following comparison can be made.

$$\frac{L_*}{L} = \left(\frac{k_1}{k_{1*}}\right) \cdot \left(\frac{R}{R_*}\right)^{1/2} \cdot \left[\frac{(1-\varepsilon_*)}{(1-\varepsilon)}\right]^{11/4} \cdot \left(\frac{\varepsilon}{\varepsilon_*}\right)^5. \tag{4}$$

With the data that we have of the brightest component of Capella, it is easy to see that

$$\frac{L_*}{L} = 0.5563 \times 10^2$$

$$\left(\frac{R}{R_*}\right)^{1/2} = 1.3386$$

$$\left[\frac{(1-\varepsilon_*)}{(1-\varepsilon)}\right]^{11/4} = 3.218 \times 10 (1-\varepsilon_*)^{11/4}$$

$$\left(\frac{\varepsilon}{\varepsilon_*}\right)^5 = \frac{0.1895}{\varepsilon_*^5};$$

in such a way that in (4) the following numerical equation for ε_* is obtained,

$$6.82\,\varepsilon_*^5 = \left(\frac{k_1}{k_{1*}}\right)(1-\varepsilon_*)^{11/4}. \tag{5}$$

Since the effective temperature is a conventional measure specifying the rate of out flow of radiant heat which a star emits per unit area, it is proposed as a work hypothesis that *in gaseous stars, the constants k_1 and k_{1*}* which appear in the numerical equation (5), *keep between them the same proportion which their respective effective temperatures have.* In other words, it will be considered that the ratios k_1/k_{1*} and T_e/T_{e*} are numerically equal; in such a way that from this viewpoint it can be stated that

$$\left(\frac{k_1}{k_{1*}}\right) = \left(\frac{T_e}{T_{e*}}\right); \tag{6}$$

where k_1 is the constant which appears in the radiation law of absorption (91.2), and in the present case refers to the brightest component of Capella. Consequently,

$$\frac{k_1}{k_{1*}} = 0.2737. \tag{7}$$

With this result substituted in (5) the following numerical equation to calculate the value of the parameter ε for the particular star studied is obtained; that is,

$$25\,\varepsilon^5 = (1-\varepsilon)^{11/4}. \tag{8}$$

Since on the other hand

$$(1-\varepsilon)^{11/4} = (1-\varepsilon)^3 \cdot (1-\varepsilon)^{-0,25},$$

in (8) the following totally equivalent relationship is obtained

$$25\varepsilon^5(1-\varepsilon)^{0.25} = (1-\varepsilon)^3;$$

where the asterisk was suppressed in order not to drag it along the following calculus. The numerical solution of the former equation can be obtained with the help of Newton´s method, as follows. In the left hand side an expansion in series is made as far as terms of the first order to obtain the following result

$$25\varepsilon^5 - 6.25\varepsilon^6 = 1 - 3\varepsilon + 3\varepsilon^2 - \varepsilon^3.$$

Next, all is divided by the coefficient of ε^6, which in this case is 6.25 and the resultant equation is rearranged to obtain the following

$$\varepsilon^6 - 4\varepsilon^5 - 0.16\varepsilon^3 + 0.48\varepsilon^2 - 0.48\varepsilon + 0.16 = 0. \tag{9}$$

Since ε is a decimal number, in a first approach and in comparison with itself, all the powers of ε of higher order are neglected to obtain the following approximate root

$$g = 0.3333.$$

Substituting the value of g in (9) we have that

$$f(g) = 0.0323.$$

Next the derivative of (9) is calculated; that is

$$6\varepsilon^5 - 20\varepsilon^4 - 0.48\varepsilon^2 + 0.96\varepsilon - 0.48 = 0. \tag{10}$$

In this last equation the root g is also substituted to obtain the following result

$$f'(g) = -0.4354.$$

The second approximation to ε is obtained using the following formula

$$h = \frac{-f(g)}{f'(g)} = 0.0742;$$

in such a way that

$$\varepsilon = g + h = 0.408$$

and

$$1 - \varepsilon = 0.592\,.$$

With the former results and from the quadric equation (90.7) and for a mean molecular weight μ equal to 2.11, the mass is calculated; that is to say

$$0.592 = 0.1697 \times 10^{-2} (M_* / \odot)^2\,.$$

Consequently

$$M_* = 18.7\,\odot = 3.71 \times 10^{34}\,gr\,. \tag{12}$$

The volume of the star is

$$V_* = 4/3 \pi R_*^3 = 6.34 \times 10^{35}\,cm^3\,;$$

and its average mass density is

$$\rho_{m*} = \frac{M_*}{V_*} = 0.0585\,gr \cdot cm^{-3}\,. \tag{13}$$

From the polytropic gas sphere theory and by means of the use of the following relation the mass density in the center of the star is obtained

$$\rho_{c*} = 54.36\,\rho_{m*} = 3.18\,gr \cdot cm^{-3}\,. \tag{14}$$

With the former results and using the equation (92.2), the central temperature can be calculated as follows

$$\frac{T_{c*}^3}{\rho_{c*}} = \frac{3 \times 8.26 \times 10^7 \times 0.592}{7.64 \times 10^{-15} \times 2.11 \times 0.408} = 0.22 \times 10^{23}\,,$$

and

$$T_{c*} = 4.12 \times 10^{7} K . \tag{15}$$

On the other hand and because k_1=1.26×10^{27} , from (7) we have that

$$k_{1*} = 4.6 \times 10^{27} . \tag{16}$$

Substituting the results (14), (15) and (16) in the radiation law for absorption (91.2) it is obtained that the coefficient of opacity from *V Puppis* has the following value

$$k_{c*} = 32.6 cm^{2} \cdot gr^{-1} . \tag{17}$$

The absolute value of the magnetic field self-generated by the star can be valued in any region within it with the help of the relation (94.13) and with the numerical data which are obtained from the polytropic gas sphere theory; that is to say, with M'=2.015 and R'=6.901. Thus, the gravitational potential in the center of the star has the following magnitude

$$\phi_{c*} = \frac{6.66 \times 10^{-8} \times 3.71 \times 10^{34} \times 6.901}{2.015 \times 5.33 \times 10^{11}} = 1.59 \times 10^{16} ;$$

where the formula (94.8) was used. In this case, from (94.13) we have that

$$H_{c*} = \left(\frac{2 \times 3.14 \times 3.18 \times 0.592 \times 1.59 \times 10^{16}}{0.408} \right)^{1/2} = 6.79 \times 10^{8} \, gauss. \tag{18}$$

Again and according to the results of the polytrophic gas sphere theory, the point z=3.5 indicates a position which is located a little farther than half the distance from the center to the surface of the star. With that election it is easy to verify that 90 per cent of its mass is being taken into account. In that position, the gravitational potential is

$$\phi_{*} = \frac{6.66 \times 10^{-8} \times 3.71 \times 10^{34} \times 3.5}{1.8203 \times 5.33 \times 10^{11}} = 8.9 \times 10^{15} ,$$

whereas also in that location the mass density is

$$\rho_* = 0.02109\rho_{c*} = 0.0671\, gr \cdot cm^{-3}.$$

With these results we have that

$$\rho_* \phi_* = 5.97 \times 10^{14}.$$

Consequently, the self-generated magnetic field has in that region the following magnitude

$$H_* = \left(\frac{2 \times 3.14 \times 5.97 \times 10^{14} \times 0.592}{0.408} \right)^{1/2} = 7.37 \times 10^7\, gauss. \quad (19)$$

If H_* and H_{c*}, are compared again, it is easy to see that for *V Puppis*, H=0.11 H_{c*} too. With the two previous values and with the one which was obtained in (1), the following graph showing the general behaviour of the magnetic field self-generated for that star can be built

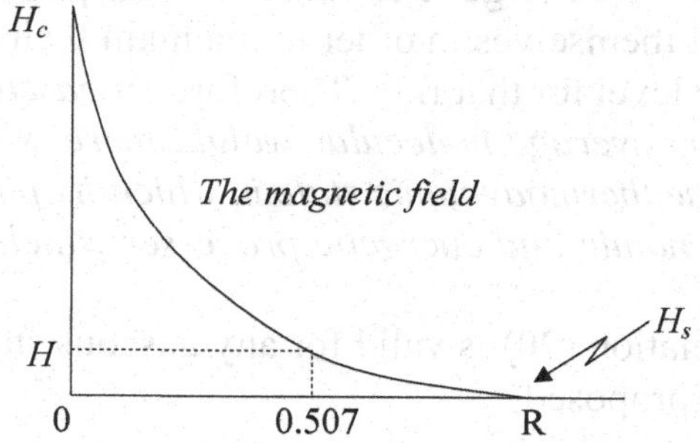

Finally, in order to estimate the distance at which *V Puppis* is found, its visual and bolometric magnitudes are used in the following expression

$$m_{vis*} - M_{bol*} = \log \frac{r}{10};$$

where r is the distance in parsecs. From direct calculus it is obtained that this system is found at 427 parsecs from the Earth; distance which is equivalent to 1,391 light-years.

The work hypothesis proposed earlier can be justified with the following argumentation. From the radiation law of absorption (91.2) and the result (92.2) by direct calculus it can be proven that

$$k_1 = constant \cdot \left(\frac{1-\varepsilon}{\varepsilon} \right) \cdot k_c T_c^{1/2}; \tag{20}$$

where k_c and T_c are the coefficient of opacity and the temperature of the star respectively, both calculated in its center itself and

$$constant = 3\mathcal{R}/a\mu, \tag{21}$$

where $\mathcal{R}$ is the gas universal constant, a Stefan´s constant, and μ the average molecular weight. According to definitions (90.5) the ratio $(1\text{-}\varepsilon)/\varepsilon$ is equal to the ratio between the pressures of radiation and the hot gases; that is to say p_r/p_g. For a particular star, its numerical value indicates its stability grade. On the other hand, the product $k_c T_c^{1/2}$ determines the magnitude of the radiant heat flow which a star emits throughout its surface, and consequently, fixes too the energetic conditions of the photosphere in such a way that they adjust themselves in order to maintain their effective temperature in the adequate level for that flux. Therefore, *in gaseous stars and for a given value of the average molecular weight there is a very close relationship between the thermodynamic state in which its photosphere is maintained and the dynamic and energetic processes which are generated inside of it.*

Since the relation (20) is valid for any gaseous star, the following comparison can be proposed

$$\frac{k_1}{k_{1*}} = \left(\frac{k_c}{k_{c*}} \right) \cdot \left(\frac{T_c}{T_{c*}} \right)^{1/2} \cdot \frac{(1-\varepsilon)\varepsilon_*}{(1-\varepsilon_*)\varepsilon}; \tag{22}$$

where the quantities withont asterisk correspond to the brightest componet of Capella considered as the measure unit; whereas the asterisc is used to indicate the parameters of any other star for which its numerical values are known. For example, Capella can be compared to *V Puppis* and therefore obtain the following results

$$\frac{k_1}{k_{1*}} = 0.2704, \text{ and } \frac{T_e}{T_{e*}} = 0.2737; \qquad (23)$$

where the first ratio is calculated from the equation (22) and the second one is obtained from their respective effective temperatures. In the case of the Sun we have that

$$\frac{k_1}{k_{1\odot}} = 0.9028, \text{ and } \frac{T_e}{T_{e\odot}} = 0.9058. \qquad (24)$$

Finally and given that *Capella* and *δ Cephei* have the same effective temperature, it is fulfilled that

$$\frac{k_1}{k_{1\delta}} = \frac{T_e}{T_{e\delta}} = 1. \qquad (25)$$

As it is easy to see, the agreement is practically complete.

The absolute magnitude

In the specialized literature it is stated that when the spectral class and the luminosity classes are simultaneously known, the position of the star in the *diagram of Hertzsprung-Russell* or *diagram H-R*, remains univocally determined and therefore its absolute magnitude is also known. That means that the factors which fix the value of the absolute magnitude of a star are its temperature and its size. In order to prove the last statement the results (88.5) and (89.5) are used to obtain the following expression

$$HR = mG^{1/2}(M/g)^{1/2}T_e^2. \qquad (1)$$

That equation is valid for any gaseous star in such a way that whichever two stars can be compared taking one of them as intermediary. What is usual to do is to consider the Sun as the measure unit and write that

$$\frac{H_* R_*}{H_\odot R_\odot} = \left(\frac{M_* \, g_\odot}{\odot g_*}\right)^{1/2} \cdot \left(\frac{T_{e*}}{T_{e\odot}}\right)^2 . \tag{2}$$

According to the definition of the stellar magnitudes given in that section and taking logarithms, the following result is obtained

$$M_{bol\odot} - M_{bol*} = 5.024 \log\left[\left(\frac{M_* \, g_\odot}{\odot \, g_*}\right)^{1/2} \cdot \left(\frac{T_{e*}}{T_{e\odot}}\right)^2\right] . \tag{3}$$

This last equation can be written as follows

$$10^{M_{bol}*} = 10^{M_{bol}\odot} \cdot \left[\left(\frac{M_* \, g_\odot}{\odot g_*}\right)^{1/2} \cdot \left(\frac{T_{e*}}{T_{e\odot}}\right)^2\right]^{-5.024} . \tag{4}$$

With the former result the fact that the absolute magnitude of any star depends on its mass and of its parameters g and T_e becomes clear. From the definition (88.5), it is easy to verify that the equation (4) can be expressed as follows

$$10^{M_{bol}} *= 10^{M_{bol}} \odot \cdot \left[\left(\frac{R_*}{R_\odot}\right) \cdot \left(\frac{T_{e*}}{T_{e\odot}}\right)^2\right]^{-5.024} . \tag{5}$$

Taking logarithms and using the numerical values of Sun´s parameters the following result is obtained

$$M_{bol*} = 97.1405 - 5.024 \log\left(R_* T_{e*}^2\right) . \tag{6}$$

If in (5) the definition (89.9) is used it can verify that

$$M_{bol*} = 89.2467 - 2.512 \log L_* \tag{7}$$

From this alternative analytical form, the fact that the absolute magnitude of a star depends on the heavenly body size and on its effective temperature becomes evident; that is to say, depends on its luminosity. In fact, it is known that the pressure differences in the atmosphere of stars of different sizes results in slightly different degrees of ionization for a given effective temperature. In 1913 Adams and Kohlschütter, first observed the slight differences in the degrees to which different elements are ionized in stars of the same spectral class but different luminosities and therefore different sizes. Suppose, for example, that a star is known to be of the spectral type G2 found in the Main-Sequence. In this case, its absolute magnitude could then be read off the *H-R* diagram at once and it would be about +5. Nevertheless, spectral type is not enough to fix without ambiguities the absolute magnitude of the star, due to the fact that the reffered star G2 could also have been a giant of absolute magnitude 0,or a supergiant of still higher luminosity and therefore of negative absolute magnitude.

With the discovery of those astronomers, it is now possible to classify a star by its spectrum, not only according to its effective temperature which gives the spectral class, as well as according to its size or luminosity. The most widely used system of classifying stars according to their luminosities or their sizes is that of W.W. Morgan and his associates. Morgan′s system divides stars of a same spectral class into as many as six categories, called luminosity classes. Those luminosity classes are the following

- Ia. *Brightest supergiants*
- Ib. *Less luminous supergiants*
- II. *Bright giants*
- III. *Giants*
- IV. *Subgiant, which are stars of intermediate size between giants and Main-Sequence stars*
- V. *Main-Sequence stars.*

A small number of stars called subdwarfs exists, lying below the normal Main-Sequence. The white dwarfs are much fainter stars. Main-sequence stars (luminosity class V) are often termed dwarfs to distinguish them from giants. The term dwarf is even applied to Main-Sequence stars of high luminosity, which may have diameters several times as great as the Sun′s. When that term is applied to a Main-Sequence star, it should not be confused with its use as applied to a white dwarf. The full specification of a star, including its luminosity class, would be, for example, for a spectral class F3 Main-Sequence star, F3V. For a spectral class M2 giant, the full specification would be M2III. Once known the absolute magnitud of a star its distance can be calculated from its visual magnitude. Distances determined this way, are said to be obtained from the method of spectroscopic parallaxes.

The magnetic field self-generated by gaseous stars

Let us consider the case of any gaseous star of radius R_*, composed by a viscous and conducting compressible gaseous fluid, which from the thermodynamic viewpoint behaves as an ideal gas; isolated in space and at very high temperature and pressure conditions which remains together by its own gravitational attraction and at dynamic equilibrium with the force produced by the sum of the pressures of radiation and the hot gases. The astronomical observations, seem to indicate that in general, the star is under the influence of a poloidal magnetic field $\boldsymbol{B}_*(x,t)$. Moreover, the star revolves on its own axis with a velocity $\boldsymbol{v}_* = \boldsymbol{v}_*(\circ_*, t)$, with $\circ_*$, the stellar latitude.

In order to determine the dynamic state of a viscous and conducting compressible gasesous fluid, which moves with velocity $\boldsymbol{v}(x,t)$ in some region of space where a magnetic field $\boldsymbol{H}(x,t)$ exists, the equations (44.1) and (48.6) are used.

In any gaseous star, the mass density can be considered as a function of time t and the radius R_*, and then $\rho_*=\rho_*(R_*,t)$, in such a way that the equation (49.6) takes te following equivalent form

$$\frac{\partial \boldsymbol{v}}{\partial t} + (\boldsymbol{v} \cdot \boldsymbol{grad})\boldsymbol{v} + \frac{1}{\rho_*} \boldsymbol{grad}(p + p_r) = \frac{1}{\rho_*} \frac{\partial}{\partial xj} \left[\sigma_{ij}{}' + \frac{H_i H_j}{4\pi} \right]; \qquad (1)$$

where and according to (35.7), σ_{ij}' are the components of the viscosity stress tensor, the equation (48.5) was taken into account, an integration by parts was made, and the relation of equivalence (89.3) was used.

Suppose that the star revolves so that the flux is steady. Then $\partial \boldsymbol{v}_* / \partial t = 0$. Next, it fulfills that $(\boldsymbol{v}_* \cdot \boldsymbol{grad}) \boldsymbol{v}_* = \boldsymbol{grad}(v_*^2/2) - \boldsymbol{v}_* \times \boldsymbol{rot}\, \boldsymbol{v}_* = 0$; because $\boldsymbol{v}_*$ is independent of x; so that from (1) we have that

$$H_*^2 = \frac{4\pi\rho_* \mathcal{R} T_*}{\mu} + \frac{8\pi a}{3} T_{*_e}^4; \qquad (2)$$

because $div\, \boldsymbol{v}_*=0$. Moreover, the relationships (88.9), (89.4), and (92.1) were used. The former result is valid for any gaseous star with radius R_* and mass density ρ_*, which revolves with a rotational velocity $\boldsymbol{v}_*(\circ_*, t)$. From (2), it is easy to see that in the inner regions of a star, the first term of the right hand side is greater than the second one, and therefore that term can be neglected comparing it to the first. Consequently and for any inner region of the star, the magnitude of the magnetic field can be calculated from the following relationship

$$H_* = \left[\frac{4\pi\rho_* \mathcal{R} T_*}{\mu} \right]^{1/2} . \tag{3}$$

On the other hand, at the surface of the star, the mass density is practically zero. And then, the magnitude of the magnetic field can be obtained from the next relation

$$H_{*_s} = m T_{*_e}^2 ,$$

which is the equation (89.5) of the text; with m given in (89.6) and its numerical value in (89.7). Suppose that H_* is the magnitude of the self-generated magnetic field by gaseous stars. It is easy to see that in the inner regions the self-generated magnetic field varies like the square root of the product of the mass density and the absolute temperature; both quantities calculated at those regions; whereas at the surface, H_{*s}, depends on the square of the effective temperature. That is, the self-generated magnetic field fulfils different laws perfectly established on the suface as well as in the stellar inward.

In the theoretical frame of MHD and because the magnetic permeability of the media differs only slightly from the unit, being that difference unimportant, it is considered that $\boldsymbol{H}_*=\boldsymbol{B}_*$ and then, the self-generated magnetic field fulfils the basic laws of magnetostatic; which in their differential form are the condition (44.1), *div* $\boldsymbol{B}_*=0$, and the following relation

$$\boldsymbol{rot}\, \boldsymbol{B}_* = \frac{4\pi}{c} \boldsymbol{j} ; \tag{4}$$

where $\boldsymbol{j}$ is the steady-state electrical current distribution localized in some region of the *convective zone* proposed in paragraph 87, and c is the velocity of light in empty space. According to (44.1), $\boldsymbol{B}_*(\boldsymbol{x})$ must be the curl of some vector field $\boldsymbol{A}_*(\boldsymbol{x})$, called the *vector potential*; that is

$$\boldsymbol{B}_*(\boldsymbol{x}) = \boldsymbol{rot}\,\boldsymbol{A}_*(\boldsymbol{x}). \tag{5}$$

For a steady-state current distribution localized in a relative small region of space, the vector potential is given by the next expression

$$\boldsymbol{A}_*(\boldsymbol{x}) = \frac{1}{c}\int \frac{\boldsymbol{j}(\boldsymbol{x}')}{|\boldsymbol{x}-\boldsymbol{x}'|}\,d^3x'\,; \tag{6}$$

where $\boldsymbol{x}'$ is a distance measured relative to a suitable origin in the localized current distribution and $\boldsymbol{x}$ is the coordinate of a point at a great distance of the localized current distribution.

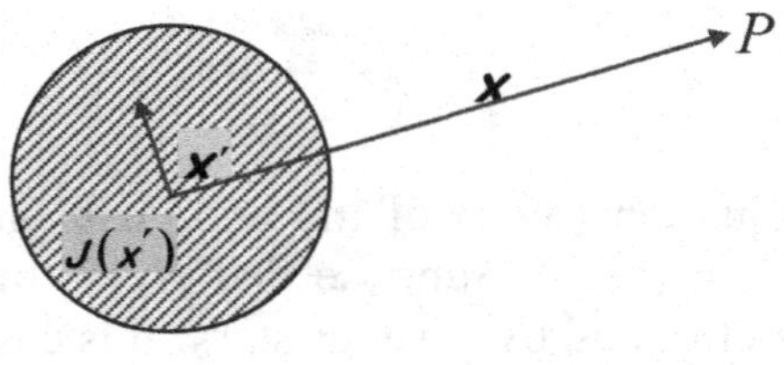

What we do is the following: in the equation (6) the denominator is expanded in power of $\boldsymbol{x}'$ as far as the lowest order of approximation. Then, for a given component of the vector potential we would have the next expansion

$$A_{*_i}(\boldsymbol{x}) = \frac{1}{cx}\int j_i(\boldsymbol{x}')d^3x' + \frac{\boldsymbol{x}}{cx^3}\cdot\int j_i(\boldsymbol{x}')\boldsymbol{x}'d^3x' + \ldots. \tag{7}$$

Since $\boldsymbol{j}(\boldsymbol{x}')$ does not depend on $\boldsymbol{x}$, it is localized divergenceless current distribution and then, this allows simplification and transformation of expansion (7). Let $f(\boldsymbol{x}')$ and $g(\boldsymbol{x}')$ be well-behaved functions of $\boldsymbol{x}'$. Then, if $\boldsymbol{j}(\boldsymbol{x}')$ is localized and has zero divergence, it is fulfilled that

$$\int\left(f\boldsymbol{j}\cdot\boldsymbol{grad}'g + g\,\boldsymbol{j}\cdot\boldsymbol{grad}'f\right)d^3x' = 0\,.$$

The latter can be established by an integration by parts of the second term, followed by expansion of the expression $f\,div'\cdot(g\boldsymbol{j})$. With $f=1$ and $g=x'$, from the former relation we obtain that

$$\int j_i(\boldsymbol{x}')d^3x' = 0.$$

Thus, the first term in (7) corresponding to the monopole term in the electrostatic expansion, is therefore absent. The integral in the second term of (7) can be transformed if it is considered an average magnetic field produced by a system of charges each one of them with charge q, in steady motion and calculated at a great distance from the system. Let $\boldsymbol{x}'$ be the vector radius of any charge and $\boldsymbol{x}$ the vector radius of the point where the field is calculated. Besides note that the average velocity of each charge is $\boldsymbol{v}=d\boldsymbol{x}/dt$ and that $\boldsymbol{x}$ is a constant vector. In this case we have that

$$\sum q(\boldsymbol{x}'\cdot\boldsymbol{x})\boldsymbol{v} = \frac{d}{dt}\sum q\boldsymbol{x}(\boldsymbol{x}'\cdot\boldsymbol{x}) - \sum q\boldsymbol{x}(\boldsymbol{x}'\cdot\boldsymbol{v});$$

where an integration by parts was made. Now, dividing the first term of the right hand side into two halfs and expanding one of them, we see that

$$\frac{1}{2}\frac{d}{dt}\sum q\boldsymbol{x}(\boldsymbol{x}'\cdot\boldsymbol{x}) + \frac{1}{2}\sum q\boldsymbol{x}(\boldsymbol{x}'\cdot\boldsymbol{v}) + \frac{1}{2}\sum q\boldsymbol{v}(\boldsymbol{x}'\cdot\boldsymbol{x}) - \sum q\boldsymbol{x}(\boldsymbol{x}'\cdot\boldsymbol{v}) =$$
$$\frac{1}{2}\frac{d}{dt}\sum q\boldsymbol{x}(\boldsymbol{x}'\cdot\boldsymbol{x}) + \frac{1}{2}\sum q[\boldsymbol{v}(\boldsymbol{x}'\cdot\boldsymbol{x}) - \boldsymbol{x}(\boldsymbol{v}\cdot\boldsymbol{x}')].$$

In this case on average, we have that

$$\overline{\sum q(\boldsymbol{x}'\cdot\boldsymbol{x})\boldsymbol{v}} = \frac{1}{2}\frac{d}{dt}\overline{\sum q\boldsymbol{x}(\boldsymbol{x}'\cdot\boldsymbol{x})} + \frac{1}{2}\overline{\sum q[\boldsymbol{v}(\boldsymbol{x}\cdot\boldsymbol{x}') - \boldsymbol{x}(\boldsymbol{x}'\cdot\boldsymbol{v})]};$$

where the dash indicates the average. Nevertheless, the average of the first term of the right hand side which contains the time derivative is zero, because it is the average of the temporary derivative of a quantity which changes within a finite interval. Integrating the remainder over all the space occupied by the system of charges and due to the fact that $\boldsymbol{j}(\boldsymbol{x}') = q\boldsymbol{v}$, finally the following result is obtained

$$\boldsymbol{x}\cdot\int \boldsymbol{x}' j_i\, d^3x' = -\frac{1}{2}\left[\boldsymbol{x}\times\int(\boldsymbol{x}'\times\boldsymbol{j})d^3x'\right]_i ;$$

where the following vector analysis formula was used

$$\boldsymbol{a} \times (\boldsymbol{b} \times \boldsymbol{c}) = (\boldsymbol{a} \cdot \boldsymbol{c})\boldsymbol{b} - (\boldsymbol{a} \cdot \boldsymbol{b})\boldsymbol{c}\,.$$

It is customary to define the *magnetic moment density* of *magnetization* as follows

$$\mathcal{M}(\boldsymbol{x}) = \frac{1}{2c}[\boldsymbol{x} \times \boldsymbol{j}(\boldsymbol{x})], \tag{8}$$

and its integral as the *magnetic moment* $\boldsymbol{m}$; that is

$$\boldsymbol{m}(\boldsymbol{x}') = \frac{1}{2c}\int \boldsymbol{x}' \times \boldsymbol{j}(\boldsymbol{x}')d^3x'\,. \tag{9}$$

Then, the vector potential from the second term in (7) is the magnetic dipole vector potential

$$\boldsymbol{A}_*(\boldsymbol{x}) = \frac{\boldsymbol{m} \times \boldsymbol{x}}{x^3}\,. \tag{10}$$

This is the lowest non-vanishing term in the expansion of $\boldsymbol{A}_*$ for a localized steady-state electrical current distribution. The corresponding magnetic induction $\boldsymbol{B}_*$ can be calculated directly by evaluating the curl of the last equation, that is,

$$\boldsymbol{B}_* = \boldsymbol{rot}\frac{\boldsymbol{m} \times \boldsymbol{x}}{x^3}\,.$$

If the following vector analysis formula is used

$$\boldsymbol{rot}\,(\boldsymbol{a} \times \boldsymbol{b}) = (\boldsymbol{b} \cdot \boldsymbol{grad})\boldsymbol{a} - (\boldsymbol{a} \cdot \boldsymbol{grad})\boldsymbol{b} + \boldsymbol{a}\,div\,\boldsymbol{b} - \boldsymbol{b}\,div\,\boldsymbol{a}\,,$$

and it is kept in mind that the derivatives are with respect to $\boldsymbol{x}$ and that $\boldsymbol{m}$ does not depend on that variable, we have that

$$\boldsymbol{rot}\frac{\boldsymbol{m} \times \boldsymbol{x}}{x^3} = \boldsymbol{m}\,div\frac{\boldsymbol{x}}{x^3} - \boldsymbol{m} \cdot \boldsymbol{grad}\left(\frac{\boldsymbol{x}}{x^3}\right).$$

Then,

$$div\frac{\boldsymbol{x}}{x^3} = \boldsymbol{x}\cdot \boldsymbol{grad}\frac{1}{x^3} + \frac{1}{x^3}\,div\,\boldsymbol{x} = 0\ ;$$

and

$$(\boldsymbol{m}\cdot \boldsymbol{grad})\frac{\boldsymbol{x}}{x^3} = \frac{1}{x^3}(\boldsymbol{m}\cdot \boldsymbol{grad})\boldsymbol{x} + \boldsymbol{x}(\boldsymbol{m}\cdot \boldsymbol{grad})\frac{1}{x^3} = \frac{\boldsymbol{m}}{x^3} - \frac{3\boldsymbol{x}(\boldsymbol{m}\cdot \boldsymbol{x})}{x^5}.$$

Now, let $\boldsymbol{x}=\boldsymbol{n}x$ be; with $\boldsymbol{n}$ a unit vector along $\boldsymbol{x}$. Then,

$$\boldsymbol{B_*} = \frac{3\boldsymbol{n}(\boldsymbol{n}\cdot \boldsymbol{m}) - \boldsymbol{m}}{x^3}. \qquad (11)$$

The magnetic induction $\boldsymbol{B_*}$ has exactly the form of the field of a dipole. Far away from any localized steady-state electrical current distribution, the magnetic induction is that of a magnetic dipole of dipole momentum given by the equation (9). Additionally, from equation (1) it can be shown that

$$\boldsymbol{T} = \frac{1}{c}\int (\boldsymbol{x}'\times[\boldsymbol{j}\times \boldsymbol{B_*}])d^3x'\ ; \qquad (12)$$

where $\boldsymbol{T}$ is the *total torque* and only the magnetic part of the force was used. Besides taking into account the equation (4) from this last result and writing out the triple vector product, it can be directly obtained that

$$\boldsymbol{T} = \frac{1}{c}\int \left[(\boldsymbol{x}'\cdot \boldsymbol{B_*})\boldsymbol{j} - (\boldsymbol{x}'\cdot \boldsymbol{j})\boldsymbol{B_*}\right]d^3x'. \qquad (13)$$

The second integral vanishes for a localized steady-state current distribution, as can be easily seen from the equation immediately before (7) with $f=g=x'$. Then, the first integral is therefore

$$\boldsymbol{T} = \boldsymbol{m}\times \boldsymbol{B_*}. \qquad (14)$$

This is the familiar expression for the torque on a dipole. It is one of the ways of defining the magnitude and direction of the magnetic induction $\boldsymbol{B}_*$. According to what was said before, it is easy to see that the self-generated magnetic field of gaseous stars is produced by some special kind of mechanism. In fact, according to density and temperature conditions, some region within the convective zone can exist which has a maximum of ionization. What occurs is that the electrically charged particles are moved by the convective streams through that region, making their contribution to the localized steady-state electrical current distribution and carry on, being continuously replaced by other particles. That is an effect similar in form to the one presented in the region where rainbow is produced. The drops of water move across the region where the rainbow is produced, making their contribution to the phenomenon and continuing with their fall; being constantly replaced by other drops. It is for that effect that the rainbow is a steady-state and localized optical phenomenon, just as the proposed steady-state electrical current distribution localized is too. Since this current distribution is produced by the high ionization of the region and the process of ionization depends on density and temperature conditions of the region, the magnitude of the magnetic field self-generated by of gaseous stars is a function of these variables, as it is easy to see from equation (3).

The self-generated geomagnetic field

In order to properly pose the problem on the origin and structure of the *Geomagnetic Field* and according to the argumentation that will be made next, it is proposed that the origin of that field may be located in the terrestrial *Outer Core* and not in the terrestrial *Inner Core*. According to the specialized literature, the Earth is structurated by the *Crust*, the *Mantle*, and the *Central Core*. The *Crust* is the thin outer region of the Earth. If it is compared to and apple, the *Crust* will be the thin envelope. On the other hand, the *Mantle* is a region that goes from the crust to the *Central Core*. It is totally solid, although throughout geological times it can behave as a plastic material. Whilst the *Central Core* is usually divided into an outer region and another inner region. There is enough evidence to prove that the outer core is a fluid and the inner core is solid. This latter region resembles a sphere composed almost totally by high temperature molten iron, possibly constant along all its volume, in such a way that it can be assumed that it is about a huge mass of isothermal molten iron. Due to the fact that it is under great pressure its mass density is so big that its value is near that of solid iron. It is very hard to assure that under such conditions the necessary temperature differences for the production to convective currents can be reached in this place. On the other hand, the outer region is subjected to pressure, density and temperature conditions such that the generation of convective currents is propitious, and therefore the outer core is proposed to be the *convective zone* of the planet; so that it is in this place where the origin of the *Geomagnetic Field* can be located. As in the case of gaseous stars, it is assumed that the *Geomagnetic Field* is generated by a steady-state current distribution localized in some region inside the convective zone, which is produced by a process of maximum ionization. Due to the fact that the outer core is considered as a very viscous, compressible and conducting fluid, which is dragged along by the Earth's rotation motion, it is proposed that it revolves with a differential rotational velocity $\boldsymbol{v}(\circ,t)$, with $\circ$ the latitude and t the time. Also, it can be considered that it is under the influence of a magnetic field $\boldsymbol{H}(\boldsymbol{x},t)$, where x is a quantity measured from any internal part of the *Outer Core* to the center of the Earth.

In order to determine the dynamic state of a mass of a very viscous, compressible and conducting fluid, isolated in space, subjected to an extremal condition of pressure, density and temperature, which remains together by its own gravitational attraction, revoling around its own axis with rortational velocity $\boldsymbol{v}(x,t)$ and being under the influence of a magnetic field $\boldsymbol{H}(x,t)$, the following form of the equation (48.6). is used

$$\rho\frac{d\boldsymbol{v}}{dt} = div\,\tilde{\sigma}^o + \rho\boldsymbol{f}\,; \tag{1}$$

where $\boldsymbol{f}$ is the force per unit mass, $\rho(x,t)$ the mass density, and $\tilde{\sigma}^o(x,t)$ the generalized stress tensor, which in components is given in (44.5).

To study the dynamic behaviour of the outer core the equation (1) will be used, where some simplifications can be made. In fact, since the rotational velocity is not an explicit function of the variable x, because $\boldsymbol{v}=\boldsymbol{v}(\circ,t)$, the term of (1) that contains the velocity becomes zero because of the following. We assume that the outer core revolves in such a way that it can be said that the regime is steady. Besides, is is proposed that $\boldsymbol{H}(x,t)$ is the self-generated magnetic field of the Earth. With those conditions taken into account, from the equation (1) the following result

$$\frac{\partial}{\partial x^i}\left(p+\frac{H^2}{8\pi}\right) = \frac{\partial}{\partial x^j}\left[\sigma'_{ij}+\frac{H_iH_j}{4\pi}\right]+\rho\boldsymbol{g} \tag{2}$$

is obtained because in that case the body force per unit mass is the $\boldsymbol{g}$ acceleration of terrestrial gravity. The form of the viscosity stress tensor in components σ'_{ij} is given in the equation (35.7). Since $\partial/\partial x^j = \partial/\partial x^i \delta_{ij}$, with δ_{ij} the components of Kronecker's delta, from (2) the following result is obtained

$$\frac{\partial}{\partial x^i}\left(p - \frac{H^2}{8\pi}\right) = \rho \boldsymbol{g} ;$$

due to the fact that $\sigma'_{ii} = 3\zeta \; div \; \boldsymbol{v}(\circ,t) = 0$. The last relationship is the equation (88.1) which governs the state of equilibrium of the Earth's central core, which in this case is magneto mechanical. Integrating the former equation and taking into account that $\rho = \rho(R, t)$, with R the central core radius, we have that

$$p - \frac{H^2}{8\pi} = \rho \boldsymbol{g} \cdot \boldsymbol{x} + constant . \tag{3}$$

From the specialized literature one obtains that the pressure is given by the next expression

$$p = p_o + \gamma C_V \rho T ; \tag{4}$$

where C_V is the specific heat at constant volume, T the temperature, γ the thermodynamic Grüneisen parameter, and p_o is an initial ambient pressure such that it is possible to assume that the hydrostatic equation (89.2) is fulfilled; that is,

$$p_o(x,t) = \rho \boldsymbol{g} \cdot \boldsymbol{x} + constant .$$

According to (4) and (89.2), from (3) we obtain that

$$H = constant\,(\rho T)^{1/2} ; \tag{5}$$

where

$$constant = (8\pi\gamma C_V)^{1/2}. \tag{6}$$

Consequently, the magnitude of the *Geomagnetic Field* in the inner regions of the Earth varies like the square root of the product of the mass density and the temperature; both calculated in those regions. Since in this case, the magnetic field self-generated by the Earth is also produced by a steady-state current distribution localized in some region relatively small inside the convective zone of the planet, it fulfils the basic laws of magnetostatic; in such a way that all the calculus which were made in the former *Selected Topic* are valid, and then they will not be repeated here. Consequently, the self-generated magnetic field by the Earth is dipolar as it is easy to see from the expression (11) of the former *Selected Topic*; that is to say

$$\boldsymbol{B} = \frac{3\boldsymbol{n}(\boldsymbol{n} \cdot \boldsymbol{m}) - \boldsymbol{m}}{x^3};$$

with

$$\boldsymbol{m} = \frac{1}{2c}\int \boldsymbol{x}' \times \boldsymbol{j}(\boldsymbol{x}')d^3x',$$

the *magnetic moment* given in the equation (9) of *Selected Topic* previously mentioned.

The strength of the dipolar field at some point of the outer core surface, can be obtained from the former expression for $\boldsymbol{B}$ and using the following data

$$m = 7.94 \times 10^{25}\, gauss \cdot cm^3$$
$$r_{oc} = 2.947 \times 10^8 cm.$$

Then, the corresponding strength of the magnetic induction is

$$B_{oc} = \frac{2m}{r_{oc}^3} = \frac{2 \times 7.94}{2.56} = 6.2\, gauss.$$

On the other hand, the value of C_V can be calculated from (5) and (6) and with the help of the following data

$$\gamma = 1.4$$
$$\rho = 11\, gr.cm^{-3}$$
$$T = 4.75 \times 10^3 K;$$

to obtain that

$$C_V = 2.1 \times 10^{-5} ergs.gr^{-1}.K^{-1}.$$

The strength of the magnetic induction at *Earth's Equator* can be estimated, by calculating B_c in the *Crust* with the help of the equation (6), from the one that corresponds to ***B***, and using the following data

$$\gamma = 0.25$$
$$\rho = 2.8\, gr.cm^{-3}$$
$$T = 3 \times 10^2 K.$$

Then, in the *Crust* and near Eath's surface we have that

$$B_c = 0.333\, gauss.$$

It is well known that the strength of magnetic induction measured on Earth's surface and at the Equator is equal to 0.307 gausss, so that the theoretical calculation and the direct measurement are practically equal.

References

1. Abell, G. "Exploration of the Universe". Holt, Rinehart and Winston (1964).
2. Chandrasekhar S. "An Introduction to Study of Stelar Structure". Dover Publications, Inc. New York (1958).
3. Eddington. A.S."The Internal Constitution of Stars". Cambridge University Press (1988).
4. Fierros. A. "The effective temperature and the absolute magnitude of the stars". To be publishing.
5. Fierros. A. "The magnetic field in the stability of the stars". To be publishing.
6. Fierros. A. " The Sunspots". To be publishing.
7. Fierros. A."El Principio tipo Hamilton en la Dinámica de los Fluidos". 2ª. Edición. Mc Graw-Hill-México (1999).
8. Rosseland. S. "Pulsation Theory of Variable Stars". Clarendon Press, Oxford (1949).
9. Schwarzschild, M. "Struture and Evolution of the Stars". Dover Publications, Inc. New York (1965).
10. Fierros. A. "The Geomagnetic Field". To be publishing (2003).

Index

SpringerPhysics

Roman U. Sexl, Helmuth K. Urbantke

Relativity, Groups, Particles

Special Relativity and Relativistic Symmetry in Field and Particle Physics

Revised and translated from the German by H. K. Urbantke.

2001. XII, 388 pages. 56 figures and 1 frontispiece.

Softcover **EUR 54,95**

(Recommended retail price)

Net-price subject to local VAT.

ISBN 3-211-83443-5

This textbook attempts to bridge the gap that exists between the two levels on which relativistic symmetry is usually presented – the level of introductory courses on mechanics and electrodynamics and the level of application in high-energy physics and quantum field theory: in both cases, too many other topics are more important and hardly leave time for a deepening of the idea of relativistic symmetry. So after explaining the postulates that lead to the Lorentz transformation and after going through the main points special relativity has to make in classical mechanics and electrodynamics, the authors gradually lead the reader up to a more abstract point of view on relativistic symmetry – always illustrating it by physical examples – until finally motivating and developing Wigner's classification of the unitary irreducible representations of the inhomogeneous Lorentz group. Numerous historical and mathematical asides contribute to conceptual clarification.